Clemens Posten
Integrated Bioprocess Engineering

Also of Interest

Microalgal Biotechnology
Posten, 2025
ISBN 978-3-11-066782-0, e-ISBN 978-3-11-066783-7

Bioprocess Intensification
Holtmann (Ed.), 2024
ISBN 978-3-11-076032-3, e-ISBN 978-3-11-076033-0

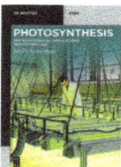

Photosynthesis.
Biotechnological Applications with Microalgae
Rögner, 2021
ISBN 978-3-11-071691-7, e-ISBN 978-3-11-071697-9

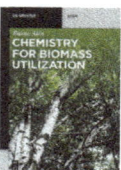

Chemistry for Biomass Utilization
Alén, 2023
ISBN 978-3-11-060834-2, e-ISBN 978-3-11-060836-6

Downstream Processing in Biotechnology
Beschkov, Yankov (Eds.), 2023
ISBN 978-3-11-057395-4, e-ISBN 978-3-11-057411-1

Clemens Posten

Integrated Bioprocess Engineering

2nd, revised and extended edition

DE GRUYTER

Author
Prof. Dr.-Ing. Clemens Posten
Senior Professor
Institute of Process Engineering in Life Sciences
Karlsruhe Institute of Technology (KIT)
Karlsruhe
Germany
and
Consultant
bio-compete
www.bio-compete.com

ISBN 978-3-11-077334-7
e-ISBN (PDF) 978-3-11-077335-4
e-ISBN (EPUB) 978-3-11-077394-1

Library of Congress Control Number: 2024934020

Bibliographic information published by the Deutsche Nationalbibliothek
The Deutsche Nationalbibliothek lists this publication in the Deutsche Nationalbibliografie;
detailed bibliographic data are available on the Internet at http://dnb.dnb.de.

© 2024 Walter de Gruyter GmbH, Berlin/Boston
Cover image: © Anthea Oestreicher, Interaction Design ZHdK
Typesetting: Integra Software Services Pvt. Ltd.

www.degruyter.com

Preface to 2nd ed – a small vote for this book

The world is facing major challenges regarding increasing world population and climate change. Bioprocess engineering must contribute to the supply of products ranging from bioplastics over fuels to food by replacing fossil raw materials and using sustainable approaches in large quantities. Even in existing fields such as pharmaceuticals or cosmetics, the challenges are growing in the face of increasing allergies or resistances. Modern biotechnology has made tremendous and amazing progress. Driving force is the desire delivering products for the mentioned urgent societal needs. This development has been rightly respected and has found its echo in public perception in general and in teaching especially. However, technical processes are necessary to make these benefits a reality. Bioprocess engineering has stayed a bit in the shade, reason enough to write a book not only as floral tour of possibilities but with didactic claim for professional university education. Already at the first leaf through, you will find elaborated verbal explanations and formula as link between the process and its physical, chemical, physiological, or molecular-biological basis, data and simulation plots to give a feeling for real production values and numerous pictures to underpin a visual understanding. The intention to write the book was to do a bit more on the didactical level as usually can be found in textbooks on bioprocess engineering and certainly as the fragmentary information from the internet can deliver. Bioprocess engineering stands in the tension between biology, technical feasibility, and societal demands. New concepts like the "cell factory" are answers to this challenge.

Dear reader of this preface. Your most urgent question now may be, whether this book is suitable and enlightening for you. Actually, it is designed for graduate students of bioengineering. Previous knowledge in reaction engineering and/or microbiology is helpful, but not urgently necessary. The book is therefore also aimed at newcomers from biology and chemical engineering. Missing background knowledge can easily be completed from other information sources. Dear student, the book is organized along a thematic thread and goes sometimes into depth, which does not open up intuitively on Bachelor level. Don't be unsettled but feel encourage to study these aspects again during your Master courses. The book wants to be a faithful companion through the whole study. Even during a doctoral thesis in bioprocess engineering, you will find mental approaches and examples that can be applied to your own questions. Why not reading the book as teacher, journalist, politician or interested layman? You will get a dense background in the fast-emerging field of bioprocess engineering to support your own opinion and qualified decisions. But also, the natural scientist or engineer already in the profession will find many new facts and approaches. Although the content of the book is not intended as reference book for "experts", it gives the framework to mentally integrate own experiences, structure own knowledge and become a more effective and creative professional. Even during a doctoral thesis in bioprocess engineering, you will find mental approaches and examples that can be applied to your own questions.

https://doi.org/10.1515/9783110773354-202

What is the content of the book, how is it organized and further, what can be a personal access to the book? The chapters follow the workflow of process development from strain selection and media design to reactor operation and process control policy. Some chapters cover cross-cutting issues like measurement and modeling. The chapters are built on each other, so working through the book in this consecutive order is strongly recommended. Cornerstones of the material are process examples from flow sheets down to evaluated data. The selection of the example processes reflects mainly the topic in the respective chapter. Yeast production is a classic and well understood process. Just this circumstance allows an accurate and instructive representation from microbiology to process management and product design. So, it is chosen as running example through the chapters. Phototrophic microalgae production as new dynamically developing field has been dedicated to being another "running example" through the workflow-oriented chapters. There are many different processes in practical use, so that it is not possible to show even the most important of them. The book is not as inclusive as possible. But concepts of fast and effective approaching new processes in order to understand, get a feeling for and improve them, is what the book tries to convey. Besides the read thread of process development, the mediation of these concepts is the real didactic structure, forming braces between the chapters.

To make such concepts of structured thinking visible, they are summarized and highlighted including different aspects of integration. The bioprocess engineer has to integrate these aspects into the design of modern and successful production processes. He must be prepared and sensitized for it, making the programmatic title "Integrated Bioprocess Engineering" sensible. This includes not only the classic integration of the various process stages, but also the merging of different sciences to arrive holistic concepts. Several new approaches are included in this second edition. Ultimately, the engineer must keep an eye on society, for whose needs the process is being developed, which gives regulatory and financial constraints and in which positive perception must take place. The engineer is supported in this by communicators, several examples from art & design are another special feature of this book. A few examples from prehistoric times are intended to round things off.

If there are questions unanswered, if you have ideas for other process examples, or even you found ambiguous formulations, pleased don't hesitate to contact me. Now enjoy reading, accompanied by a quote of Louis Pasteur, one of the greatest forefathers of biotechnology: *"Fortune favors the prepared mind"*. My hope is that this book will help the esteemed reader a little bit to follow this spirit.

Clemens Posten, Karlsruhe, Germany, Easter 2024

Contents

Chapter 3
Media – supplying microorganisms with a comfortable environment and
building blocks for growth —— 60

Chapter 4
Kinetics – finding quantities for bioprocess reactions —— 90

Chapter 5
Bioreactors – designing a home for the bioreaction —— 118

Chapter 1
Introduction – a thread through this book

Biotechnology is a fascinating science and an invaluable weapon in the battle against hunger and diseases. It uses the capabilities of enzymes, microorganisms, and isolated cells that produce complex molecules to bring comfort to our daily lives. Success stories about groundbreaking techniques in genetic engineering and the discovery of new bioactive compounds are at the center of public interest. Integrated bioprocess engineering provides the means to go from ideas to actually achievable products for the benefit of our society. This is not an easy path. The challenge can only be approached by sound application of engineering sciences and smart use of biological capabilities.

The chapters of this book follow the workflow during usual bioprocess design and operation, from lab work through fermentation to the final product. On the way, standard unit operations such as bioreactor design including mixing and aeration are addressed. This is achieved by fundamental chapters giving descriptions of state-of-the-art techniques and leading the reader through the jungle of technical calculations. The activities happening inside a reactor are decisive to understand a process; how to handle the technical operations then comes easily. Beyond this "hard-core" process engineering, concluding paragraphs raise additional questions and suggestions meant to strengthen the ability to take the broader view, which is essential for further success.

In several case studies called "running examples," special processes will be introduced. Different aspects of the interplay between cell physiology and the process conditions become clear through these examples. Here, know-how is complemented by know-why for these special cases. This is meant to allow for knowledge transfer from basic approaches to the huge variety of other already existing processes and to allow the readers to find ideas for their own projects. Further examples are given in smaller formats as an overview of different solutions for a given problem.

Integration takes place on different levels from direct coupling of process steps to societal aspects. The examples for integrative aspects given in the previously mentioned chapters are summarized and generalized in specifically dedicated paragraphs. Frameworks and concepts of the current way of thinking in science and industry are highlighted and outlined. Finally, exercises provide an opportunity to reinforce and train the teaching contents as well as to encourage further investigations.

1.1 Motivation – window shopping for biotechnological products

People encounter more and more biotechnological products in everyday life, whether they know it or not. Dairy products like yoghurt or kefir are traditional examples. Fermented foods are found in many cultures all over the world. Wine and beer have been

https://doi.org/10.1515/9783110773354-001

consumed for millennia thanks to the fermentative activity of yeast. Furthermore, yeast lets dough rise. Acetic acid has been known since antiquity and used in the acidification and conservation of foods. Since the Middle Ages it has been produced by the so-called Orleans method, an example of surface fermentation. This provides the acid bacteria for oxidation of ethanol (wine) to acetate. Later it was intensified (by increasing the air/water interface) by letting the suspension trickle over beechwood chips. For technical use acetic acid production even by modern fermentation processes is not competitive with chemical production. That does not hold for vinegar, a good example of integration into society, where public perception is an important issue especially in the food area (Fig. 1.1).

(a)   (b)

Fig. 1.1: Traditional food: (a) vinegar (aceto balsamico), one of the oldest biotechnological foods, presented here in modern lifestyle appearance (© Meine Pestoria); and (b) cheese is a traditional food as well. Scrubbing, essentially inoculation with selected microorganisms, supports the formation of the rind during cheese ripening (© Schönegger Käse-Alm).

Convenience food is commercially processed food that includes biotechnological steps or ingredients to optimize taste, smell, or ease of consumption. The flavor enhancer, glutamate, is the most prominent example and is present in many different packaged foods. It has been produced in a direct fermentation process since the 1950s by the bacterium *Corynebacterium glutamicum*. Today, total world production of this amino acid is estimated to be two million tons per year. Less well known is the polysaccharide, xanthan. It is present as food thickener and gelling agent in many soups, dressings, and ice creams. It is produced by the bacterium *Xanthomonas campestris*. Citric

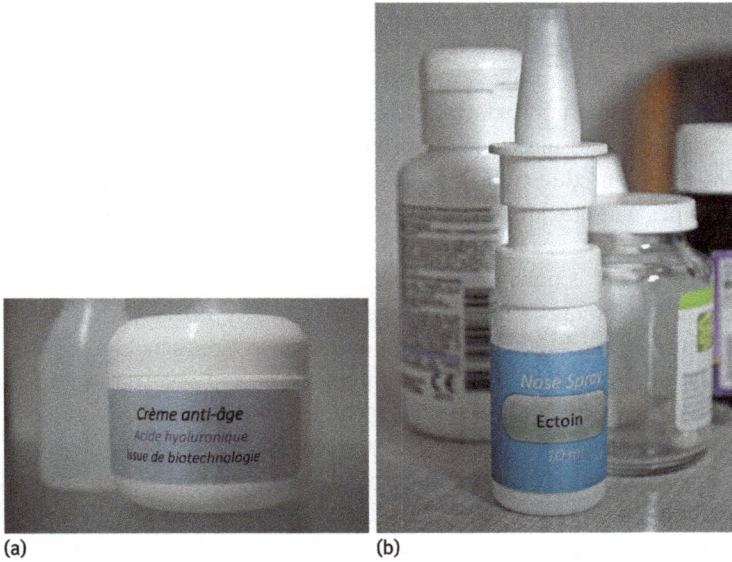

Fig. 1.4: Cosmetics: (a) product for skincare based on hyaluronic acid written as "bio-hyaluronic acid" when produced by microorganisms; and (b) nasal spray with ectoine to prevent drying out of the nasal mucous membrane in winter when heated air in the living room is very dry.

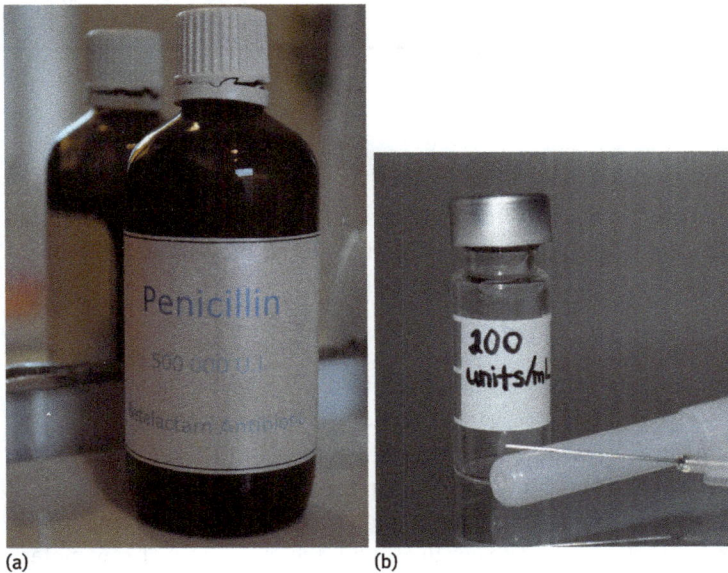

Fig. 1.5: Pharmaceutical products: (a) penicillin is still the most powerful weapon against bacterial infections; and (b) insulin was the first medical drug made by genetically engineered bacteria. Since then, it has helped millions of people live with diabetes.

(surfactants). Biotechnologically produced and biodegradable surfactants (e.g., rhamnolipids) are labeled as "biotensides" and are produced by microbial fermentation or enzymatic catalysis of plant-derived oils. Especially for outdoor applications they are more environmentally friendly than surfactants based on fossil oils (Fig. 1.6).

Beyond food, cosmetics, and medicine, bioproducts have acquired their place in technical applications. Sodium gluconate is used as aggregate in concrete mixes and acts as set retarder to prevent cracking during fast curing. Metal surface treatment is another application. Great hopes are placed on bio-based, biodegradable, and biocompatible plastics. A success story is polylactides (polylactic acid, PLA), for which the lactic acid is produced by fermentation. PLA-blends are already in use in different specific applications such as packaging (especially deep-drawn products like yoghurt pots and coffee pods), small parts of technical appliances, or as fibers for technical fabrics. However, the most significant biotechnological product in terms of volume is ethanol. Even at the petrol station we can find it. With bioethanol, biotechnology entered the energy market. The huge need for sugarcane in competition with its use as food however makes this process controversial (Fig. 1.7).

Bioprocesses will reach our daily life more and more, directly as commercial products or indirectly via alternative production methods.

---

i In future, more and more products will be manufactured with the help of biological systems. Traditional processes, which are based on fossil raw materials and often only work with many complex catalysts and energy-intensive processes, are to be replaced. New products can also be expected outside the pharmaceutical and food sectors. This trend is known as the bio-economy.

1.2 Bioprocesses – what they are and what they are able to do

Aren't cells just catalytic particles? What is so special about bioprocesses that you need a special book as also special courses of study and job descriptions? The next paragraphs will shed some initial light on these questions.

1.2.1 Bioprocess engineering – attempt at a definition

The bioprocess engineer deals with living material employed in technical processes. This basic view states two characteristics as the heart of the activity field. However, it is not intuitively clear exactly what "bio" means and exactly what kind of "process" is the target. The process or chemical engineer deals with the processing of raw materials that are being converted into added value products employing physical, chemical, and biological means. The characteristic view is to see the materials as a more or less unstructured flow of substances. This delimits the process engineer from mechanical

Fig. 1.6: Products for use in the household: (a) washing powder contains different enzymes like proteases and (b) a relatively new product is biotensides in environmentally friendly household cleaners.

Fig. 1.7: Technical products: (a) polylactide is one of the current bio-based plastics and a major material for 3D printing, here employed to form a globe; and (b) vehicles can be fueled in many countries with pure bioethanol or with ethanol as a fuel additive.

engineering handling of single workpieces. It also excludes medical engineering or agricultural technologies. Manipulating single plants or cells is therefore not a typical task of a process engineer. However, during a process, molecules, particles, or cells are definitely and intentionally altered.

The term "bio" assigns a role to the bioengineer in the larger field of biotechnology. The *Organization for Economic Co-operation and Development* (OECD) gives a definition of biotechnology (OECD iLibrary, 2018):

> "The application of science and technology to living organisms, as well as parts, products and models thereof, to alter living or non-living materials for the production of knowledge, goods and services."

This definition can be analyzed further. In an older version only microorganisms were mentioned, whereas now plants and animals are increasingly in the focus. Under "parts of cells" things such as enzymes could be understood. It is also worth stating that the "production of knowledge" as such is a kind of technology as long as it is a targeted process using biological and engineering principles. This is how bioprocess engineering is not only "running the process" but also has to apply the tools and paradigms of engineering to understand and design processes and not only production processes.

This definition of the OECD covers all modern biotechnology but also many traditional or borderline activities. To specify bioprocess engineering further, the definition is accompanied by a list-based definition, where item IV states:

> "Process biotechnology techniques: Fermentation using bioreactors, bioprocessing, bioleaching, biopulping, bioleaching, bio-desulphurization, bioremediation, bio-filtration and phyco-remediation."
>
> This definition unfortunately focusses on a transitory list of applications. Modern definitions, most of them valid only in particular countries, aim more on the application of different sciences and products.

This is a collection of technical means (here only the bioreactor) and different fields of application. A collection of things cannot be a reliable definition. For that, we need a general understanding on the level of meanings. In this book we will discuss specific processes and general methods going beyond these strict technical aspects. In bioengineering or more precisely bioprocess engineering the work is not limited to large scale operations as suggested by the definition. Zooming into a cell shows "processes" as well, which can be addressed and investigated by an engineer's way of thinking and their specific methods. This includes transport and reaction phenomena to be described by kinetics, balances, and thermodynamics familiar to chemical engineering at the large scale. In fact, chemical engineers were amongst the first drivers of metabolic flux analysis. Nevertheless, designing of bioprocesses is the core activity of the bioprocess engineer.

1.2.2 Yeast production – a classic but instructive process

To get a first insight into the structure of bioprocesses a closer look at yeast production will help. The demand for baker's yeast is directly associated with the need for bread, making the production a rapidly increasing industry. Growth rates are estimated to be more than 1% in developed countries and 10% in developing countries. Nevertheless, yeast is a cheap product compared to other biotechnological products, making an optimized process necessary with respect to cost efficiency.

A flow chart is given in Fig. 1.8. The workflow starts in the lab where single yeast cultures are stored in frozen vials or agar slants. Yeast companies hold several strains in their strain collection to be employed for different regions and different types of bread. For sugar bread, where the dough may rise for several days, the yeast cells have to be robust against high osmotic pressure, in contrast, e.g., to baguette, where short-term high fermentative activity is needed. The strains originate in best cases from one single cell. Specific institutes, the yeast banks, host such cell lines for many breweries and baker's yeast production companies. The selected strain is further propagated by transfer into a shaking flask. This can be done by a platinum loop or "eye", from which this operation has its name "inoculum" (from the Latin *oculus*, *meaning* eye). Further propagation goes through vessels with increasing size up to a final 30-L cultivation vessel. The purpose of this part of the lab work, the seed production, is to provide biomass as inoculum for production, commercial fermenters, or bioreactors. By the way, the word "fermenter" should no longer be used for bioreactors. This sequence of cultivations is in industry often called the "seed train." "Seed yeast" is produced under sterile and/or aseptic conditions. Seed yeast cream is separated and stored in seed yeast cream tanks, also called stock fermenters, and inoculated in portions into several production fermenters.

Other work items in the lab include analysis and preparation of the medium to supply the cells with the necessary amounts of sugar and other nutrients. Molasses, a sugar containing waste stream from sugar production, is used in yeast production for this purpose; see Chapter 3 concerning media design. Molasses preparation is a major step in the upstream process of yeast production. In order, to have a pumpable medium, molasses have to be diluted with hot water. Diluted molasses will be sterilized continuously under pressure, stored in clean or sterilized molasses tanks, and fed into fermenters during yeast propagation. In this case, water is treated with chlorine. In some factories diluted molasses will be acidified, sedimentation occurs, and molasses will be clarified continuously by decanters. The choice of these operation steps depends on the quality and source of the molasses (beat or cane). With the final contamination check the work in the lab is completed. This part of the production process delivering the seed is generally for bioprocesses called "upstream" processing.

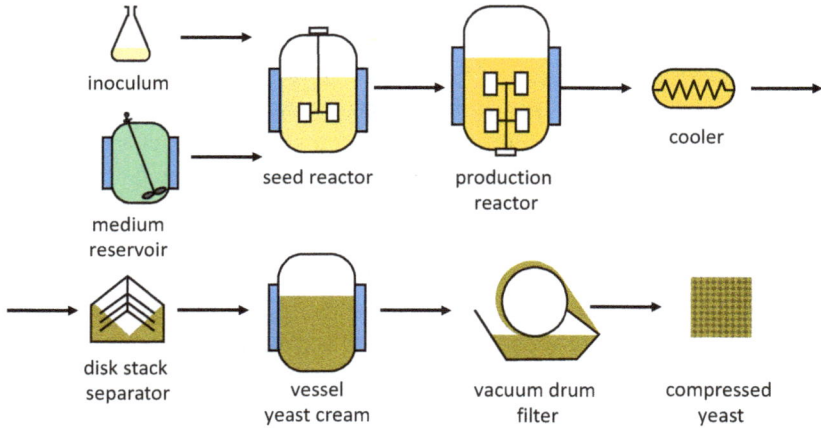

Fig. 1.8: Flow chart of a yeast production process. The pure strain in the inoculum is mixed with the medium to start the seed reactor. Stepwise scale-up leads to the final production reactor. After cooling, the yeast is concentrated in a special centrifuge, the disk stack separator, giving yeast cream as the first product. Production of compressed yeast for the supermarket requires an additional filtration step.

The next steps take place in bioreactors located in the production area. Further yeast propagation occurs in bioreactors of progressively larger volumes V_R by a factor of about 10. The index R means the working volume of the reactor. These large bioreactors are called "fermenters" in technical jargon. Temperature, pH value, and medium concentration are strictly controlled. The pH value is kept at 4.5–5.5, which is quite acidic compared to the optimum for bacteria. Together with the potential of yeast to produce ethanol, this is considered as one of the reasons why humans were able to produce yeast despite the contamination risks. The temperature is kept between 30 and 35 °C.

The first steps, called "seed fermentation," are kept under aseptic conditions, excluding other microorganisms from the reactor. The first product is the pure yeast that can be either sold or used to inoculate the next stage. The amount of yeast transferred from a small bioreactor into a larger one is still called inoculum, although it has nothing to do with a loop or "eye" anymore. Prior to inoculation there can be a concentration step by centrifugation. These seed fermentations end in 30 m³ scale. In the final step named "trade or commercial" fermentation, which occurs in large scale up to $V_R = 300$ m³ the medium is not sterilized due to cost reasons, so the next production steps are not necessarily free from contamination, e.g., by "wild yeasts." Such contaminations are not expected to propagate for the duration of one or two trade fermentations. Nevertheless, cleaning of equipment, steaming of pipes and tanks, as well as air filtering is practiced ensuring aseptic conditions. These central steps during production, where basically biomass is propagated, form the "bioreaction" stage.

At this point we have to think about measuring the amount of yeast produced. This can be done by taking a sample and counting the cells or filtering the sample and weighing the "wet cells." Both methods are problematic as cells are not all the same size and contain different amounts of water, between 80% and 90%. In addition, there is gusset water between the cells. For engineering and selling purposes a reliable measure in terms of mass is needed. This is achieved by sampling and drying the samples according to a standardized protocol. This includes filtration or centrifugation of a sample and drying it at 100 °C for up to 24 h. After drying, the samples consist only of the solid parts of the cells and do not contain free or physically bound water. This measure is referred to as "cell dry mass m_X [g]." The amount of yeast in the bioreactor in terms of concentration is then given by cell dry mass per liquid volume c_X [g/L]. In shaking flasks typical values are around 5 g/L, while in the production reactors 50–60 g/L are reached before harvesting.

Yeast out of the trade reactors has a cell dry mass concentration c_X of about 6% which corresponds to 60 g/L. During the next processing step, conditioning for the market, the yeast cream is further concentrated by several washing, centrifugation, and filtration steps. A disk stack centrifuge delivers a suspension of about 180 g/L, which is still pumpable. This "yeast cream" can be sold to industrial bakeries. The vacuum drum filtration leads to a filter cake with 30% solids content. This final product, the compressed yeast, is sold in big blocks for bakeries or in small (traditionally 42 g) yeast cubes in supermarkets. Product formulation includes application of additives like oils for better handling. These operations of a bioprocess are in general referred to as "downstream processing." "Vinasse", the liquid phase from the concentration steps, is still rich in some compounds and can be sold as a by-product after dehydration.

Each cultivation steps occurs within 24 h inclusive of reactor cleaning and the whole chain from the seed to the product occurs in one week. This is to keep a standard schedule every day and every week, a tribute to human working conditions and an example for operational integration of bioprocesses into the social environment.

1.2.3 The three columns of bioprocess engineering – what bioprocesses have in common

A generic bioprocess is shown in Fig. 1.9. It contains the most important operations already known from the yeast production process. These are essentially present in most biotechnological production processes. The chapters of this book follow this generic process from upstream through bioreaction to the downstream stage. Consequently, we start by looking at the upstream stage, beginning with microorganisms, the unique living entities that make a process a bioprocess. Which of them are practically employed in bioprocesses, and what are their specific needs and capabilities? In a broader sense, we also have to deal with the biosystems of cells of higher organisms and enzymes. In the next step, we look at media like molasses in the case of yeast. Media designs can be straightforward using balance equations and other rational considerations. Nevertheless, a lot of empirical knowledge is still behind the usual recipes. The thread then runs to the bioreaction stage.

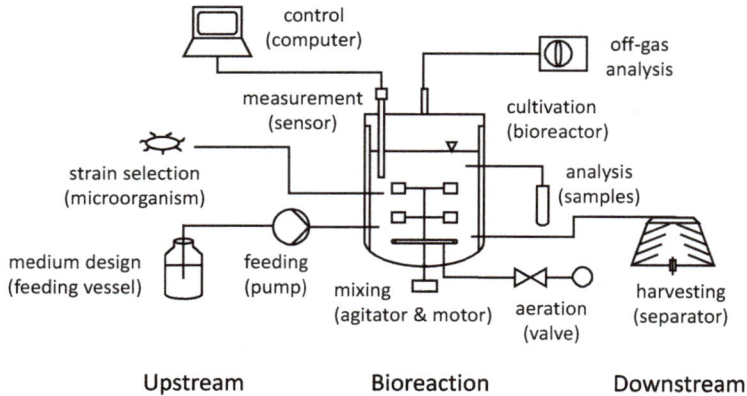

Fig. 1.9: Flowsheet of a generic bioprocess showing all major parts of a production line. The single elements are generic insofar as the related function can be taken on by other hardware components from case to case.

The core element is of course the bioreactor. The functions of the bioreactor in the bioreaction stage are analyzed and we show how the demands are represented in the technical design. The functions include supplying the organisms with appropriate environmental conditions with respect to temperature, nutrient, and oxygen concentrations as well as protection against undesired microorganisms. However, humans working with the reactor too have their claims concerning manageability and safety. Technical means are, for example, agitation, aeration, and heat transfer. However, value creation happens by growth and product formation. Kinetics and process dynamics are means to describe and control these processes. Especially, interaction of the microorganisms with media has to be reconsidered for optimal design of the reaction process. To control dynamics of a bioprocess regardless of the highly complicated interrelations, specific process control policies have been developed for bioprocesses. These will be deduced step by step. Nobody wants to run cultivation in blind flight, so the next important point is measurement, a topic that has made much progress in recent years. Based on respective data, mathematical process modeling is state of the art in science and industry. The basic approach to modeling will be given separately (Chapter 12).

Finally, the thread reaches the downstream stage. In many companies this section is even located in an area spatially separated from the cultivation facilities. The centrifuge is a symbol for harvesting, the first step of the downstream processes. Further steps follow but are not the subject of this book. In case studies the interplay between the three stages of bioprocesses will become clear. In fact, the different operations can be directly integrated on the process level overcoming the concept of unit operations. On the level of process development, integration means the constructive cooperation of biologists, chemists, process engineers, and basically all other sciences as depicted in Fig. 1.10.

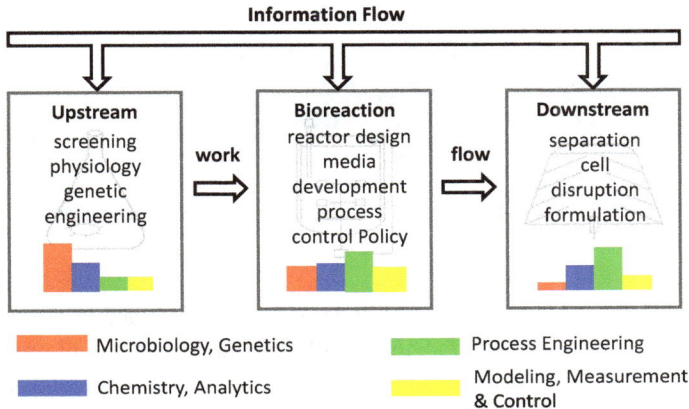

Fig. 1.10: The three columns of a bioprocess, namely upstream, bioreaction, and downstream. The basic steps present in nearly all bioprocesses are also indicated.

The most important duty of the process engineer is to become integrated into the development team and to integrate knowledge and skills from people in the upstream and downstream area into their work and decisions. Biochemical engineering science as well as practical implementation needs an interdisciplinary mind set. While the workflow goes from upstream to downstream, information flow in all directions is a necessity. For a tongue-in-cheek example: What a mess it would be if a highly productive strain was isolated and developed in the lab, and then years later the strain was tested in commercial media in large scale where strong foaming is defeated with surface active antifoam agents, finally leaving the membrane separation unit with a mission impossible?

1.2.4 Microalgae – biomass of the "third generation"

Today humans gain sun energy by photovoltaics, while wind and waves are driven by the sun as well. However, the overwhelming gift from the sun is food and feed for terrestrial plants. While sun energy as such is not limited, arable land cannot be further enlarged mainly due to water shortage. This is the point where microalgae come in. Microalgae have been recognized as biomass of the "third generation." This view comprises the use as food, feed, or fuel. In this running example we investigate the necessary technology to produce microalgae as alternative biomass (Fig. 1.11). These photobiotechnological processes are so different from heterotrophic bioprocesses, the anticipated production volumes are so huge, and the expectations so far-reaching that microalgae deserve a separate detailed investigation.

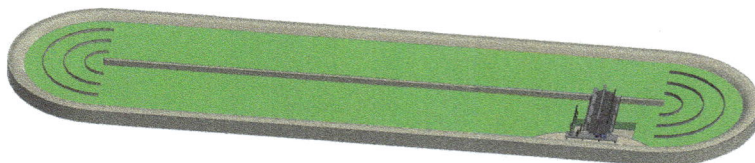

Fig. 1.11: To expose a large amount of the microalgae suspension to light, the cultures are kept in open tanks facing the sun. A paddle wheel circulates the suspension and mixes it.

The potential of microalgae has already found public interest. Barack Obama's clean power plan speaks 2015 about *"A Huge Win for Algae."* Different media are full of presentations of large cultivation facilities and small start-ups dealing with microalgae. Here we collect some of the repeated statements often made and amend some spontaneous comments.

– Microalgae are the only way to renewable biomass! Of course, microalgae offer big opportunities. Other options are macroalgae or salt water-resistant vascular plants. With respect to fuels, power-2-fuel technology is a competitor.
– Microalgae grow five times as fast as terrestrial plants! A single algae cell cannot divide faster than the shoot cells of higher plants. What is meant by this statement is growth of a culture at given light conditions in an annual average. Plants let light pass by chance to the soil, suffer from temperature changes, dryness, and different seasons. This statement of higher productivity will be true only if the microalgae are protected and optimally cultivated, taking advantage of their unicellular habit and growth in suspension.
– Production is possible in arid areas where no other claims exist, and water demand is minimal! This may be true in closed photobioreactors, but cultivation in open ponds is accompanied by evaporation.
– Growth takes place in sweet and salt water (in contrast to most agricultural crops). That is obviously true. Salt water is available in excess. Large production facilities will be close to coasts, where salt or brackish water is available.
– The whole cells can be used as products without persistent residues! Of course, algae do not have a stem with lignin that is difficult to recycle. Nevertheless, they have a cell wall partly consisting of persistent polysaccharides. In other cases, we are interested only in specific compounds.

– Microalgae take up a lot of carbon dioxide! In this point they are not different from all other plants. What sounds as a big chance can turn out to be a technical challenge.

In the next parts of the running example, we go into engineering details and try to understand how the sun, the cells, the reactors, and all other necessary items interact, how effective microalgal bioprocess engineering is, and how current limitations could possibly be overcome.

1.3 Integration – goals, methods, ways of thinking

The bioprocess engineer first does what all process engineers do, which is to make a production process work. Such processes essentially consist of conversion processes of a material from raw material to product. The characteristic feature of this is that it involves "unformed" materials, e.g., powders made of particles, liquids containing active solutes, gases, and energy. The processing of a single workpiece is therefore not "process engineering." The special feature of bioprocess engineering is that the material and feedstocks are living or even bioactive organic substances. There are particular difficulties, but also potential. That is what this book is about.

All bioprocesses need organic material, e.g., from plants as raw material/reactant. This should be converted into product as effectively as possible. Furthermore, no material or energetic side streams should arise. At least it should be possible to transfer them into other processes. The product should also be recyclable after use, at least be biodegradable, and not accumulate uncontrollably in the environment. These requirements are summarized under the term sustainability. Biological processes usually take place at ambient temperature. Furthermore, they can catalyze chemical conversions very specifically. These are undoubtedly huge advantages. Nevertheless, compared to purely chemical processes, they are very slow and process a lot of water. Furthermore, they can be sensitive to disturbances and often require specialized knowledge. In this tense field, the bioprocess engineer must nevertheless ensure high and stable turnovers in the apparatus involved, a goal called process intensification. The following two paragraphs discuss in more detail two features of bioprocesses that require special thought to be successful.

1.3.1 Integration of sciences – acquiring knowledge on demand

First and foremost, bioprocess engineering is the application of process engineering means to biological systems. The easiest option is to cherry-pick from the pool of unit operations and devices in process engineering. Here we meet aspects of mechanical and chemical engineering as well as biotechnology and informatics. Calculations are in most cases given as correlations based on natural laws and experiences. We have to carefully check the constraints of the validity of the equations given in references or the suitability of standard equipment. By definition, process engineering deals with "unformed" materials like powders or suspensions and not with direct manipulation of single objects. Nevertheless, the focus of bioengineering is to look at biological systems. Microorganisms of course have their own special qualities, so that adjustment is necessary to meet the specific demands and to harness their full potential. Biological knowledge is available in descriptions of the physiology of organisms including cell morphology or metabolic pathways. The task of the bioengineer here is to find reliable interfaces between biology and process conditions. These are in best cases kinetics and stoichiometry but could include complex cell adaptation as response to changing environmental parameters.

To do so, a standardized approach may be followed based on thinking in hierarchical levels and problem decomposition. On the process level, upstream, bioreaction, and downstream must be investigated as shown above. On the reactor and cultivation level we can virtually separate gas, liquid, and the bio-phase and understand their interactions. Here we meet all fields of mechanical and chemical engineering. Design criteria for reactors seem to be well developed and quite clear. Nevertheless, the bioengineer must make specific adaptations in geometry and material to meet the cells' needs, e.g., for oxygen supply coupled with low shear stress. On the cell level the task is to understand cells' needs and to go further into detail with respect to metabolism. Step-by-step, we come closer to basic physical or chemical laws. At the end of all our attempts additional knowledge is necessary and we must design targeted experiments to acquire data, e.g., for kinetics, material, or reaction constants. All the small bricks collected in this way are interconnected. Integration of sciences means to close the brackets between the single details and to go upward again to finally reach a strong pyramid.

In fact, all applied sciences are based on natural sciences, namely physics, chemistry, and parts of biology, where mathematics is the common language for description. This can be regarded as a pyramid of different layers. On each level, specific paradigms, which are fundamental ways of thinking, have already been developed. This makes things easier and prevents us going over details that may be or may be not necessary to consider each time. Terms like energy transfer, diffusion or viscosity have been employed successfully for decades and are still useful for most of the applications. One task in understanding and designing processes is the formulation of given natural laws or commonly employed correlations for given conditions. These include geometrical situations, given materials, or finally, specific qualities of the cells. With further progress, deeper understanding is necessarily gained by looking at intermolecular forces or complex intracellular metabolic networks. Breakthroughs in engineering sciences have often been gained by scrutinizing commonly accepted paradigms. Nevertheless, the art of engineering includes the art of simplification. Things should be as complex as necessary and as simple as possible for a given purpose. Monod formulated his famous kinetics using only two a priori unknown parameters. This turned out to describe growth curves fairly precisely, despite the thousands of enzymatic steps in the bacterial metabolism.

Bioprocess engineering can also give something back to the understanding of biological systems. Controlled cultivation in bioreactors keeps the cells in an adapted and defined state and allows us to precisely measure metabolic fluxes and to identify intracellular bottlenecks.

1.3.2 Bringing microorganisms to work – the concept of "cell factory"

Bioprocesses consist of timely and spatially highly ordered steps of transport, biochemical reactions, mechanical and thermal separation, product formulation, and delivery. Most of these steps require a special unit operation with special equipment. The process itself may be part of a factory, where basically also a pattern of transport and conversion steps applies on a higher level. Means are of course different, transport is down by conveyer belts or trucks, central control is the administration, and so on. Looking downward to the smaller unit of a bioprocess – the cells – again this ordered structure of basic transport and conversion steps applies again with different means. This has brought up the idea of the cell factory as shown in Fig. 1.12.

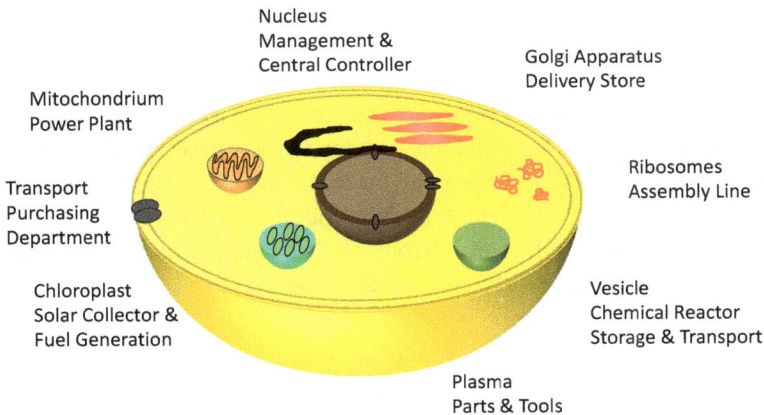

Fig. 1.12: Visualization of the idea of a "cell factory"; the different organelles or sites in the cell can roughly be matched to the unit operations and devices in a chemical factory.

This analogy can teach us several things. All steps must be balanced in terms of turnover. It does not help to build more assembly belts, if not enough raw material is provided. This has to be ensured along the supply chain from an external deliverer, over receiving controls, processing steps, and finally loading onto the belt. The product will be packed, targeted, and mailed to the customer. For cell engineering it means that overexpression of a product must be accompanied by balancing the whole cell with respect to all other production steps. The "management" should be convinced that the product that is, for example, a recombinant protein is worth being made without corrupting the central control, which is necessary to balance the other intracellular processing steps.

The role of bioprocess engineering lies mainly in the supply chain by maintaining optimal conditions in the reactor, supplying the necessary parts, and receiving the product. However, these steps are not trivial. Building blocks, which may help the cell doing its job, have to be identified and provided in the medium; it must be ensured that the cell can take up these parts and use it properly. Product formation shall not overcharge metabolism in general. Separating the product means isolating, e.g., a protein, from thousands of others. While during downstream processing this is a serious problem the cell can master it to perfection. Coordination between molecular and process engineering can lead to a successful decision on which unit operations are done in the cell factory and which of them must be done on the macroscopic process level. A prominent example is targeting recombinant proteins to outside the cell, which makes macroscopic separation much easier. On behalf of the scientists involved in discovering the molecular basis for intracellular protein targeting, the name of Guenther Blobel, who is Nobel Prize laureate (1999) for the discovery that "proteins have intrinsic signals that govern their transport and localization in the cell" may be recalled here.

1.3.3 Characteristics of living cells – unique selling points from engineering view

Bioprocess engineering delivers platform technologies for a spectrum of microorganisms and classes of products to reduce development time for new processes. Nevertheless, for specific products adaptations have to be made. In any case, the employment of a bioprocess has to be justified against the background of costs and societal needs. The discussion should start with a look at microorganisms and at what makes them different from inorganic catalysts. Basic points are shown in Fig. 1.13.

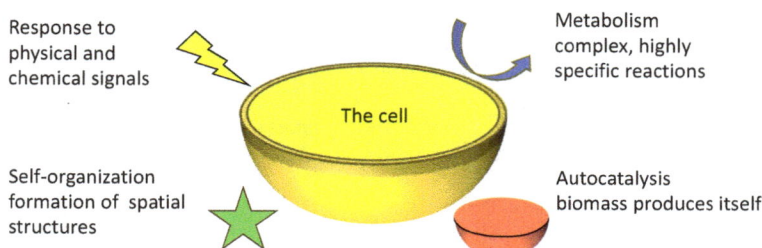

Response to physical and chemical signals

Metabolism complex, highly specific reactions

The cell

Self-organization formation of spatial structures

Autocatalysis biomass produces itself

Fig. 1.13: Unique selling points of microorganisms as catalysts: these include self-reproduction, synthesis of complex molecules, autonomous formation of spatial structures, and the response to environmental signals.

The most fantastic feature of living cells is self-reproduction being the basis for life in general. No classic catalyst has this ability. Employing microorganisms for production purposes means to employ self-reproduction. A great diversity of microorganisms is available as biocatalysts for different purposes. As engineers we have to provide a suitable environment to enable this amazing ability in a closed containment: the bioreactor.

Microorganisms form complex molecules, which can be produced in chemical processes only with great effort in many consecutive inefficient conversion steps. Complexity in this context means sequential order of monomers in polymers, and regiospecific and enantioselective conversions. For many pharmaceuticals no other production routes exist apart from production by bioprocesses. This is indeed a very strong unique selling point for their employment.

Cells are in no way ideally mixed micro-reactors. They maintain a highly complex compartmentation and spatial order from organelles over enzyme complexes, facilitating ordered metabolite flux down to transport of molecules to distinct locations. Such spatial structures are not only necessary to decouple the hundreds of enzymatic steps but can be also employed directly for production purposes. Long chains of separated chemical reactors would be necessary to mimic this unique feature.

Cells react to physical and chemical signals by adaption of their physiology and their metabolic pattern to the environment. Chemical signals include media concentrations leading to overshoot metabolism on the one hand or starvation on the other. Specific chemicals induce specific reactions like excretion of specific products or are employed as inhibitors. Physical signals include temperature, mechanical stress, or light. Such signals are also employed during bioprocesses, e.g., to switch the cellular metabolisms from growth to production mode. For all these unique selling points examples will be given in the following chapters.

1.3.4 Integration into society – the final meaning of bioprocesses

Bioprocesses are linked to human society in many ways. Engineers must understand and consciously shape these interactions (Fig. 1.14). People will only accept processes if they have direct benefit as is the case with food or medicines. Of course, the costs must also be commensurate with the benefits. These aspects will be addressed in various examples in this book.

On the other hand, the process itself must not pose any danger. Appropriate precautions and regulations should be in place to ensure the safety of products, processes, employees, and customers. Sustainability is a task on a global scale. Decarbonization must be forced through the production of alternative fuels or foodstuffs. These aspects of integration into society should also be taken into account.

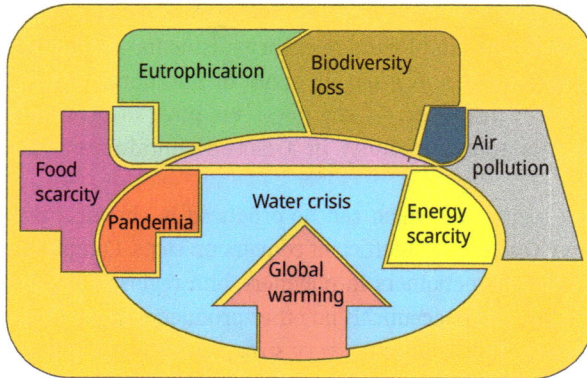

Fig. 1.14: The current problems facing mankind also pose a major challenge for bioprocess technology. The big puzzle: Which potentials of bioprocesses can be tackled and how?

1.4 Exercises, questions, and suggestions

1. Make a shopping tour and look at the list of ingredients on the cans, packets, and bottles. Try to find out which of the items are biotechnologically derived.
2. Compare the list in Fig. 1.3b with the ingredients of a medium for microorganisms (Chapter 3).
3. What are the "parts and model" of living organisms in the sense of the OECD definition?
4. Which unit operations in the yeast production process are employed in the field of mechanical, chemical, or thermal process engineering?
5. Bioproduction takes place for many fields of application: food production, agriculture, technical additives, environmental and energetic processes, pharmaceuticals, commodity conversion, etc. Collect examples of processes and products from this book or other resources.
6. Do you have a better idea for a definition of "bioprocess engineering"?
7. During reading of the book try to make your own opinion about the future significance of microalgae.

Chapter 2
Biosystems – microorganisms and other biocatalysts

Omne vivum ex vivo (Latin: All life is from life). Up to the nineteenth century this insight was a subject of discussion until Louis Pasteur and other researchers brought it to the point. Nowadays we accept this law of biogenesis as a matter of course. Strain development including directed evolution and genetic/metabolic engineering is an integrated step in upstream process development.

Usually process engineers get production strains from the upstream department. Nevertheless, communication between biologists, biotechnologists, and bioengineers is necessary to consider the strengths, weaknesses, opportunities, and threats of different production scenarios including strain selection in the spirit of integrating the different sciences. This early decision has far-reaching consequences on bioreaction and the downstream steps, and finally commercial success. According to the definition of biotechnology biosystems include microorganisms, plant, and animal cells, but also parts of it. Single cells in suspension culture do not represent the only form in which organisms are used in bioprocess; different kinds of structures are employed. An overview is given in Fig. 2.1 and briefly discussed below. It is not the aim of this chapter to give a comprehensive treatise of the biology of microorganisms, but to give ideas about the diversity of possible production systems and to sensitize for opportunities, risks, and technical constraints in bioprocess design.

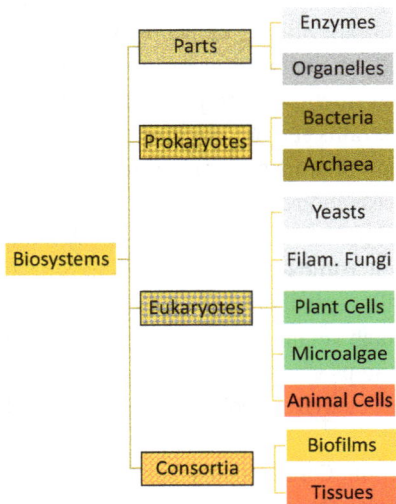

Fig. 2.1: Overview of different biosystems including prokaryotic and eukaryotic microorganisms, plant and animal cells, parts of them like enzymes, and their structures like biofilms.

https://doi.org/10.1515/9783110773354-002

2.1 Motivation – a first bioengineering view to microbiology

Agricultural crops and domestic animals have been farmed and bred for centuries and have adapted to the artificial environment provided by people. This concept of domestication can also be transferred to microorganisms. Some of them are employed in bioprocesses traditionally or have been in use in modern biotechnology for decades. For these strains, elaborate cultivation procedures are established and process design follows standard rules. Undesired side effects like by-product formation, sticking to reactor walls, flocculation, or strong foaming are reduced by long-term selection. These are arguments to stay with such strains as production platform for different products. For newly screened strains, behavior in the reactor must be tested for a selection of the best strains coming out of the biological tests to avoid disappointing results during fermentation and cell harvesting.

Process safety is a major concern in biotechnology and is influenced by the choice of the production organism. The simplest case is of microorganisms generally regarded as safe (GRAS status). Cultivation of genetically engineered organisms requires specific measures for the production plant, the reactors, as also for handling and documentation. This holds especially for pharma products. Potentially pathogenic organisms demand respectable safety regulations as well. The decision is made for each strain – even some *E. coli* strains are classified as potentially pathogen. Consequently, process engineering is facing many complicated detail problems and cost issues.

Also, in connection with genetically engineered organisms, process stability has to be investigated. Loss of plasmids, for example, can slowly reduce productivity in the long term during continuous cultivations. But wild strains can be subjected to spontaneous mutations or at least phenotypical changes. Microorganisms running through a life cycle like sporulation need special attention in process control.

Size, cell wall, and cell shape determine mechanical stability of cells in the agitated reactor. Turbulence causes the formation of eddies of many different length scales (Kolmogorov length scale), where the smallest structures are in the range of 50 μm in bioreactors. Bacteria and yeasts being much smaller are more or less on the safe side. Furthermore, efficiency of downstream steps especially cell separation and cell disruption depend on cell size. Filtration and centrifugation become more and more difficult for biomass with decreasing cells size.

Strain selection determines medium composition. Consequently, it includes tests not only for growth on technical cheap media but also for consumption of most of the carbon sources present in the medium. This holds especially for products of low value. Resistance against osmotic pressure or occasionally occurring inhibiting chemicals in the medium are further demands on microorganisms for production. Tolerance against low pH values can be an issue and helps suppressing contaminations.

2.2 Diversity of biosystems – appearance in technical environments

Microorganisms are prevalent in incredible diversity in nature. Screening programs look for new strains with special properties in all areas of the environment, may be soil, deep sea, or symbionts in other organisms. Axenic cultivation is a big problem. For example, complex cultivations of sponges were carried out whose bioactive effects are actually produced by bacteria living in them. On the other hand, standard expression systems are required to avoid customization work. One idea being pursued is to isolate genes from specific species and experiment in standard organisms. But even this is not always easy. For example, carotenoids with a strong antioxidant effect would lead to damage in heterotrophic cells. In other cases proteins need different intracellular environments.

The term "biosystem" covers all levels of organization from bioactive macromolecules to pro- and eukaryotic microorganisms, plant and animal cells, symbioses, and other communities. Even a short overview is practically impossible in the framework of this book. The following paragraphs list some technically important groups and emphasize the different aspects for bioprocess engineering.

2.2.1 Enzymes – the universal toolbox for biocatalysis

Biocatalysis is the chemical process through which biological catalysts, which could be one or more enzymes (or even cells), perform chemical reactions between organic components. Thereby the catalyst can significantly lower the activation energy of a reaction, so that the speed of the reaction is increased, and the catalyst itself is not used up. Generally, protein-based enzymes are responsible for catalytic reactions in all living organisms and are considered as the common catalytic units that form the basis for biotechnological transformations processes, since they can increase the reaction speed up to a factor of 10^8–10^{20}. Classical fields of application can be found in food and drink processing, where the production of wine, beer, cheese etc. is dependent on the effects of microorganisms or enzymes. In fact, biocatalysis underpins some of the oldest chemical transformations known to humans with the oldest records of brewing done by Sumerians about 6,000 years ago. The Incas used salvia (probably without deeper understanding of the mechanisms acting behind) as a source of the enzyme amylase to breakdown starch from corn to produce of a sort of beer called "chicha." The identification of enzymes as the catalyst of a specific reaction, however, was first successfully described in the nineteenth century when a German physiologist Wilhelm Kühne defined in 1877 the term *"enzyme,"* which was later used to describe the chemical activity of nonliving substances. Since then, a lot of efforts were made to gain knowledge of the working mechanisms of the enzymatic catalysis and therefore, enzymes play a key role also in the modern biotechnological applications (e.g., cheese thickening was one of the first processes that used recombinant enzymes, see Fig. 2.2).

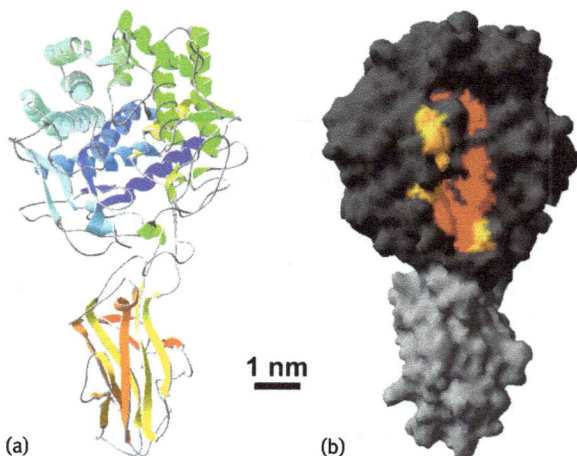

1 nm

(a)   (b)

Fig. 2.2: Structural visualization of cellulase enzyme from bacteria *Thermomonospora fusca* (TfCelE4). (©Olga Blifernez-Klassen): (a) ribbon structure model with artificially colorized α-helices and β-sheets shown. (b) A surface plot model where some functional groups are colorized (substrate binding domain (gray), catalytic domain (dark gray), and conserved amino acids within the catalytic domain (red, yellow)).

Protein/enzymes mass is usually given by dalton (Da) or kilodalton (kDa) and is typically in the range of 3 kDa to more than 1,000 kDa. The unit Da corresponds to atom mass u ($u = 1/12$ of ^{12}C carbon atom mass) and can be used for calculations on mol basis (1 mol ^{12}C = 12 g). Another important factor for enzyme applications is the enzymatic activity. Usually, the activity of enzymes is given in units (U) and stays for the amount that can catalyze μmol substrate within one minute. For example, if hexokinase (first step of glycolysis) activity is given by 350 U/mg enzyme, it means that 1 mg of the enzyme (hexokinase) will catalyze the phosphorylation of 350 μmol substrate (glucose) within one min under optimal conditions (which are normally also stated by the supplier). Another way to express maximal possible enzyme activity is the turnover number (also termed k_{cat}), which specifies maximum number of catalytically substrate conversions per enzyme in one second.

Enzymes often require for function cofactors; some cofactors like metal ions (Cu, Fe, and Mg,) do not change during the reaction. Other cofactors like ATP or NADH(P)$^+$+H$^+$ are used up by the catalytic reaction and can be designated as co-substrate. These co-substrates have to be regenerated with another coupled enzymatic system. This needs to catalyze an exothermic reaction, from which the substrate is lost. Ongoing research develops such processes robust enough for industrial application. A commonly accepted top level classification based on the reaction mechanism lists oxidoreductases (oxidation/reduction), transferases (transfer a functional group, e.g., phosphate), hydrolases (hydrolysis of various bonds), lyases (cleave various bonds other than hydrolysis and oxidation), isomerases (catalyze isomerization), and ligases (join two molecules with covalent bonds). A selection of commercially applied enzymes covering different reaction groups and application fields is given in Tab. 2.1.

Tab. 2.1: Selected list of common technical enzymes.

Scientific name	Function	Application	Comments
Protease	Protein degradation	Washing powder	*Bacillus* sp.
Amylase	Starch degradation	Food industry, bioethanol industry, textile industry, bakery industry	*E. coli, B. subtilis, B. amyloliquefaciens, B. licheniformis*
Chymosin	Proteolytic	Cheese thickening, substitution of rennin	*Aspergillus* sp., *E. coli*
Phytase	Hydrolysis of phytic acids	Additive in animal food	*Bacillus* sp., *Lactobacillus plantarum*
Cellulase	Cellulose degradation	Laundry detergent, pulp and paper industry	*Clostridium* sp., *Cellulomonas* sp.

Features for favorable application of enzymes can be identified from this list. First of all, the ability to distinguish between similar substrate molecules is one of the most important advantages of a biocatalyst. This selectivity is often functional group specific (chemoselective) or the enzymes may distinguish between functional groups which are chemically situated in different regions of the substrate molecule (regioselective or stereospecific). This high selectivity may offer several benefits to the overall process like high catalytic efficiencies and mild operational conditions, minimized side reactions, easier separation, and fewer environmental problems. However, this also covers a unique selling proposition making chemical alternatives with unspecific catalysts difficult if not impossible. Applications in food industry take advantage from the fact that enzymes can be left inside the product. Actually, it would be impossible to remove them in many cases like cheese. For fluid and molecular disperse systems different methods have been developed for enzyme immobilization. That allows for avoiding mixture into the product and enzyme losses during the process. Due to their relatively simple applicability and low costs, employment of enzymes outside classical biotechnological production is an emerging field.

2.2.2 Bacteria – organisms for all seasons

Bacteria are present ubiquitously in the environment and are key players in global organic and inorganic material cycles. Traditionally they are involved in food fermentation including different dairy products. On the other hand, they are part of the microbiota and residue on our skin and inside the intestine. Humans have been suffering from bacterial diseases for millennia and are still suffering from them. Therefore, bacteria are targeted by biotechnological research with respect to finding antibiotic activities. Antonie Leeuwenhoek (1632–1723) was the first to watch bacteria in his strongly improved microscope. The year 1674 is regarded as the beginning of microbiology. It is noteworthy to mention that Leeuwenhoek was not only an enthusiastic scientist but also a good communicator and made the society aware of his findings. A modern electron microscopy image (Fig. 2.3) shows the typical appearance of bacterial cells.

Bacteria species can be – besides their phylogenetic classification – divided into two groups based on chemical and physical properties of their cell wall measurable by so-called Gram staining. Both bacteria groups possess rigid polysaccharide (peptidoglycan also known as murein) layer within the cell wall in order to compensate the hydrostatic pressure (~ 2 bars). The peptidoglycan layer is more pronounced in gram-positive bacteria, since gram-negative bacteria possess an additional (to the cytoplasm membrane) outer membrane, on top of the thin peptidoglycan layer. These differences in the cell wall structure are responsible for positive or negative staining with crystal violet dye, commonly used for discrimination of gram-positive and gram-negative bacteria.

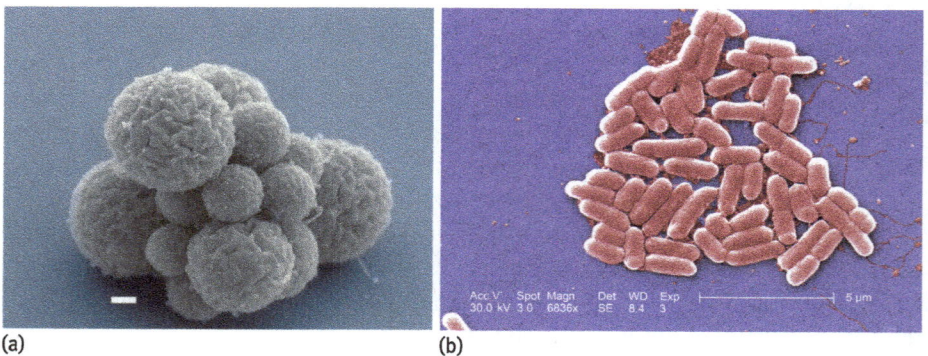

(a)   (b)

Fig. 2.3: Bacteria cells appearance under scanning electron microscope (SEM): (a) a cluster of JCVI-syn3.0 cells, artificial cells with minimal bacterial genome (Craig Venter, ©Science) showing spherical structures of varying sizes; (b) group of gram-negative *Escherichia coli* bacteria (here strain O157:H7), artificially colorized for better visualization (©CDC).

The size of technically relevant bacteria is about 1 μm diameter $*$ 2 μm length making a volume of about 1μm^3. These dimensions make filtration and centrifugation difficult without pretreatment of the suspension.

The huge diversity of metabolic patterns is consequently represented in technical processes and products. A list covering some technical relevant strains and fields of application is shown in Tab. 2.2.

Tab. 2.2: Selected list of bacteria of industrial importance.

Scientific name	Biology	Application	Comments
Escherichia coli	Intestinal	Recombinant proteins	Insulin production since 1982, model for genetic engineering
Corynebacterium glutamicum	Soil bacterium	Amino acids	Since 1957 as glutamate producer in Japan, model for metabolic engineering
Bacillus subtilis	Soil bacterium	Extracellular technical enzymes	Proteases, lipases, etc., for washing powder
Lactobacillus	Milk products and intestinal	Lactic acids, dairy products	Starter cultures in food industry
Streptomyces	Filamentous soil bacterium	Streptomycin	Many industrially produced antibiotic agents, streptomycin (1952), ivermectin (against worm infections, 2015)

Largest product examples with respect to quantity are small metabolites. Anaerobic processes are operated for solvent (acetone/butanol) production as well as for bulk products for the chemical industry. A prominent example is 1,3-propanediol that is used as precursor for polymerization. Also, the organic acid lactate is an important example not only for food industry but recently discussed for polylactate production as bioplastics. Aerobic production of organic acids is employed for amino acids where glutamate is an ingredient in many consumer foods. *Bacillus* is a natural producer of extracellular proteins making it a preferred candidate for technical enzymes such as hydrolases like protease, amylase, or cellulase. Recombinant proteins are formed by *E. coli* and intracellularly stored as inclusion bodies, the most prominent example being insulin. But production of antibody fragments is an important application, as well. Bacteria can target proteins into the cytoplasm, the periplasm, or to the extracellular environment. This has decisive consequences for product recovery. Other applications of bacterial metabolic performance cover wastewater treatment.

2.2.3 Archaea – the underestimated extremists

Archaea are prokaryotes with a cell size ranging from 0.1 to 15 μm in diameter and do not contain cell nucleus or any other membrane-bound organelles. Despite their general morphological similarity to bacteria, they represent a fundamentally different biochemical and phylogenetic group compared to the bacteria kingdom and possess genes and several metabolic pathways that are more closely related to those of eukaryotes. Cell walls of archaea do not contain peptidoglycan and are instead composed of different polysaccharides, proteins, and glycoproteins. Some methanogenic archaea contain a polysaccharide (pseudomurein), which is like the peptidoglycan of the bacteria. This difference in the structure is sufficient to prevent cell wall degradation by lysozyme, which is efficiently used for bacteriolysis. However, most common archaea are enclosed only by so-called S-Layer (consisting only of protein and glycoproteins), which seems to be sufficient to prevent osmotic lysis of the cells. The name archaea means "ancient things" and points towards extreme milieus where these organisms can often be found (temperature > 80 °C, pH of 0.7, high salt concentrations, and anaerobic conditions), which is similar to conditions on the early earth. The ability to proliferate and be metabolically active under various extreme conditions makes these organisms very interesting for biotechnological applications. Overexpression of archaea enzymes, which are resistant either to heat and/or acidity and/or alkalinity in established mesophilic overproducing hosts (bacteria, yeast), *Pyrococcus furiosus*, have found many applications in modern biotechnology. Thermostable DNA polymerase (key enzyme in polymerase chain reaction) gene is originally from hyper thermophile archaea and is produced for commercial purposes in *E. coli*. Industry is using recombinant enzymes (amylases and galactosidases) from other *Pyrococcus* species for food processing at high temperatures, for example, in the production of low lactose milk.

Despite the versatile use of recombinant archaeal enzymes, the use of archaea cultures seems to be less variable. Although only a few applications of living archaea exist, they are all applied in larger dimensions. For example, the biological production of methane (biomethane production) in various anaerobic processes is performed by archaea. Although the process of anaerobic digestion is mainly driven by a versatile microbial community of facultative and obligate anaerobe bacteria, the final step (methanogenesis, paragraph on biogas) is performed exclusively by archaea (so far known). Other thermophile archaea (e.g., *Sulfolobus metallicus* or *Metallosphaera sedula*) are used industrial bioleaching processes, where metals are extracted from the ores (paragraph on bioleaching). Furthermore, industrial production of PHB, PHBV, and different exopolysaccharides can be performed with extreme halophile archaea species (Tab. 2.3).

Tab. 2.3: Archaea strains used for industrial product formation.

Scientific name	Biology	Commercial products	Comments
Haloferax mediterranei	Extreme halophilic	Exopolysaccharides poly(3-hydroxy-butyrate-co-3-hydroxy-valerate) (PHBV)	Can be cultured not only on glucose but also on extruded cornstarch, rice bran, or other hydrolyzed organic wastes
Haloarcula sp. *IRU1*	Extreme halophilic	Poly(3-hydroxybutyrate) (PHB)	Can be cultured on petrochemical wastewater, high PHB yield

2.2.4 Yeast – the working horses of bioprocesses

Baker's yeast is probably the microorganism that is most closely related to human culture serving as leavening agent in bread making and providing fermentation activity for brewing and winemaking. Yeasts grow at quite low pH values of 4–5 suppressing most of the potentially competing bacteria. This "self-sterilization" property may be one reason why humans could manage dough maturing without specific precautions. Ethanol formation during brewing also supports self-sterilization.

In contrast to bacteria, yeasts do not propagate by cell fission but by budding, an asymmetric division process. During unequal budding a "daughter cell" is formed on the "parent cell" and grows. It separates from the parent cell before it reaches the same size. In the meantime, new buds can be formed, leading to small primary aggregates. Light- and electron-microscopic pictures of unequally budding yeast are shown in Fig. 2.4.

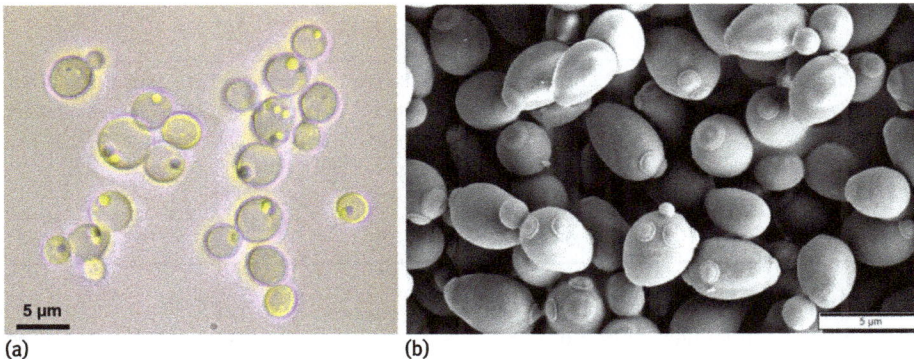

5 µm

(a)   (b)

Fig. 2.4: Examples of yeast cell appearance: (a) appearance of yeast cells (*Pichia pastoris*) under light microscope (© Viktor Klassen) and (b) culture of yeast cells (*Saccharomyces cerevisiae*) depicted by SEM. Both are proliferating by unequal budding (©Ramasamy Patchamuthu).

With the size of about 5-μm diameter, the flocs being accordingly bigger, yeast are assessable by centrifugation in technical scale. This is a good example for a process where strain selection and growth conditions have a direct influence on downstream processing.

This process is of technical relevance as parent cells exhibit a higher fermentative activity and stability in the dough, while the buds grow potentially faster during production. Filtration after yeast production and sedimentation or floatation during fermentation is largely influenced by the size of the flocs. Yeast can be regarded as microorganisms with a high degree of "domestication." For 200 years brewers and bakers have been conducting strain selection more or less intuitively to bring up process-relevant features characterizing today's production strains. A segregated view on the cell population helps understand yeast culture. This concerns not only parent and daughter cells but also cell size. Mother cells show different numbers of bud scars. Hence, we can speak in a generalized sense from cell age differently from the "immortal" bacteria.

But yeast has more abilities than making ethanol and carbon dioxide. Some other strains and their respective applications in modern bioprocess engineering are given in Tab. 2.4.

Tab. 2.4: Industrial applications of common yeast stains.

Scientific name	Biology	Commercial products	Comments
Saccharomyces cerevisiae	Unequal budding, aerobic and anaerobic	Yeast for baking and brewing	Since millennia for alcoholic beverages
Pichia pastoris (*Komagataella phaffii*)	Methylotrophic, methanol for induction	Recombinant proteins for pharma/food industry	Strong natural promotor
Ogataea angusta (*Hansenula polymorpha*)	Methylotrophic, thermotolerant (50 °C)	Enzymes (phytase, hexose oxidase, lipase) as food/beverage additive Proteins/peptides (hepatitis B surface antigen, hirudin, insulin-like growth factor for biopharmaceutical use	Tightly regulated native promoters
Candida		Single cell protein (SCP)	SCP also *Pichia*

Yeasts have been subject to one of the earliest regulations in history. In the so-called "Reinheitsgebot" (German Beer Purity Law, since 1516 Bavarian law based on earlier regulation), it was stated that, "Furthermore, we wish to emphasize that in future in all cities, markets and in the country, the only ingredients used for the brewing of beer must be Barley, Hops and Water." Note that yeast is not mentioned in this statement as the biological principle of fermentation was not known at that time. Yeasts came into the beer from the air or from impure fermentation vessels. Until now integration of bioprocesses into society means to obey or to propose regulations that influences process design. Yeast production will be further outlined in this book as a running example.

2.2.5 Filamentous fungi – the broad-spectrum chemists

Filamentous fungi being employed in bioprocesses do not grow as suspended single cells, but in long threads consisting of a chain of cells. These hyphae grow only at the tips (apices), while the elder cells do not divide any more. But they contribute to metabolism and transport the products to the growing tip. Depending on growth conditions and length of a hypha, side branches are generated, thus forming a braid which is called mycelium. While the hyphae have a diameter of typically 5 µm, the whole organism can reach macroscopic dimensions. Under natural conditions in the soil the mycelium can cover several square meters. The humongous fungus (*Armillaria solidipes*) is known to be one of the largest living organisms. In the Malheur National Forest scientists have found a single specimen covering 8.4 km^2 (3.4 square miles). It grows underground digesting on dead wood but attacks living trees too. The weight is estimated to be 660 t and the age 2,400 years. That it is indeed a single organism is proven by DNA sequencing.

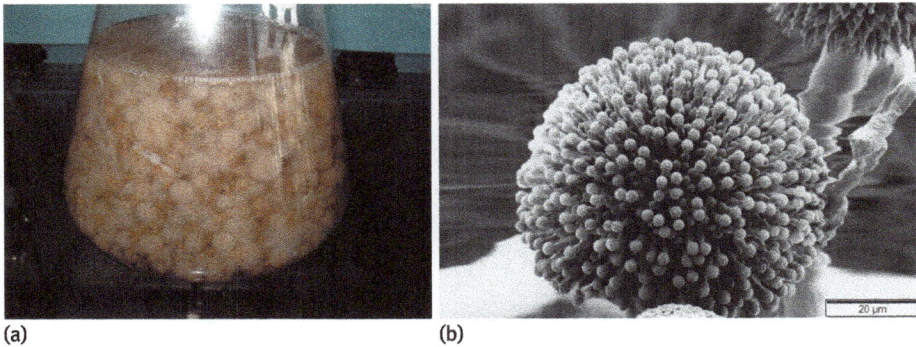

(a)   (b)

Fig. 2.5: Appearance of filamentous fungi: (a) mycelial biomass pellets of *Ganoderma lucidum* are 7–12 mm in diameter after submerged fermentation (©Marian Petre); (b) scanning electron microscope image of a fruiting body of the fungus *Aspergillus niger;* small spheres on the surface of the fruiting body are spores (reproductive cells) (©Mogana).

In bioreactors, growth habit looks different (Fig. 2.5). Growing tips experience strong shear forces bending them back to other threads or even letting them break. From the fragments new hypha can grow. Finally, a hairy ball of wool – the pellet – emerges. It can be several mm in diameter. This has severe consequences for the process. Oxygen is taken up by the outer layers of the pellet and cannot diffuse fast enough into the center. That leads to oxygen depletion and death of the affected cells. Usually, pellets are hollow. Filtration on the other hand is easier, as particles are large and the filaments quite rigid (Tab. 2.5).

Tab. 2.5: The two most technically important filamentous fungi.

Scientific name	Biology	Commercial products	Comments
Aspergillus niger	Filamentous Fungus, also known as black mold	Glucoamylase and other enzymes, citric acid, and gluconic acid	Generally recognized as safe (GRAS) Genome fully sequenced (2007)
Penicillium chrysogenum	Filamentous fungus	β-Lactam antibiotics (e.g., penicillin)	Genome fully sequenced (2008)

The discovery of penicillin by Alexander Fleming (1881–1955) is widely known and features in many history films. It is a prime example for integration of science into society. After the first observations and isolation of penicillin in 1928, it took 10 years until the world took notice of this great discovery; process development started in 1939 and penicillin was produced on a large scale, being available for soldiers in 1944. The background was the urgent need for a wound healing therapy during the Second World War. Today's discussion is still about "technology push versus market pull" or "fast applied research versus basic research."

This paragraph concludes with a quote from Vera Meyer (see Chapter 12): "Fungi form a kingdom of their own, the kingdom of Funga. An estimated six million fungi exist on earth. They range from unicellular to multicellular, from visible to invisible. And they have crazy metabolic activities that we can use in biotechnology and the bioeconomy."

2.2.6 Plant cells – slow but resourceful

Since Neolithic Age people used plants to cure diseases. From corpse discoveries it is known that moss cushions were carried during travelling, probably to put on wounds for healing because of their (of course not recognized) antimicrobial activity. Since Middle Age the use of Salicin from the bark of the willow tree (*Salix* sp.) against headache is documented – today in a chemically produced modified form as acetylsalicylic acid. The indigenous population in North America used purple coneflower *Echinacea* against inflammation of sores; nowadays extracts are commercially available in each drugstore as immune stimulant. Direct cultivation of the respective plants is subject to seasonal changes in availability and drug content. Furthermore, it is difficult to get approval as a drug for complex natural mixtures of active compounds, so pharmaceutical companies isolate active substances and produced them chemically or used them as lead structures in drug discovery.

Nevertheless, there is ongoing research to employ plant cells cultures for production as callus (unorganized proliferative mass of cells), meristem (from apical shoot) culture, or as hairy root (induced by *Agrobacterium*) in mostly heterotroph suspension or surface culture (Fig. 2.6).

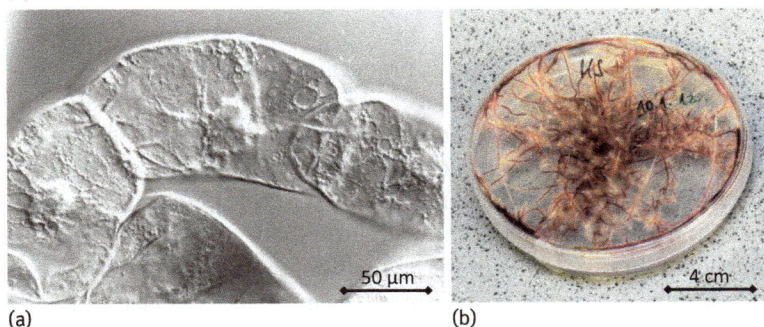

(a)   (b)

Fig. 2.6: Examples of plant cells/tissues. (a) Cells in a filament of *Echinacea purpurea* in heterotrophic suspension culture. The biggest part of the cell is filled with the vacuole. The kernel is connected with the cell wall via plasma strings. (b) Hairy roots of *Beta vulgaris* (beet) grown in a Petri dish (© E. Steingröver).

Some examples with current technical applications or systems under research are given in Tab. 2.7.

Tab. 2.6: Important plant cell-based production systems.

Scientific name	Biology	Commercial products	Comments
Taxus brevifolia	Plant cell from tejo tree	Paclitaxel anticarcinogenic	Production since 1993 in Germany, up to 75,000 L scale, 1 bill €/$ market volume (2016)
Lithospermum erythrorhizon	Purple gromwell	Shikonin, antibiotic properties (external ulcers), red color for cosmetics and textiles	Since 1983 in Japan (purpura red circle in flag)

Tab. 2.6 (continued)

Scientific name	Biology	Commercial products	Comments
Physcomitrella patens	Spreading earth moss, phototrophic cultivation	Human hormones, identical glycosilation	Research and diagnostics, first clinical phase, approval for diagnostics
Sphagnum palustre	Wet habitats, peat bogs	Decoration, gardening	Will be grown in large amounts as seed for peat bog rewetting

According to their natural function plant cells produce bioactive secondary metabolites. Although 50,000 plant species are used for medical applications worldwide, only very few plant cell applications have reached commercial scale. Some companies use special reactors, where the tissues are sprinkled with medium to produce a variety of secondary products for pharma and cosmetics. The two first examples in the table represent the main applications. The reason for limited distribution is the high cost due to low productivities. In case of *Taxus*, the agricultural production as the competing approach is obviously not feasible, what seems to be the case for *Lithospermum* as well. In other cases, the bioactive compound can be synthetized chemically.

Phototrophic cultivation can claim some advantages to produce recombinant proteins as in the case of the moss *Physcomitrella* ("pharming" = production of recombinant proteins by farming of genetically engineered terrestrial plants). Mosses are one of the oldest land plants on earth, dating back 450 million years. Cultivation is based on pure mineral medium (with plant hormones) that makes cell separation easy. The absence of organic C-sources prevents contamination. Another intrinsic product safety aspect (compared to mammalian cell culture) is that no viruses are known to be affecting plants and humans. Products from plant cell culture others than proteins are limited to compounds, which are difficult to express in other cell systems. Plants have abilities for post-translational modifications and can also excrete large proteins. Recent development is aiming at production of human identical proteins with respect to glycosylation and could therefore lead to a production platform as an alternative to animal cell culture.

2.2.7 Animal cells – masters of protein decoration

Cell culture is the process by which cells are grown under controlled conditions, generally outside of their natural environment. In practice, the term "cell culture" refers to the culturing of cells derived from multicellular eukaryotes, especially animal cells, in contrast to microbial cultures. In recent years, there has been an increase in the use of mammalian cells (Fig. 2.7) as an expression system to produce biopharmaceuticals like antibodies, vaccines, hormones, and nucleic acids. The advantages are higher quality and efficiency in the production of human or humanized glycoproteins compared with nonmammalian cells (Tab. 2.6).

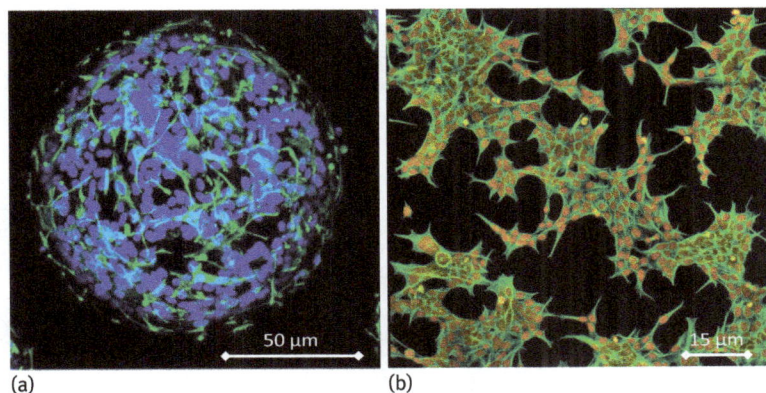

(a)   (b)

Fig. 2.7: Examples of animal cells visualized by immunofluorescence: (a) hep G2-cells, cultured in tridimensional microcavity of 3^D-KITChip (cytoceratin 18 (green), cell nuclei (blue)) (© E. Gottwald). (b) Human embryonic kidney (HEK) 293 cells (F-actin (green), 11–15 µm nuclei (red), phosphohistone H3 (yellow) (©creativecommons).

Tab. 2.7: Animal cell-based production systems.

Scientific name	Biology	Commercial products	Comments
Animal cell line Chinese hamster ovary (CHO)	Epithelial cell line derived from the Chinese hamster ovary	Therapeutic antibodies	Most important system for humanized recombinant proteins
Animal cell line hybridoma	Fusion of white blood cells (B-cells) and myeloma cells	Monoclonal antibodies (mAbs)	Permanent (immortal) cell lines due to origin from cancer cell
Insect cell line *Spodoptera frugiperda* Sf9	Isolate of ovarian tissue of fall armyworm	Various recombinant proteins	Baculovirus vectors used as transgene shuttle system

Mammalian cells have a typical diameter of 15 µm (Fig. 2.8b) – quite large compared to yeast and bacteria. They grow quite slowly (t_d about 10–20 h) and require a complex expensive growth medium. Beside sugars, they contain amino acids and animal serum prepared from blood. Mammalian cells possess a cell membrane but no rigid cell wall. That means that together with their large size, the cells are very sensitive to shear stress and cannot easily be cultivated in cell retention stirred tank reactors. HEK and CHO cell line are cultivated up to 3 m^3. In many cases bubble free aeration must be provided, see chapter aeration. For production the cells can be propagated in suspension culture but are often attached as monolayers (2D) on the surface of small spheres ($\varnothing$ 200 μm) microcarriers for high-yield culture. As technical limit up to 200 million cells/mL can be obtained; reactor sizes up to 6 m^3 are in use. For medical research cells are grown more and more in so-called 3D culture, e.g., in cavities of silicon chips (Fig. 2.8a).

2.2.8 Organized communities – making rich booty as a consortium

Not in all cases bioprocesses work with monoseptic cultures containing only one species. In nature that is a great exception. In biogas plants carbohydrates are hydrolyzed to sugars, degraded to carbonic acids, and then further metabolized to hydrogen and methane. Each of these steps is carried out by several different microorganisms specialized to do a few specific reactions. In this way the different strains support each other; the community is much more efficient than only one strain. Cutting off one brunch of the degradation pattern has turned out to be a problem, e.g., to get pure hydrogen. Such communities have, for the time being, not been fully understood, despite modern molecular methods are applied (Fig. 2.8).

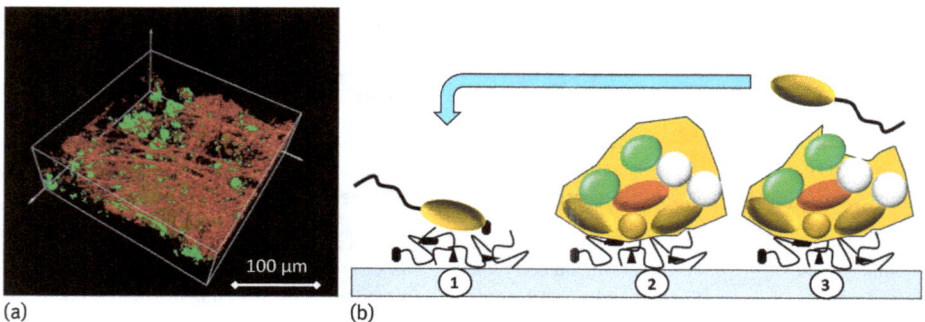

(a) (b)

Fig. 2.8: Visualization of biofilm structures: (a) 3D imaging of Confocal Laser Microscopic Characterization data presented as isosurface projection. Color allocation: red: nucleic acids, green: lectin-specific extracellular polymeric substance (EPS) glycoconjugates (©M. Wagner). (b) Simplified diagram with three stages during biofilm formation on surface, attachment by ambient molecules (1), proliferation/EPS synthesis (2), senescence/detachment (3) (©Uschi Obst).

While in bioprocesses ideally mixed cell suspensions are usually applied, in nature many microorganisms live attached to surfaces. The cells cover themselves with an extracellular matrix consisting of self-produced extracellular polymeric substances (EPSs), consisting mainly of polysaccharides and proteins as well as nucleic acids and lipopoly-saccharides in minor concentrations. This structured community finally forms the bio-film. The EPS protects the cells from drying out or from toxic substances. Even in environments with extremely low substrate concentration like in drinking water tubes the microorganisms can survive without being washed out. Inside a biofilm sometimes much higher volumetric productivities than in a bioreactor can be observed. Biofilms are much more than an ensemble of cells; a lot of chemical communication is going on orga-nizing the structure and cooperation between the different organisms. This encouraged researchers to employ biofilms in technical system. While in wastewater that happens naturally, artificially supported biofilms for production are also under investigation. Sub-strate can be provided on the biofilm side in direct contact or by diffusion through a membrane on which the film is attached. Particularly, cell retention in continuous culti-vations is one specific advantage, making a separation unit superfluous.

2.3 Cultivation conditions – environmental parameters to take care about

Across all biological groups microorganisms make high demands on chemical and physical parameters in their environment. Besides chemicals there are three other en-vironmental factors affecting microbial growth in particular: temperature, pH value, and water availability. Appropriate environmental conditions have to be guaranteed in in the bioreactor. Shifts in the environmental conditions can be an additional de-gree of freedom in process design to influence growth and product formation inten-tionally. These environmental factors are listed in the next paragraphs.

2.3.1 Primary nutritional groups

A commonly applied classification of primary nutritional groups is depicted in Fig. 2.9. The three basic criteria are the energy source, the redox source, and the carbon source of the organism. These three qualities determine later the design of a process, its efficiency, and its intensity. The most widespread processes are heterotrophic processes ("hetero-"). An organic molecule ("organo-") from the environment is taken up and used both as a C source and as an energy source. To do this, some of the substrate molecules must be transferred to an electron acceptor. This redox reaction provides the energy whose flow is then closely coupled to the redox flow. In aerobic processes, the electron acceptor is oxygen. Such aerobic processes with, e.g., glucose as carbon and energy source make ac-tive aeration of the culture necessary but lead to the most intensive bioprocesses with respect to volumetric productivity of biomass and products. About 50% of the carbon from glucose is allocated for biomass.

If no O_2 is available, many microorganisms can also use already formed metabolites as electron acceptors. However, the product must then be excreted, which means a loss of carbon. Such anaerobic processes are usually quite simple and lead only to low biomass production containing about 5% of the carbon of glucose consumed. Most of the carbon up to 95% is directed into the reduced intermediate, which is typically the desired end-product. We meet such processes in ethanol fermentation or biogas production (Chapter 9, Sections 9.2 and 9.4). Note that the term "fermentation" should only be used in connection with anaerobic bioprocesses including microbial or enzymatic food processing. Strict anaerobic bacteria need removal of oxygen from the feed medium. Facultative anaerobes can at least remove the oxygen by themselves. But also, microaerobic conditions can be applied, where oxygen respiration covers the maintenance energy demand of the cells, but product formation occurs via anaerobic metabolism.

Lithotrophic processes ("troph-", Greek trophe = nutrition) include nitrification or denitrification in wastewater treatment or bioleaching (see Section 9.5). The lithotrophic microorganisms get their energy from a redox reaction using an inorganic electron-donor and -acceptor ("litho-", Greek lithos = stone). The oxidized molecules can usually not be incorporated into biomass and is excreted thus contributing to geological processes. Apart from those mentioned, such processes are rather rare. However, at the beginning of evolution there were only chemolithotrophic organisms. Only later did heterotrophic organisms appear on the scene. Aerobic microorganisms only appeared long after the "invention" of photosynthesis (3 trillion years ago). However, many metals close to the surface oxidized, which is why the chemolithotrophic microorganisms were displaced into niches (e.g., black smokers, Fe-, Mn-, or S-containing springs from greater depths). However, heterotrophic organisms also carry out chemical conversions of mineral components (e.g., limestone) and thus continue to contribute to the geological formation of the earth's crust. Finally, the metabolic pattern, where the cell uses organic substrates but inorganic electron acceptors as, e.g., nitrate, is called anaerobic respiration.

In the case of the trophic groups mentioned so far, the gain of energy, electron acceptors, and carbon are always linked to the extraction of substances from the environment and the production of residual substances. But there is also energy gain from the nonmaterial source "light." Photosynthesis is largely carried out by higher plants and algae. These are the primary producers of organic substances on which all other organisms depend. They also complete the carbon and oxygen cycle, at least roughly. While mankind has been practicing targeted agriculture for around 10,000 years, the technical use of macro- and microalgae is only just beginning in terms of quantity. Absorption of photons ("photo-," Greek phos = light) is the primary energy source of phototrophs. But still, there must be an electron acceptor. This is intracellularly produced by water splitting at a catalytic center supplied with photon energy. The generated electrons and protons (H^+) stay inside the cell, while oxygen is excreted. Photobiotechnological processes using microalgae and cyanobacteria are under bioprocess engineering research.

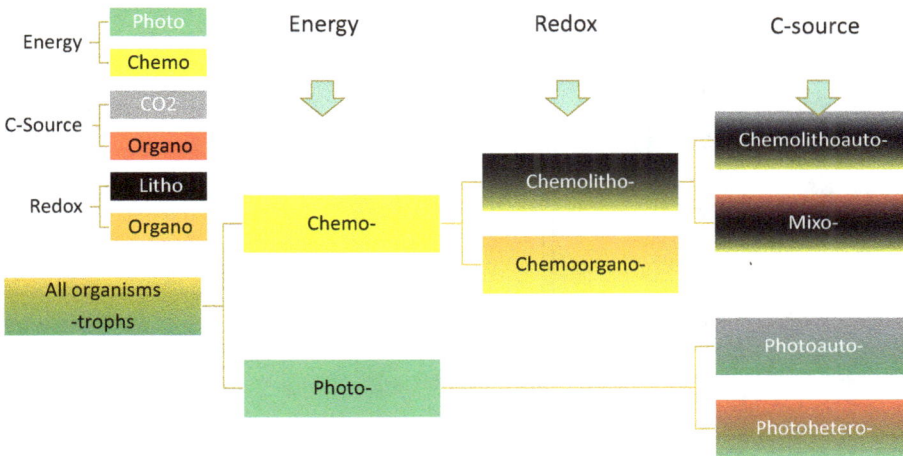

Fig. 2.9: Primary nutritional groups; the colors underpin the different aspects.

The decoupling of energy gain from carbon flux enables microalgae to use CO_2 as the sole carbon source. That makes them highly interesting for decarbonization of our economy. From a process engineering view, only water, light, and CO_2 must be provided. A comparable situation is possible, when electrons are provided and processed by the cell to fit the redox demand without metabolizing chemical compounds. The electrons are provided via electrodes and generated from sustainable sources like photovoltaic. This is the field of electro-biotechnology. It is a combination of electrochemical and biotechnological processes (Fig. 2.10). Applications range from biofuel cells for wastewater treatment, to the regeneration of cofactors in enzymatic processes, up to bioelectrical syntheses with whole cells.

The consequences of nutritional groups for bioprocess design are further discussed in Chapter 4, and in numerous examples throughout the book.

2.3.2 Temperature

Temperature dependence of chemical reaction constants k_{Comp} (mol/s) are usually approximated by the Arrhenius equation (2.1) as given by Svante Arrhenius in 1889:

$$k_{Comp} = A_{Comp} \cdot e^{\frac{-E_{a,Comp}}{RT}} \tag{2.1}$$

The two parameters A (mol/s) being a pre-factor and E_a (kJ/mol) the activation energy have to be determined experimentally for different enzymes or microorganisms. As a rule of thumb, increase of temperature by 10 °C leads to increase of the reaction rate by a factor of 2–4. That underlines the importance of a good temperature control. Physical pro-

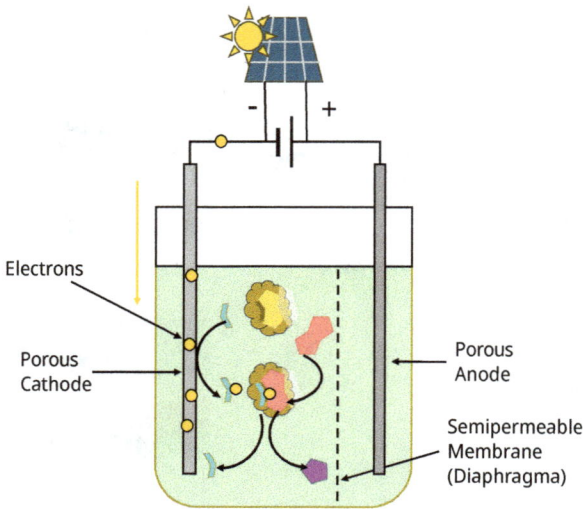

Fig. 2.10: Principle of a process in electro-biotechnology; electrons from a sustainable source regenerates the cofactor of an enzymatic step.

cesses like diffusion exhibit much weaker temperature dependencies (e.g., 20%/10 °C). In microorganisms, hundreds of chemical reactions occur in parallel coupled by transport steps like diffusion. In principle, ach single reaction step making up observable enzyme kinetics (see Chapter 4) has different values for E_a. Nevertheless, the Arrhenius equation fits quite well to describe temperature-dependent growth rates but only for temperatures below a distinct value. Above this temperature optimum reactions, rates of enzymes or growth rates of microorganisms decrease. Notwithstanding from chemical systems, in biochemical systems, the general assumptions of the Arrhenius equations being temperature-dependent collision and assignment of an activation energy to molecular species are not entirely true. Macromolecules show increasing mobility of single residues; enzymes may partially lose their spatial structure leading to decreased substrate affinity. On the cell level membrane fluidity can change or regulation of single pathways for adaption may occur.

To summarize these effects different mathematical correlations to describe temperature response have been proposed. They relate mainly to a maximum growth rate (see Chapter 4 for definition) assuming that all other influences are in a nonlimiting and noninhibiting range. The most commonly used approach is the assumption of a decay rate in which the rate constant increases with temperature following again an Arrhenius equation:

$$\mu(T)_{max} = \frac{A_{react} \cdot e^{\frac{-E_{react}}{RT}}}{1 + A_{decay} \cdot e^{\frac{-\Delta G_{decay}}{RT}}} \tag{2.2}$$

This approximation has a defined optimum (Fig. 2.11) and decays (ΔG = free enthalpy) in the direction of decreasing and increasing temperatures. The four unknown parameters have to be estimated from data. Even knowing that such a correlation is not clear with regard to reversible/irreversible effects and timely aspects for duration of temperature impact, it is practically applicable at least for moderate temperature changes in a bioprocess. In the chapter on sterilization, we will come back to the decay aspect. Other correlations claiming better intuitive access correlate cardinal growth temperatures (T_{min}, T_{opt}, T_{max}) by a smooth curve. The assumption of a temperature range outside of which no growth is possible may be more realistic than the Arrhenius formula. An example is given as follows:

$$\mu(T) = \mu_{max} \cdot \left(\frac{T_{max} - T}{T_{max} - T_{opt}}\right) \cdot \left(\frac{T - T_{min}}{T_{opt} - T_{min}}\right)^{\left(\frac{T_{opt} - T_{min}}{T_{max} - T_{opt}}\right)} \tag{2.3}$$

Besides the three cardinal temperature values, the maximum specific growth as fourth parameter has to be determined from data.

Due to the immense importance of temperature for the physiology of microorganisms and for the technical applications as well, microorganisms are classified according to their optimum temperature as listed in Tab. 2.7.

The Arrhenius equation is specific for each given strain out of this classification. Nevertheless, optimum growth rates are usually lower for extremophiles. Consequently, it is sensible that often enzymes of extremophiles are cloned into mesophiles for production. In other cases, like food industry, the respective strains are used directly. Note that temperature control is not a technical problem as such but optimum temperature for a specific strain/medium combination may not exactly be known. Temperature shifts are on the other hand applied to induce changes in the microbial metabolism. Microorganisms react to overheating by production of heat shock proteins. This is applied for production of recombinant proteins in *E. coli*, where a temperature shift induces the respective promotor. Other cells can show changes in their lipid profiles or show other potentially useful reactions.

An aspect for integration of downstream and bioreaction is to strip the product like ethanol directly during the cultivation, thus saving energy for distillation. In this case it turned out that ethanol and temperature have a negative impact on the cell wall, which made the process not feasible.

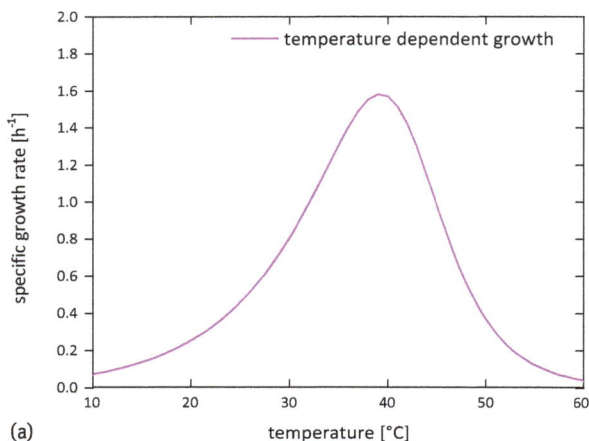

(a)

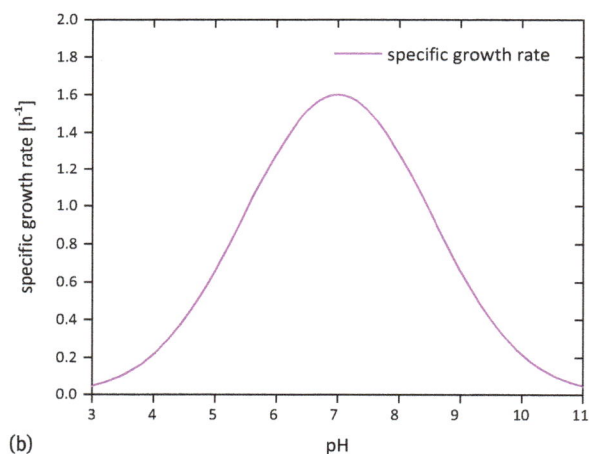

(b)

Fig. 2.11: Growth curve for *E. coli* as function of primary growth factors: (a) temperature impact according to the Arrhenius formula (eq. (2.3), $A = 6.4 \cdot 10^{14}$, $B = 1.38 \cdot 10^{48}$, $E_{react} = 86.4$, $\Delta G_{decay} = 290$). The decline from the optimum to higher temperatures is usually steeper than the one toward lower temperatures, so T_{max} is closer to T_{opt} than $T_{min;}$; (b) pH impact according to eq. (2.4) is assumed to be symmetrical ($pH_{opt} = 7.0$, $\sigma_{pH} = 1.5$).

2.3.3 Acidity and basicity

Each species shows its optimum growth at a specific pH value. Deviation from this optimum will lead to remarkable decrease in growth velocity. Technically speaking it is not a big problem to adjust the pH value in the fresh medium before it is used. During the cultivation pH shifts can be compensated by acid or base. Nevertheless, in many cases it is not reliable to do so, for example, if an organic acid is the product (Tab. 2.8).

Tab. 2.8: Definition and occurrence of habitats with different temperature levels.

Temperature T_{Opt} (°C)					
Range	−15–+15	10–30	20–45	45–80	80–113
Designation –philes (Greek = affection)	Psychro-cryo-	Psychro-tolerant	Meso-	Thermo-	Hyperthermo-
Occurrence	Ice, meltwater		Intestine animals	Warm lakes	Hot springs, volcano lakes
Remarks	≤0 °C high salt content		Standard in bioreactors	Hot water pipes	No organics present
Application (examples)	Food maturing	Volatile products	Standard production	"Thermo-zymes" protease	*Thermus aquaticus* Taq polymerase

Tab. 2.9: Designation and occurrence of habitats with different pH values.

pH [–]			
Range	< 4	6–7	> 8
Designation, – philes	Acido-	Neutro-	Alkali-
Occurrence	Geothermal areas (yellowstone), metal mining sites	Soil, freshwater habitats, and oceans	Soda lakes (Na_2CO_3), high Ca^{2+} groundwater
Remarks	Suppressing of contaminants	High organism diversity	Suppressing of contaminants
Application	Bioleaching	Standard cultivations	Enzymes, e.g., paper industry, and detergents

Like in the case of temperature it is not simply possible to find mechanistic kinetics for the impact of pH on enzymes and microorganisms. Degree of dissociation of residues (e.g., COO^- and NH_3^+) and its effect on surface charge would be only one item. In the case of enzymes, the isoelectric point (pI value) is the measurable expression of pH influences on the surface charge. Specific action on the active site alters the affinity constants. To have at least a preliminary anticipation of the impact of pH changes the correlation:

$$\mu(\mathrm{pH}) = \mu_{max} - \frac{1}{2} \cdot \left(\frac{\mathrm{pH_{opt}} - \mathrm{pH}}{\sigma \mathrm{pH}} \right)^2 \tag{2.4}$$

can be used. We see immediately in Fig. 2.11 that the representation is formally equal to a normal distribution (Gauss) where pH_{opt} is the optimum pH value and d_{pH} the standard deviation with d_{pH} being often in the range of 1.05–1.5 pH units for growth and enzyme activity. Interestingly a symmetric approach is feasible. Like in statistics, in summaries many different influences may be unknown. Validity of such models should not be overstressed but could be in the range of $\pm 2d_{pH}$ according to published data sets. For process development more specific but unfortunately also formal kinetics have to be found from case to case where relevant (Tab. 2.9).

2.3.4 Osmolality

High concentrations of toxic compounds are a potential problem. High concentrations of substrates or essential salts can be a problem. Like high salt concentrations also high concentrations of organic carbon sources cause high osmotic pressures. Osmophilic organisms can live in such environments because they protect themselves by the synthesis of osmo-protectants. These small soluble molecules let the intracellular osmotic pressure increase, thus preventing water loss by osmosis. For definitions see Tab. 2.10.

Tab. 2.10: Designation and occurrence of habitats with different salt concentrations.

Salt concentration as NaCl (mol/L)				
Range	0–1	0.2–2.0	0.4–3.5	2.0–5.2
Designation – philes	Nonhalo-	Slight halo-	Moderate halo-	Extreme halo-
Occurrence	Technical media	Seawater	Lake Qarun, Little Manitou Lake, Lake Urmia	Hyper saline lakes, Dead Sea, Lake Assal, Don Juan Pond
Remarks	Normal case	Marine aquaculture (algae)	β-Carotene production (*Dunaliella salina*)	Habitats for halophile organisms

Mainly yeasts but also some bacteria can stand high osmotic pressures (Tab. 2.11). In technical terms this is quantified by water activity a_w, the vapor pressure of the solution divided by the standard partial vapor pressure of water. The osmotic concentration is the molar concentration of molecules contributing to osmotic pressure, which are ions and all dissolved organic molecules.

Approximation of the impact of water activity on growth rate can be done pH by the Gaussian function similarly as for pH. It is superfluous to note that the mentioned environmental parameters affect the cells in mutual interactions. Some principles on

Tab. 2.11: Classification of microorganisms with respect to osmotic concentration.

Osmotic concentration (mol/kg)	0–1.0	1.0–2.0	2.0–4.0	4.0–8.0	>8.0
Water activity a_w (–)	1–0.98	0.98–0.96	0.96–0.92	0.92–0.84	<0.84
Environment	Normal medium	Seawater	Molasses	Jam	Dried fruit
Organisms	Most bacteria	Most bacteria	Osmotolerant, *Lactobacillus*	Baker's Yeast, *Zygosaccharomyces*	*Saccharomyces rouxii*

how to handle such cases will become clear in the case studies; others are given in the additional references.

2.4 How the choice of a microorganism influences the whole process

The microorganism is the soul of the bioprocess. The selection of the biological group of organisms or the strain has an impact not only on the bioreaction, the product quality, and the productivity, but also on all stages of the process including upstream and downstream. Likewise, all indicators for performance are influenced by the particular strain.

2.4.1 Criteria for selection – going through main aspects of process design

First and foremost, each product has only a limited set of possible microbial producers associated with it. Thus, simple molecules such as solvents are produced with anaerobic microorganisms, firstly because they can do it and secondly because they have a high energetic efficiency and a high carbon yield. As an example, see the ethanol production in Chapter 6. Having a specific catabolic pathway, *Zymomonas mobils* shows low biomass yield and therefore higher product yield, but high turnover rates. Both properties speak in favor of its use. Unfortunately, the pH of 7 makes it too easy for foreign organisms to contaminate the culture. The costs for axenic operation are not paid back by the advantages of the bacterium.

More complex molecules like antibiotics, antioxidants, enzymes are usually produced with aerobic systems. Since the molecules are synthesized in anabolism, the cell needs a high energy flow, which is only guaranteed in aerobic cells. The carbon yield is not so important in view of the proceeds for the product. This energy flow is often the actual limiting element of a production system. Simply increasing the copy number of the recombinant genes would have no effect, because it is on cost of the production machine as such, the other proteins. Verify quantitative aspects in Chapters 4 and 5. However, aerobic processes are technically more complex. Production processes with animal cells are the most complex; they have high media costs and the lowest volumetric productivity values. This is justified by the complexity of the products, e.g., glycosylated pharmaceutical proteins (not possible with bacteria, and the level of their benefit. For example, a single dose of an injection against the consequences of a heart attack can cost up to €/$1,000. In order to keep costs low nonetheless, process intensification is the most important criterion for process quality in the industrial environment. This includes in particular high volumetric productivities and high yields.

Every bioprocess (unfortunately) needs a justification why the product cannot be produced more simply, e.g., chemically. Direct production with plants ("pharming") is not excluded either. Chiral products are often more complex to synthesize chemically than biologically. A unique selling point for bioprocesses is therefore chirality or the ordered monomer sequence in polymers. Examples are Vitamin C for chirality (Chapter 9) or polysaccharides and proteins for their sequence. May be, it is a matter of time, when cell-free protein synthesis (of course with biological catalysts) will be operable. Other examples are carotenoids or related antioxidants, which can be produced both chemically and with the help of microorganisms, e.g., microalgae. It is a long debate whether individual chiral differences between the engineered and the natural product play any role at all for animals or for humans.

While genetic engineering methods are accepted for biotechnological production in the pharmaceutical sector, there are still reservations in the case of foodstuffs. The use of genetically engineered microorganisms to produce vitamins or organic acids is also no longer discussed. Nevertheless, only traditional microorganisms such as yeast or *Lactobacillus* are considered for the direct production or modification of food. New ingredients (after 1995) are subject to the Novel Food Regulation, which limits the strain selection. In any case, a main need to be guaranteed by the product of the bioprocess is health. With respect to social perception, health and safety aspects are most important.

Microorganisms – even from the same group – differ in their ability to utilize different media. In terms of sustainability, alternatives to glucose are being sought, particularly for residual materials. The focus is on glycerin (from biodiesel production) or starch residues, for example. One example is the fermentative production of 1,3-propanediol with *Clostridium*. Bioconversion is necessary to obtain only the 1,3-isomer. Sometimes media tend to foam. This can be countered with anti-foaming agents. However, these later block the membranes for filtration. This means you have to use a simpler medium and a different microorganism. This is just one example of many different things that can happen and must be considered for process robustness.

There are also new ideas. For example, thermotolerant microorganisms could make cooling the reactor unnecessary and enable continuous harvesting through evaporation. In this sense, strain development or selection proceeds along a domestication idea. Many more hints are included in other examples described in this book and elsewhere.

2.4.2 The good, the bad, and the ugly – microorganisms as products

Now we have nearly all intellectual means to start a bioprocess and can start thinking about products. The best unique selling point is microorganisms as there is no other technical process to deliver them for different purposes and different markets. Due to their ability for autocatalysis, they can be the commercial products by themselves. This may happen on a small scale when a lab purchases interesting strains from a strain bank or on a large scale for industrial application. Different application areas are listed in Tab. 2.12.

Tab. 2.12: Applications for which living microorganisms and cell lines are traded as products.

Purpose of employment	Organisms	Remarks
Samples from strain banks	Nearly all	
Starter culture	Yeasts, *Lactobacillus*	
Protein source, food, and feed	Yeasts, microalgae	
Agrochemicals	*Bacillus thuringiensis*	
Technical application, buildings	Fungus mycelium	Fig. 12.18
Plant propagation	Plant cells, e.g., cyclamens	
Cell based therapy	Animal cells	
Bioweapons	Anthrax	

Food and feed processing is a classic field where microorganisms are employed. The bacteria, yeasts or fungi are customarily not propagated directly in the food factory but are purchased from professional suppliers. The most popular example is bakeries, where the yeast is bought from yeast factories. The same holds for vintners and breweries. Similarly, lactobacilli (and others) are necessary for preparation of dairy products like yogurt, cheese, sourdough, or fermented vegetables. The inoculum for starting the food fermentation process is called the "starter culture" (fermentation starter, in some branches "mother"). Some of them like fungi to prepare Kombucha tea are even available for private application. To guarantee a high level of reproducibility and quality the food manufacturer relies on the experience of the deliverer of the starter culture with respect to genetic stability and physiological activity. The microorganisms are not only active as biocatalysts in the food but contribute also to taste, smell, or nutritional value.

Apart from a food matrix, microorganisms are used directly as food or food supplements. Since the 1960s single cell protein (SCP) from edible microorganisms was assumed to contribute substantially to food and feed supply for mankind. Feed and food yeasts (*Candida, Pichia*) were the focus. As substrate hydrocarbons or methanol were foreseen, as fossil oil reserves seemed to be exhaustless. Increasing oil prices and other economic and political constraints made this development invalid. The trend of producing energy from biomass with the aid of microorganisms shows that value-added chains can even be turned to the opposite direction, depending on market needs and available resources. Nevertheless, the best example for a viable SCP process for animal feed (ICI Pruteen) led to the development of continuous culture in large scales of 1,500 m³. There are ongoing activities to produce high value proteins in the form of edible microorganisms from sugar, lignocellulosic materials, or other agricultural residues. The main example for human novel food is "Quorn" (after the Leicestershire village of Quorn) with myco-protein as its main ingredient. It is produced by the *Fusarium venenatum* fungus in a continuous fermentation process. While Quorn was developed in the 1980s and has been available as retail product since 1993 the current trend towards vegetarian diets supports its prevalence (market value of over €/$100 million) as it is a perfect substitute for meat.

Treatment of food by microorganisms or producing microorganisms for food on agricultural products is highly welcome as a refinement process. Unfortunately, such processes are competitors for arable land in the bioenergy/food nexus. As food production seems to be limited and fossil carbon sources are even to be exhausted soon, production of microorganisms based on organic substrates to contribute to the exponentially growing human protein demand is hopeless. A way out of this dilemma is to consider sun energy, which is technically used to produce electrical energy and consecutively hydrogen. This route of energy production is meant to be available in excess in the future. On the chemical side, synthesis of fuels from hydrogen and carbon dioxide is called "power-to-fuel" technology. Consequently, on the biotechnological side, growing microorganisms assimilating H_2 and CO_2 to produce CH_4 following this route is a recent challenge. This is an example that others besides oxygen and carbon dioxide can serve as educt and product. In the next step, conversion of the gaseous substrate methane to protein by microorganisms may be a viable route to help ensure global food security (e.g., by company Calysta). Note, that this route is exactly diametrically opposed to the attempts to make electrical energy or hydrogen by microorganisms from organic matter.

The examples given above are meant to serve as beneficial processes fulfilling human needs, but humans have also learned to misuse microorganisms as weapons aimed at each other. The ability of microorganisms to propagate and produce extremely strong toxins is the kinetic basis of many fatal diseases. This means that since antiquity people have used microorganisms as weapons even without knowing the biological background. Even 3,000 years ago, the Hittites employed infected cattle to sap the food base of their enemies. Ancient people contaminated the drinking fountains of their enemies with decaying corpses. In the Middle Ages infested bodies killed by the plague were thrown over city walls. During the settlement of Europeans in America the indigenous population was decimated by smallpox epidemics. In North America some cases are reported where European solders are accused of infecting native tribes deliberately with smallpox, e.g., by giving them things like infested blankets. Since the nineteenth century, when biotechnology was put on a rational basis, not only the fight against diseases but also the development of biological weapons was the focus of research driven by political demands.

Bioweapons have the potential to kill people but to leave material values like buildings unaffected. Botulinum toxins, known from pharmaceutical cosmetics (Chapter 1), have been produced as bioweapons in several countries starting before the Second World War. It is quite unstable under outdoor conditions, so that an affected area can be entered just after two days. A recent study said that 1 g toxin distributed by infested packaged milk could potentially kill 100,000 people, mainly children. In contrast, in the twenty-first century research has begun around "nondeadly" weapons. The idea is to destroy specific materials, e.g., by microbes degrading fuel reserves. Plant and animal toxins are also under inspection by militaries. The military logic here is that such weapons are currently not covered by the Biological and Toxin Weapons Convention (BTWC). On the other hand, they have some inherent limitations making them incalculable. This has prevented excessive use so far. Spreading of an epidemic cannot be strictly limited to a given area once it breaks out. This causes potential danger for the instigator's own people. The development of specific sensors detecting the toxic compounds under such scenarios is the next logical step in the arms race. The onset of the impact will take some time, which enables the opponents of war to take counter measures. Furthermore, it is regarded that infected soldiers already facing death are specifically dangerous. The basic question of how the microorganisms can distinguish between their own population and the enemy's is treated on the genetic level. Specifically designed bacteria, so called target-delivery-systems, are conceivable, which attack only people carrying specific genes. The attacker's own people could be protected by specific antibiotics. Such ethnic weapons are discussed as the most appalling scenario. It is horrible to imagine bioweapons in the hand of terrorists (bioterrorists) who do not care at all for collateral damages or ethics.

The Center for Disease Control and Prevention (CDC) published a list (out of a potential 200) of the "dirty dozen" biological warfare agents (Tab. 2.13). These bacteria, viruses, or toxins are especially dangerous because of their high lethality in conjunction with easy transmission and spread.

It is highly desirable that engineers working on bioprocesses are aware of these potential dangers for mankind and actively take part in the political discussions to ban such horrible developments.

Tab. 2.13: List of biological warfare agents.

Pathogen		Disease	Release	Incubation
Bacillus anthracis	Bacterium	Inhalation anthrax	Spores	Aerosols
Yersinia pestis	Bacterium	Pneumonic plague	Vegetative cells	Aerosols, person to person
Francisella tularensis	Bacterium	Tularemia (rabbit fever)	Vegetative cells	Aerosols
Brucella suis	Bacterium	Brucellosis	Vegetative cells	Aerosols
Coxiella burnetii	Bacterium	Q fever	Vegetative cells	Aerosols
Burkholderia mallei	Bacterium	Glanders	Vegetative cells	Aerosols
Variolavirus	Virus	Smallpox	Virus	Aerosols, person to person
VEE-virus	Virus	Venezuelan equine encephalitis	Virus	Aerosols
Marburg-virus	Virus	Marburg fever	Virus	Aerosols, person to person
Clostridium botulinum	Bacterium	Botulism	Toxin (protein) (known from cosmetics)	Ingestion food/water
Ricin	Plant	Ricin intoxication	Toxin (protein)	Food/water
Staphylococcus aureus	Bacterium	SEB intoxication	Toxin (known from food poisoning)	Food/water
Trichothecene	Fungus	Trichothecene	Toxin	Food/water

2.5 Running example: microalgae – the solar cell factory

Algae provide 50% of the carbon dioxide assimilation and 50% of the oxygen production on our planet. Furthermore, they are the basis of the food chain/web in the waters and seas. This is reason enough to take a closer look at these photosynthetic creatures that are essential for our survival. This is a good reason to take a closer look at these organisms, which are indispensable for our survival. The public perception of these phototrophic organisms and their role in biotechnological research falls far behind that of their terrestrial relatives. We know the use of spirulina since ancient times and from pre-Columbian Central America. In Asia, macroalgae and microalgae are also traditionally used as food. In Europe, especially Scotland and Normandy are traditional users of macroalgae from the shelf as fuel or as fertilizer.

The use of microalgae in the sense of cultivation and production in a controlled environment started in appreciable quantities in the 1970s. Powders or tablets from spirulina and chlorella as food supplement have been the main products. Exaggerated expectations toward bulk foods or biofuels followed by disappointments have led to several waves in the activity of science in recent decades. However, microalgae production has reached a remarkable market level and are recognized as "biomass 3.0" after freshly grown terrestrial crop plants as the primary source for feed and energy and use of treated residual biomass as the second option. Like biomass from higher plants, microalgae biomass can be processed to different products in different fields besides food and feed also bioplastics. The scheme of a typical phototrophic (green algae) cell is shown in Fig. 2.12.

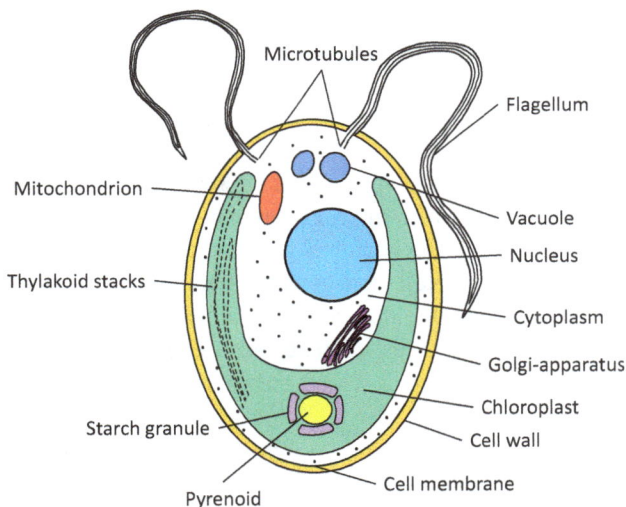

Fig. 2.12: Morphological structure and principal cell organelles of a eukaryotic algal cell; the cup-shaped chloroplast is dominant with the thylakoids, where the actual photosynthesis takes place.

The most obvious difference from the heterotrophic cell is the large chloroplast in which light is captured. Then its energy is metabolically harnessed along an electron chain and finally used to bind CO_2 to glucose.

2.5.1 Choosing from diversity – widely employed microalgae for commercial bioprocesses

Microalgae are unicellular photosynthetic organisms without roots, stems or leaves (Latin "alga" = beach, seaweed). These photosynthetic microorganisms are ubiquitous (aquatic, salt, fresh water and/or terrestrial). However, individual species occupy specific habitats so that some of them are motile like animals, whereas others are suspended in water bodies or located on surfaces. This evolutionary specialization results in further morphological diversification of microalgae strains, which may be unicellular, multicellular, round, oval, or filamentous. Microalgae are not a defined taxonomic group but belong to several taxa. The term "microalga" describes more a habit like bush or tree than a phylogenetically uniform classification. Major groups like the prokaryotic cyanobacteria ("blue-green algae") and the eukaryotic microalgae, where the commonly known classes Chlorophyceae ("green algae"), Porphyridiacea ("red algae"), and Bacillariophyceae (e.g., diatoms) have to be distinguished. Actually, these classes have a basically different cell morphology, based on which they are classified in domains. Some of them have gone through endosymbiosis even two or more times. While cyanobacteria (blue-green algae) have a cell structure very similar to the prokaryotic bacteria, the eukaryotic microalgae cells are comparable to other unicellular microorganisms but with one or more chloroplasts. Independent from cyanobacterial or eukaryotic origin and multi- or single-cell structure, the size of individual cells is usually in the range of 3–20 μm.

Considering the large biological diversity and the new developments in the field of genetic engineering, microalgae are foreseen as one of the most promising sources for new products and solutions. Between 200,000 and several million species of microalgae are expected to exist on Earth, from which approximately between 40,000 and 60,000 species have been identified so far. The chemical composition of only a few hundred species has been investigated in detail, and nowadays less than 15 strains are used for cultivation on an industrial scale. Biological fundamentals of the most relevant microalgae used for commercial purposes will be briefly described in the following paragraphs showing/explaining the reasons why these microalgae can be successfully cultivated on a commercial scale.

Cyanobacteria represent the oldest algae (at least 3.8 billion years old), and thus they are presumably responsible for oxygen evolution on Earth. These prokaryotes with gram-negative cell walls are extraordinarily robust and efficient photosynthetic organisms. Nevertheless, cyanobacteria are very diverse, so for example some species produce toxins that can taint and poison drinking water (e.g., *microcystis*). Others are harmless or even ecologically indispensable (e.g., oceanic picoplankton *Prochlorococcus* is the most abundant organism on the planet). Industrially, the most important and well-known species are *Arthrospira* (spirulina), *Synechocooccus*, and *Synechocystis*. *Nostoc*, visible as a slime carpet, has lately gained increasing interest.

"Spirulina" (Fig. 2.13) is commonly used as a commercial name for the two industrial relevant strains *Arthrospira platensis* and *Arthrospira maxima*, which are both multicellular, filamentous cyanobacteria (blue-green algae) composed of cylindrical cells arranged in helicoidal trichome helices. *Arthrospira* has a very broad physiological plasticity when it comes to medium/water salinity conditions, so it can live in freshwater (salinity <2.5 g/L) but is the dominant flora in water bodies with high carbonate and bicarbonate alkalinity and high pH (9–10.5). Generally, temperatures in the range of 35–38 °C are regarded as optimal for growth and photosynthetic perfor-

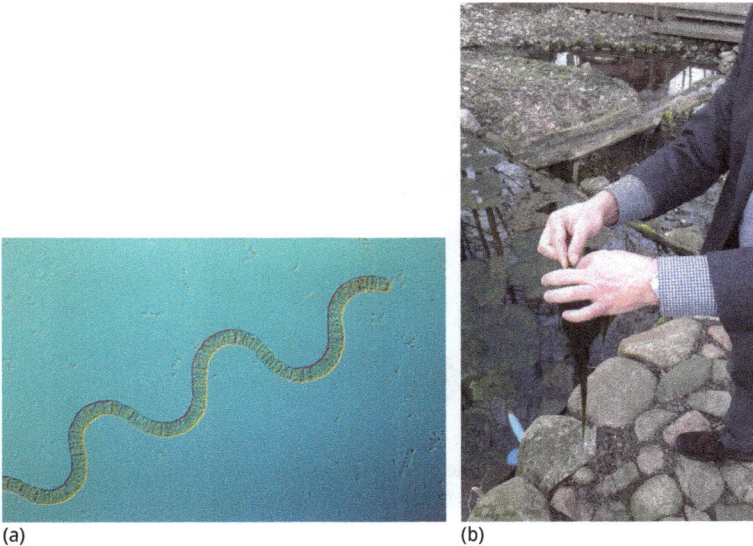

(a)   (b)

Fig. 2.13: Different appearances of *Arthrospira*: (a) *spirulina* picture under light microscope
(©Viktor Klassen); filamentous mat during manual sample taking by Prof. Kruse.

mance of spirulina. Its ability to grow rapidly under natural conditions make it a
promising source of food and food supplements. The natural blue colorant phycocyanin, along with other biomass components, is an important product of spirulina. This
and the fact that spirulina has an easily digestible cell wall, have led to the rapid development of industrial plants for large scale biomass production.

From the wide variety of eukaryotic microalgae, biotechnology currently focuses
on the green algae (Chlorophyta), red algae, and brown algae as well as some diatoms.
The green microalgae are phylogenetically the closest to the higher plants and include
a wide range of organisms with very different morphologies ranging from micro- to
macroscopic forms. The robust cell wall of most green algae improves the mechanical
stability of the cell, while reducing the shear sensitivity to mixing or centrifugation,
but also complicates the purification of the inner cell products and digestibility when
used as food. *Chlorella vulgaris* is a unicellular freshwater microalgae, naturally living
in both aquatic and terrestrial habitats. *C. vulgaris* cells are round, nonmotile, and
comparably small with the size ranging from 4–10 μm in diameter. The reproduction
(optimal temperature 30–34 °C) is asexual by autospore (2–16) production from one
mother cell, which is divided into three stages: growth (increase in cell size), ripening
(mitosis), and division (release of daughter cells). *Chlorella* species possess very rigid
cell walls, the structure of which can vary greatly among species. Many *Chlorella* species are capable of mixotrophic or even heterotrophic growth on organic substances
such as glucose, acetate, and glycerol.

Some algae species are listed in Tab. 2.14. A closer look can be found in the de Gruyter textbook *Microalgae Biotechnology* by Posten/Griehl.

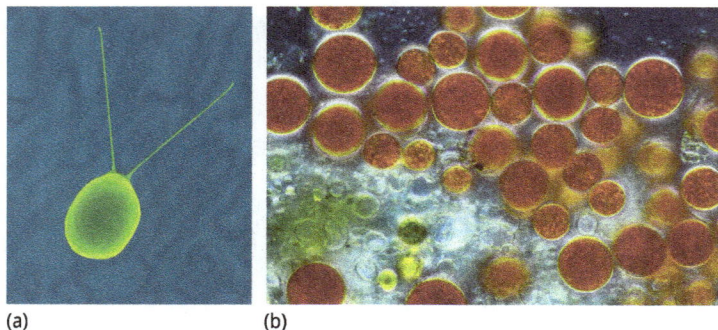

(a)   (b)

Fig. 2.14: Microscopic picture of *Haematococcus*: (a) the green mobile stage (macrozooid) and (b) production stage of astaxanthin (metatocyst) (© Both pictures Science Photo).

Tab. 2.14: Species of microalgae with some technical or commercial importance.

Scientific name	Biology	Commercial products	Comments
Arthrospira platensis	Cyanobacteria, forms helical, multicellular filaments	"Spirulina" (trade name), food-, feed-supplement (esp. aquaculture)	Phycocyanin is a blue dye, popular in the fitness industry
Chlorella vulgaris	Green algae	Food supplement, vitamins, micronutrients	Biomass as powder or tablet; cell wall has to be disrupted before use
Dunaliella salina	Green algae, red color during production, hypersaline	β-Carotene, perspective for glycerol	Grown in open salt lakes, e.g., lake Tyrrell, Australia, due to salinity no contamination problem
Haematococcus pluvialis (Fig. 2.14)	Green algae, appears in red during production	Astaxanthin, the most effective antioxidant known	Increasingly produced in closed photobioreactors, causes common vernacular "blood rain" or "blood snow"
Nanochloropsis		Feed for aquaculture	
Chlamodymonas reinhardtii	Prototype of green algae		Scientific model organism

2.5.2 Production of microalgae – a means to tackle ongoing challenges

Microalga biomass is generally composed – like most microbial biomass – from the main compounds protein, carbohydrates, (polyunsaturated) lipids, and as a specialty, of pigments (Fig. 2.15). Their nutritional value can be equated with fruits or seeds of higher plants. The difference with respect to production of terrestrial plants is clearly that the cells in the suspension represent completely usable biomass and not only a part of the harvest. Furthermore, the fraction of the different compounds can be shifted in wide ranges according to process strategy. Furthermore, microalgae biomass represents a valuable source of a balanced mineral content (e.g., Na, K, Ca, Mg, Fe, Zn, and trace minerals) and all essential vitamins (e.g., A, B1, B2, B6, B12, C, E, nicotinate, biotin, folic acid, and pantothenic acid). This makes microalgae as whole cell preparations, keeping the high value of the balanced constituents, and making them interesting for commercialization as nutraceuticals. Specific extracts are commercialized for their different bioactivity, e.g., as antioxidant.

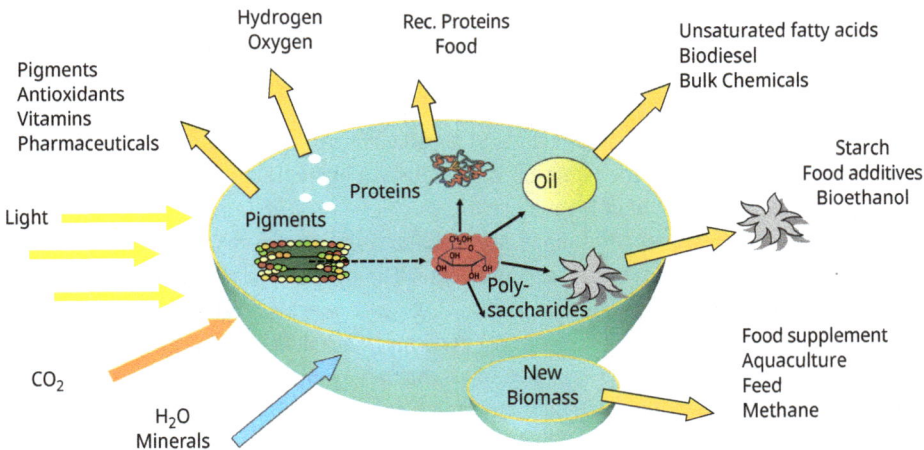

Fig. 2.15: The product palette of microalgae.

Chlorella and *Spirulina* (*Arthrospira platensis*), both possessing the GRAS status, are traditionally used as food supplement and currently dominate the fast developing market of microalgae-based health food production. The global Spirulina market is valued at US $542.52 million in 2022 and is projected to reach a value of US $1,129.55 million by 2030 (Vantage Market Research). However, other microalgae species appear also to arise slowly within the food and feed market as they run slowly through regulation procedures. One limitation in bringing interesting new species into the market is lack of scientific insights relating nutritional value or unknown side effects, making a sophisticated approval process necessary, while *Spirulina* and *Chlorella* already possess the

GRAS status, because they are traditionally used. Another cyanobacterium, *Aphanizomenon flos-aquae* with brand names like AFA, has currently made a sensation in the USA. A general concern in blue-green alga is their ability to produce toxins. Microalgae for food use are usually produced in raceway ponds. Particularly, *Chlorella* has a strong cell wall, which has to be disrupted before further processing to increase bioavailability. Whole cells as dietary supplements are commonly marketed as dry powder, compressed pastilles (green or orange color), or as capsules. Research is working on reducing the market price to such an extent that microalgae can become a real alternative to fruit and vegetables with a significant contribution to the daily diet.

But many of the high value ingredients are available as extracts. Lipids are essential for all living organisms primarily because all cell membranes are built up from polar lipids. Polar lipids consist mainly of phospholipids and glycolipids. Algae are no exception but have a unique advantage insofar as they possess large amounts of membranes in the thylakoids and form especially high amounts of polyunsaturated fatty acids ("PUFAs"). Humans lack the ability to introduce double bonds in fatty acids beyond carbons 9. Linoleic acid, α-linolenic acid, eicosapentaenoic acid (EPA, 20:5 ω-3), and docosahexaenoic acid (DHA, 22:6 ω-3) are very valuable as functional food. EPA and DHA can be found predominately in the polar lipid fraction of especially marine microalgae species and are important for human health. Enhanced DHA intake, e.g., has positive cardioprotective effect on adults and may improve infant cognitive performance and enhance visual acuity. Essential fatty acids are sold predominately as oil capsules or are worked into functional food. The natural supply is via some plant oils or via see fish (or extracts from fish liver), again a hint to the marine food chain from algae to fish.

The most striking feature of algae are of course the nicely colored pigments. They are gained by lipophilic extraction, e.g., n-hexane. The lipid fraction contains a high variety of pigments such as chlorophylls, carotenoids, tocopherols, and sterols. Chlorophyll amounts are usually about 0.5–1.5% of DW and can be used as food and pharmaceutical colorants. Carotenoids in microalgae possess primarily photoprotective (antioxidant) functions. Correspondingly, they are deployed as sources of antioxidant activity in human and animal diets. Despite the high variety (>600 carotenoids known) in nature only a few were used commercially. Despite the current possibility of chemical synthesis of β-carotene (all-trans isomers), there are still several β-carotene production plants from *D. salina* running nowadays (e.g., Australia), because cis-isomers of ß-carotene can only be produced naturally. Astaxanthin (AXT) represents another carotenoid, which is produced on commercial scale as fish feed (accountable for red color, e.g., of salmon fillet) in *Haematococcus pluvialis*. It is the strongest natural antioxidant. In some countries it is also available as food supplement. It is also produced on commercial scale as fish feed (accountable for red color, e.g., of salmon fillet). Chemical synthesis is possible with the debate whether it is equivalent to biologically produced material with different isomers. Moreover, other lipid-soluble carotenoids such as lutein, zeaxanthin, lycopene and fucoxanthin are used in animal feeds, pharmaceuticals, cosmetics, and food colorings, however, to a lesser extent.

Besides the lipophilic pigments, cyanobacteria and Rhodophyta contain also water-soluble fluorescent pigments called phycobiliproteins, where tetrapyrrole chromophoric prosthetic groups, named phycobilins, are linked to a protein backbone. For commercial production of red colorings phycobiliprotein the rhodophyte *Porphyridium* is most used. Phycocyanin is the best blue color for food applications, e.g., in chewing gums, candies, dairy products, jellies, ice creams, and soft drinks. It is also available as blue-colored drink for general health support. In *Arthrospira*, it is a light collecting pigment produced up to 15% in cell mass.

Another group of lipids to be found in microalgae are neutral lipids (e.g., triacylglycerol, "TAG"), which accumulate in the cells up to 60% DW as storage lipids. TAGs are interesting as future solar fuels, however, comparable high cultivation, harvesting, and downstream costs prevent their application on industrial level. The other side of the medal is that fossil fuels are still unbeatably cheap, which is not in the sense of sustainability.

In general, protein represent with 50–60%, the most abundant compound in microalgae biomass, containing all essential amino acids making it equivalent to egg protein. Nutrition with whole cells is not only a protein alternative but simultaneously a source for the valuable compounds mentioned above. Many areas on our planet like parts of Africa or Europe are protein-deficit areas. As long as people can afford it, they buy soya beans, e.g., in South America at the cost of rain forest and the local population. This is reason enough to think about microalgae as a protein and vitamin source not only as food supplement but on a large scale. Soy protein is traded with about US \$1,500/ton, while microalgae protein is available not under US \$5,000/ton. This is a strong incentive for further technology development also with a view to applicability in countries with limited infrastructure avoiding male nutrition.

A further promising application for high value microalgae-based products might be represented by the synthesis of recombinant proteins for medical use. Although, mammalian cell culture (e.g., CHO) and bacterial expression systems are currently dominating the market for therapeutic proteins, microalgae could represent an alternative promising platform. Microalgae possess all necessities for production of complex therapeutic proteins like other eukaryotes including posttranslational modifications. Phototrophic production systems furthermore offer intrinsic security as no viruses can infect the culture and no other organic material is in the medium simplifying product separation. These qualities could make production on lower cost possible, thus enabling the disposability of affordable quantities of vaccines for developing countries.

Furthermore, microalgae are important feed additives for aquaculture animals (fish, shrimps, mollusks, and/or their larvae) accounting for about 75% of the algae market. Microalgae are the basis of the marine food chain. They are the feed for abovementioned animals and especially for krill in the oceans. Small fish feed on small animals, big fish chase small fish, and finally humans are the predators of big fish. Edible fish are usually carnivores and get PUFA and carotenoids from this food chain. Otherwise, fish will not be the healthy food as it used to be. In the meantime, aquaculture produces 30% of the consumed fish worldwide.

2.5.3 What else to learn from microalgae – awareness and sustainability

Microalgae are masters of symbiosis and living together. This concerns, firstly, the formation of macroalgae, known as seaweed. They form large thalli, precursor to the tissues of higher plants. Seaweeds are historically and nowadays harvested in much larger quantities than microalgae in more or less extensively used seminatural stands. But cell differentiation occurs as well. In *Nostoc*, for example, some cells specialize in nitrogen fixation and provide neighbors with precursors for protein synthesis. *Volvox* forms hollow spheres that can move in a coordinated manner. Small, young colonies grow up inside the sphere in a protected area. Upon release, the parent colony perishes. This process is interpreted as one of the earliest and few examples of programmed death in the field of microorganisms. A *Volvox*-like microalga is postulated to be the ancestor of all green plants. Symbiosis with species other than microalgae began billions of years ago. Cyanobacteria were taken up by heterotrophic microorganisms by endocytosis. They were integrated there genetically and metabolically and thus survived. According to the endosymbiont theory, the eukaryotic microalgae arose from this symbiosis. While this theory was already postulated in the late nineteenth century, it was first systematically investigated and published in 1967 by the American biologist Lynn Margulis (1938–2011). On a higher level, microalgae and filamentous fungi have entered into a symbiotic relationship with lichens. The microalgae give the fungus substrates and gets therefore protection against water losses. That allowed the algae also to spread into terrestrial environments. Another example of this organizational level are the microalgae in symbiosis with coral polyps.

Unlike the attention we pay to terrestrial plants, known to us as forest, meadow, or crop plantation, little attention is paid to aquatic plankton. Warming up of the oceans has serious effect on the planktonic drifters. They depend on situations where ocean currents meet, and nutrient-rich waters rise. This is the case, for example, at the Arctic and Antarctic shelf ice. There, krill, and subsequently larger predatory fish, gather and close the food web. The shift of the ice edge has already led to dramatic changes, which are equivalent to desertification of the terrestrial flora. The frightening news of the last years has currently led to a strong increase in research and company foundations in the field of microalgae. This brings us closer to Barack Obama's clean power plan talking about "a huge win for algae." However, the overall problems have not been understood completely. The artist Anthea Oestreicher has made this her theme. Besides fluid performances also object art has been created (see Fig. 2.16).

This work is an example of immersing oneself in a jumble, being in touch with each other through relations and materials, together with nonhuman actors (microalgae) and the interactions with sun and material agencies. The works of the multidisciplinary de-

Fig. 2.16: This way the portrait of Lynn Margulis (as a pioneer of the concept of endosymbiosis) becomes, not only "Means-to-Purpose," but a unique object with its own aura. Rather than using animal gelatin and silver halogenides, the artist combines less damaging cyanotype chemicals (potassium ferricyanide and ferric ammonium citrate) on a body of algal biopolymer (agar)-based bioplastics for sun-derived, sustainable photographs (© Object & Photo: Anthea Oestreicher).

signer and artist extend the biological concept of symbiosis and cohabitation to an onto-logical dimension. Merging knowledge and modes of being through art creates narratives of living and thinking symbiotically. This example shows that art and design can help promote social awareness and thus ensure acceptance and sustainability.

2.6 Questions and suggestion

In this paragraph some aspects of integration of microorganism and process have been mentioned. Collect issues and examples from the other chapters and try to set up a detailed overview, see also the supplementary material.

By the way: recently an industrial process has started producing DHA with *Schizochytrium*. This orgnism is cultivated heterotrophically and is not really an algae. Try to find out the truth!

Chapter 3
Media – supplying microorganisms with a comfortable environment and building blocks for growth

A medium must be provided apart from a choice of the production strain. The spectrum of media ranges from synthetically composed media to complex natural media, depending on the purpose and scale of cultivation. Media design is based on biological information regarding stoichiometry and kinetics and can be straightforward using the respective calculation. This approach goes only up to a certain point, after which strain-specific information and data based on experiences have to be respected. For further optimization, modern biotechnological techniques can be applied. In this chapter, basic ideas for media design are provided and specified by calculations and data.

3.1 Media – the chemical basis of the process

The main aspects of media design are listed in this paragraph. Specific examples will be given in the case studies.

A culture medium is, by definition, a chemical matrix of substances designed to support growth of microorganisms. Usually, it is an aqueous solution of some organic substances and inorganic dissolved salts.

The medium has to provide all the necessary nutrients for energy supply (catabolism), and components for biomass formation (anabolism) and reductants in connection with energy generation. The nutrients have to be present in the form and concentration that allow for rapid uptake and utilization by the cells. However, it is not always easy to find out the specific needs of the specific species or strains. Furthermore, the medium determines the environmental conditions with respect to acidity and buffer effectiveness, osmotic pressure, viscosity, and salinity. In some cases, the related ingredients are not used by the cell but they have to be present.

For practical purposes, media can be classified according to different aspects. Minimal media or synthetic media are based on known chemicals, and contain only necessary components. Complex media can contain many auxiliary materials to support the best growth of the organisms under investigation. Often, they contain more or less well-defined ingredients, e.g., from plant origin. The choice of a medium is an important item in cost calculation, especially for low value products. Technical media, therefore, often have a composition different from lab media. They may be derived from residuals, e.g., from food processing.

https://doi.org/10.1515/9783110773354-003

3.2 Media design – starting from scratch

Before looking up recipes from handbooks, we try to translate basic knowledge about the need of microorganisms into quantitative terms. Microorganisms need medium compounds to build up their own cell material and an energy/redox carrier to perform the anabolic steps accordingly. Microorganisms consist of a couple of macromolecules, the sum of which makes up almost the entire cell mass. These bioactive compounds are polymers, namely, nucleic acids, proteins, carbohydrates, lipids, and additional compounds in minor amounts like pigments, vitamins, etc. The amount inside a given cell mass is quantified by quota q, which is the relative mass fraction. Typical values of q_{nuc}, q_{prot}, q_{carbs}, and q_{lipid} can be found in Fig. 3.1. In practice, the sum of all q is in most cases only 90%, as dissolved salts in plasma or monomers are not recorded. The quotas are less variable than you might think. Exceptions are storage substances, which can range from 0% to 50%, for example. Cells need low-molecular building blocks such as the monomers to build polymers. Fortunately, this is often not the case, but the chemical elements involved must always be present in the medium. This usually enables the organisms to form metabolites by themselves.

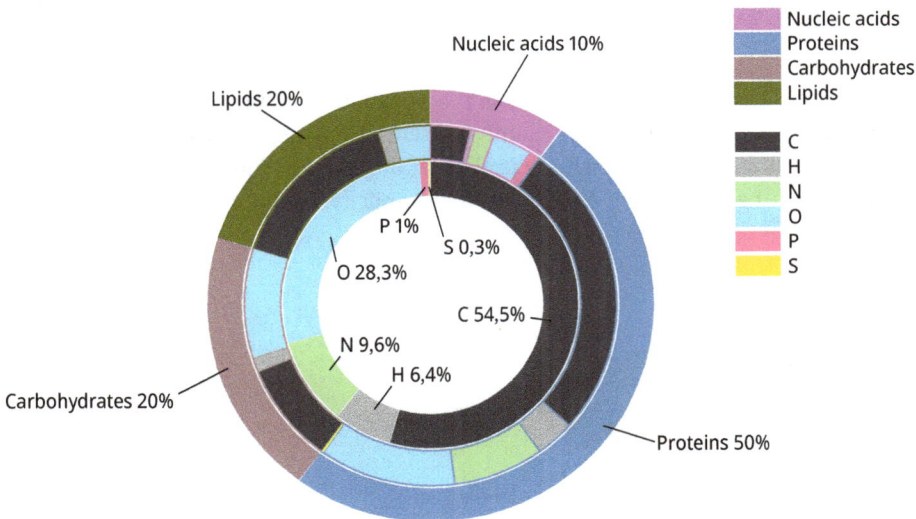

Fig. 3.1: The cell consists of a couple of macromolecules, the sum of which makes up almost the entire cell mass, the outer ring. These in turn have an elemental composition (middle ring) that is determined by the monomers. Finally, the cell as a whole has an elemental composition, the inner ring.

In biochemical engineering, reaction processes are quantitatively described by material balance equations for the cell or for the reactor, considering each chemical compound appearing in the process. This requires knowledge of the stoichiometry of each reaction inside the cell. However, this molecular species balance is too complicated at the moment, as we cannot follow all reactions inside the cell. Since the number of atoms of each element is a conserved variable, the atomic species balance, often called elemental balance, can be applied independently from specific reaction schemes. In the following paragraph, this idea is outlined for aerobic cultivation of heterotrophic microorganisms as the most common case.

We start by drawing a system boundary around the biomass. Figure 3.2 represents a single cell. The molar amounts of substances "Comp" n_{Comp} (mol) going into or out of the biomass per given time interval Δt has to now be considered. Molar amounts, as such, are not a conserved quantity, as they may change by conversion, but they allow us to set up the elemental balances. Therefore, all atoms forming the nutrients or products have to be added up according to their relative amount in the compound "Comp." The stoichiometric coefficients $s_{E,Comp}$ are defined as the number of atoms of element "E" in the formula of compound "Comp"; for example, for C in glucose, it would be 6. In practice, mass balances are easier to handle. The mass fractions of the single elements $e_{E,Comp}$ in a compound can be determined from the stoichiometric factors and the molar masses M_{Comp} are measured directly as given in eq. (3.1). For the biomass itself, elemental fractions can be found in Fig. 3.1. They can be measured or found in databases:

$$q_{Comp} = m_{Comp}/m_X; \quad e_{E,Comp} = s_{E,Comp} \cdot M_E/M_{Comp} \tag{3.1}$$

The interval Δt is defined in this case as the duration of a cultivation process. Compounds entering are the nutrients mainly present at the beginning of the cultivation, while products and biomass itself are present at the end. The amount taken up or produced during the cultivation, Δm_{Comp}, is, for simplicity, called in the following equations as m_{Comp}.

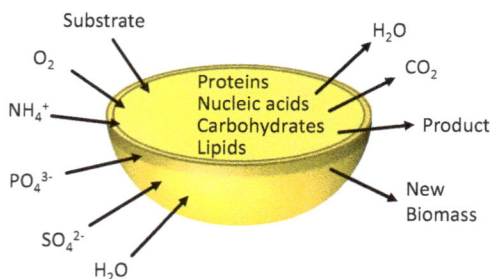

Fig. 3.2: The cell as catalyst, given that compounds in the medium are converted to products and new biomass. The exact metabolic pathways are not considered in this view.

Writing down the mass balance equation for carbon (C) and the substances involved, namely, the substrate, e.g., glucose, carbon dioxide, product, and biomass yields:

$$e_{C,S} \cdot m_{Subs} - s_{C,CO_2} \cdot m_{CO_2} - s_{C,P} \cdot m_{Prod} - s_{C,X} \cdot m_X = 0 \tag{3.2}$$

This also holds for concentrations as long as the volume is constant. Of course, gases have also to be considered, which are usually only O_2 and CO_2. Note that the substances going out of the system are counted negatively. As we do not consider any substance to accumulate inside the boundary (new cells are counted as being outside the boundary), the balance must be zero.

The same holds for the nitrogen (N), phosphorus (P), and sulfur (S) balance:

$$e_{N, NH_3} \cdot c_{NH_3} - e_{N,P} \cdot c_{Prod} - e_{N,X} \cdot c_X = 0 \tag{3.3}$$

Although nitrogen is present in the medium as NH_4^+, here it is calculated as NH_3 as the cell excretes H^+ to maintain the intracellular charge balance. The medium then becomes more acidic. When nitrate is absorbed, it becomes more alkaline:

$$e_{P,PO_4} \cdot c_{P, PO_4} - e_{Prod, X} \cdot c_X = 0 \tag{3.4}$$

This holds, assuming the product does not contain phosphorus:

$$e_{S, SO_4} \cdot c_{S,SO_4} - e_{S,P} \cdot c_{Prod} - e_{S,X} \cdot c_X = 0 \tag{3.5}$$

To have a more generalized and compact notation, it is convenient to sum up the fractions of the element E over all n_{Comp} compounds:

$$\sum_{i=1}^{n_{Comp}} e_{Comp,i} \cdot c_i = 0 \tag{3.6}$$

The different concentrations are lumped in the vector c:

$$\sum_{i=1}^{nq} e_{C,i} \cdot c_i \tag{3.7}$$

For a better overview, this sum can be split into terms representing the concentrations at the start of fermentation and the concentrations at the end:

$$\sum_{i=1}^{n_{Comp,in}} e_{C,i} \cdot c_i - \sum_{i=1}^{n_{Comp,out}} e_{C,i} \cdot c_i = 0 \tag{3.8}$$

Oxygen usually cannot be used for this scheme. The molecular O_2 absorbed does not end up in the biomass but is excreted again as CO_2. The oxygen in the biomass comes from the substrates or many hydration reactions. The same applies to hydrogen.

These equations can be used to calculate the initial values for a medium. It is important to understand that there are two directions of thought. How much biomass do I get for a given medium or how much of each component do I need to achieve a certain biomass? Examples are given further down in this paragraph.

But there remains another problem. Even assuming that we know all the mass fractions, $e_{E,Comp}$, there is still missing information. Counting in Fig. 3.2, the unknown compounds and the available equations reveal that there are more parameters necessary to know. Some degrees of freedom in the system are left, as we do not know how much carbon is directed into the biomass and how much into carbon dioxide and the product. The same holds for nitrogen and its partitioning into product or biomass.

Now, a point is reached where additional stoichiometric data need to be included. From batch experiments, we know that the yield $Y_{X,S}$ is often close to 0.5 g/g for aerobic growth on glucose (without product formation). The background lies in ATP generation by respiration and the subsequent use of ATP for growth. This will be further outlined in Chapter 4. We can now use this observation to get a preliminary empirical value. Knowing this yield, we can fill in the first value in Tab. 3.1 to design a generic medium. All information given in the balance equations represent relative relationships. The most important decision now is to decide on the targeted amount of biomass. As an example, for a new biomass with $c_X = 50$ g/L, a substrate of $c_S = 100$ g/L is needed.

> The basic structure of a culture medium can be determined from the stoichiometric coefficients for the turnover rates. These result from elemental balances and from known stoichiometries of the metabolic pathways.

Up until now, we have looked at the cell as being a molecular species with a constant stoichiometry. That is of course not true. Furthermore, we do not know a priori the carbon content of the biomass $e_{C,X}$, nor the molecular mass of the biomass, which is obviously a virtual number. Nevertheless, the elemental composition of the biomass can be measured and normalized to one carbon atom, giving a virtual molecular formula and molecular mass as basis for engineering considerations.

> A calculation example for an average microorganism is given in Tab. 3.1. It contains the elements being covalently bonded in macromolecules. Try to follow the logic behind the numerical results.

Tab. 3.1: Elemental composition and molecular formula for an average generic bacterium "*Virtuella generica*"; values are similar to *E. coli*.

Species	Mass fractions $e_{E,X}$ (g/g)						Empirical formula	Molar mass (g/mol)
	C	O	N	H	P	S		
Virtuella generica	0.47	0.31	0.11	0.07	0.025	0.009	$CH_{1.8}O_{0.5}N_{0.2}P_{0.02}S_{0.007}$	25.47

With these data, we can further complete our medium composition under the specification to reach the biomass concentration of $c_X = 50$ g/L, as listed in Tab. 3.2. The determination of the final biomass concentration is necessary as all equations derived so far are relative to biomass formation, neglecting, for now, product formation. From the N balance, the ammonia concentration can be calculated:

$$e_{N,NH_4} \cdot c_{NH_4} - e_{N,x} \cdot c_X = 0 \rightarrow c_{NH_3} = \frac{e_{N,x} \cdot c_X}{e_{N,H_3}} = 6.68 \, \frac{g}{L} \tag{3.9}$$

The phosphorus and sulfur balances are revealed accordingly:

$$e_{P,PO_4} \cdot c_{PO_4} - e_{P,x} \cdot c_X = 0 \rightarrow c_{PO_4} = \frac{e_{P,x} \cdot c_X}{e_{P,PO_4}} = 3.75 \, \frac{g}{L} \tag{3.10}$$

$$e_{S,SO_4} \cdot c_{SO_4} - e_{S,x} \cdot c_X = 0 \rightarrow c_{SO_4} = \frac{e_{S,x} \cdot c_X}{e_{S,SO_4} \cdot c_S} = 1.35 \, \frac{g}{L} \tag{3.11}$$

$K^{+,}$ as the main cellular inorganic cation does not appear in the molecular formula but is present in the cells with a fraction of about 0.01; according to mass balance, it has to be 0.5 g/L in the medium.

Tab. 3.2: Generic medium to produce 50 g/L biomass without product formation.

Compound/ ion	Purpose	Rule	Side effects	Concentration, c (g/L)	Amount n (mol/L)	Comments
$C_6H_{12}O_6$	C-source, energy source	Yield, ATP balance	Osmotic pressure	100	0.556	1)
NH_4^+	N-source	Elemental balance	pH	7.07	0.392	2)
PO_4^{3-}	P-source	Elemental balance	Precipitation	3.72	0.0392	3)
SO_4^{2-}	S-source	Elemental balance		1.32	0.0137	4)
K^+	K-source	Meas. mass fraction		0.5	0.013	5)
Cl^-	Counter ion	Charge balance		8.68	0.245	6)

1) Yield depends on the C-source; 2) NO_3^- alternative; 3) often limited in nature; 4) –; 5) potentially Na^+; and 6) alternatives.

Elemental balances have been adapted from plant fertilization. Carl Sprenger and the chemist Justus von Liebig observed in 1828 that plants needed balanced amounts of N and P. In soil with high nitrate content, only phosphate could bring better yields and vice versa. This finding was called the Liebig's "law of minimum." It states that growth is controlled by the scarcest resource; today it is called the limiting factor. Later, this rule was popularized by a graphical model (Fig. 3.3), showing a barrel with staves of unequal length, where the shortest one determines the holding capacity.

Fig. 3.3: Graphical representation of Liebig's law, where the staves are the relative amount of the respective element. The maximum filling volume is given by the smallest one, which, in an unbalanced medium, may not be the highest value.

The next step is caring for the charge balance. Of course, the net charge has to be zero. This approach allows us to find a combination of dry salts to be weighed in. Any excess of charges in the medium is physically impossible. When using already defined recipes, the charge balance is automatically zero as we weigh in neutral substances. The charge balance is set up as

$$\sum_{i=1}^{n_{Comp}} z_{C,i} \cdot n_i = 0 \tag{3.12}$$

where z is the charge number of the respective compounds, e.g., $z_{PO4} = -3$. In our self-designed medium, the total charge is highly positive, so we have to find negatively charged counter ions. A good candidate is chloride, Cl^-. It is abundant in nature but not directly a building block for cell growth and is present in the cells only in small amounts. In the intracellular space, proteins and polysaccharides also act as negatively charged counter ions. The charge balance also has to stay zero during cultivation. In our case, it is not guaranteed that the cell takes up positively charged ions with the same velocity as negatively charged ions. The most important example is NH_4^+. Cells take up the uncharged NH_3, leaving a proton H^+ in the medium. Ammonia uptake happens, consequent to acidification during the cultivation. This has to be compensated by titration, e.g., with NaOH, again with a drawback, namely increasing salinity.

All considerations so far are based on constant biomass composition and constant stoichiometry of the substrate uptake and product formation. In addition, practically unhindered uptake of the nutrients is assumed. This concept is called balanced growth. Nevertheless, further optimization is necessary with respect to yield or growth velocity, a long and empirical process.

The elements mentioned so far are covalently bonded in the major cellular components. Together with the cations K^+, Mg^{2+}, Ca^{2+}, and Fe^{3+}, they are called macronutrients, as they are required in considerable amounts. To get an idea of the quantities to be fed, a look at the biological function is helpful. The quantitative contribution to cell mass is given by a real analysis of *Escherichia coli*, as listed in Tab. 3.3.

Tab. 3.3: Necessary mineral elements, and their biological function and content in *E. coli*.

Ion species	Involvement in reactions and structures	*E. coli* content $e_{I,X}$ (mg/g)	Comments
K^+	Cofactor for certain enzymes (e.g., pyruvate kinases)	0.12	Principle inorganic cation
Mg^{2+}	Cofactor for certain enzymes (e.g., kinases); contained in ribosomes and phosphate esters	0.08	Cell membrane
Ca^{2+}	Cofactor for certain enzymes, especially exoenzymes (proteases)	12.74	Cell membrane
Fe^{3+}	Component of certain enzymes, cytochromes, and DNA synthesis	0.02	Reduction to Fe^{2+} is possible
Na^+	Involved in membrane transport	0.07	
Cl^-	Inorganic anion		Essential
Trace elements		(µg/g)	<100 µg/g
Co^{2+}	Part of vitamin B_{12}	0.019	Synthesis of B_{12}; solely bacteria
Cu^{2+}	Cofactor for metalloenzymes (e.g., cytochrome oxidase)	0.038	
Mn^{2+}	Cofactor for enzymes (e.g., PEP carboxylase and recitrate synthase)	0.018	Essential for all microorganisms
Zn^{2+}	Cofactor for metalloenzymes, especially RNA and DNA polymerases	0.059	Essential for all microorganisms
MoO_4^{2-}	Present in dehydrogenase form	0.006	

Micronutrients, also called trace elements, are related to specific metabolic functions like cofactors for specific enzymes or to maintain protein or cell membrane structure. The dosage is highly species-dependent. The given sample analysis may not necessarily reflect the real need of the cell or the availability in the medium during the experiment. Many companies have their own recipes with which they obtain good results for a given purpose. Sophisticated use of micronutrients is also a means to control metabolic fluxes to increase product formation. Examples are given in other chapters.

3.3 Practical application – a discussion of recipes from references

After setting up a theoretical minimal media, a look at a commercial minimal medium is advisable. Such media are used in labs to study microbial physiology. A popular medium for use in labs is the M9 medium, as given in Tab. 3.4. The recipe is given in the form of six different stock solutions. They are to be mixed to create a ready-to-use medium. Preparing different stocks beforehand makes handling easier, but it is primarily a means to prevent mutual interactions between the medium constituents during storage and sterilization – such as precipitation or chemical reactions. Another point to take care is that glucose solutions have higher density than water (c_S = 100 g/L → ρ = 1.04 kg/L; c_S = 200 g/L → ρ = 1.08 kg/L). The recipe is formulated on a mass per volume basis, in the sense that the chemicals of each stock solution are not to be solved in, e.g., 1 L of water, but have to be solved in a smaller amount of water and then filled up to yield 1 L of solution. The same holds for the other stock solutions. This approach is a tribute to easy pipetting, and leads to pre-calculated glucose and salt concentrations in the final medium. An additional point is that 1 L of medium contains less than 1 kg of water. During cultivation, the density of the medium changes due to glucose uptake by the cells – it has to be considered in feeding and sampling for precise balancing on a mass per mass basis. The amounts of stock solutions are recommendations and do not influence the final medium concentration.

Tab. 3.4a: Composition of M9 medium (Cold Spring Harbor), modified for $c_{Gluc,4}$ = 4 g/L glucose in the final medium.

Stock "glucose" 100 mL	$m_{Gluc,4}$ (g)	M (g/mol)	$n_{Gluc,4}$ (mol)	Remarks
Glucose	20.0	180.2	1.109	
Solve in 90 mL warm water bath, fill up to 100 mL				
Use autoclave for sterilization				

Stock "salts" 1,000 mL	$m_{Salt,4}$ (g)	M (g/mol)	$n_{Salt,4}$ (mol)	Remarks
"5×"				Multiple of final concentration
$Na_2HPO_4 \cdot 2H_2O$	42.5	178.0	0.239	P-source, basic salt, buffer
KH_2PO_4	15.0	136.1	0.110	Acid salt, used as buffer
NaCl	2.5	58.4	0.043	Adjusts salinity
NH_4Cl	5.0	53.5	0.093	N-source, acid salt
Dissolve in 800 mL, adjust to pH 7 using 4 M NaOH, and fill up to 1 L				
Filter, sterilize, and store in the dark; autoclaving leads to ammonia evaporation				

Tab. 3.4a (continued)

Stock "magnesium" 100 mL	$m_{Mg,4}$ (g)	M (g/mol)	$n_{Mg,4}$ (mol)	Remarks
"1 M"				Round molarity
$MgSO_4 \cdot 7H_2O$	24.65	246.5	1.00	
Dissolve in 80 mL and fill up to 100 mL				
Autoclave and store at room temperature				

Stock "calcium" 100 mL	$m_{Ca,4}$ (g)	M (g/mol)	$n_{Ca,4}$ (mol)	Remarks
"100 mM"				Round molarity
$CaCl_2 \cdot 2H_2O$	1.47	147.0	0.01	
Dissolve in 80 mL and fill up to 100 mL				
Autoclave and store at room temperature				

Stock "iron" 100 mL	$m_{Fe,4}$ (g)	M (g/mol)	$n_{Fe,4}$ (mol)	Remarks
"50 mM/100 mM"				Round molarity
$FeCl_3 \cdot 6H_2O$	1.35	270.3	0.05	Unstable
Citric acid $\cdot 1H_2O$	2.11	210.1	0.10	For complexation
Dissolve in 80 mL and fill up to 100 mL				
Filter, sterilize, and store in the dark at 4 °C				

Stock "trace elements"1,000 mL	$m_{Stock,4}$ (mg)	M (g/mol)	$n_{Stock,4}$ (mmol)	Remarks
"100×"				Multiple of final concentration
$MnCl_2 \cdot 4H_2O$	100	197.91	0.505	
$ZnCl_2$	170	136.29	1.247	
$CuCl_2 \cdot 2H_2O$	43	170.48	0.252	
$CoCl_2 \cdot 6H_2O$	60	237.93	0.252	
$Na_2MoO_4 \cdot 2H_2O$	60	241.95	0.248	
Dissolve in 800 mL, adjust to pH7, fill up to 1,000 mL				
Filter and sterilize; more than needed to reduce errors from weighing accuracy				

Tab. 3.4b: Dosing scheme for 1 L M9 medium.

Glucose (mL)	Salts (mL)	Magnesium (mL)	Calcium (mL)	Iron (mL)	Trace elements (mL)	H_2O
20.0	200.0	1.0	1.0	1.0	10.0	Add to 1 L

Tab. 3.4c: Final concentration in the medium.

Organics or ion	$n_{Medium,4}$ (mmol/L)	$c_{Comp,Medium}$ (g/L)	g Compound/g biomass	Fraction from theoretical	Remarks
Glucose	22.2	4.00	2.0	1.00	intentional
NH_4^+	18.7	0.34	0.17	1.18	Reasonable
$HPO_4^{2-}/H_2PO_4^-$	69.8	6.7	3.35	44.29	Buffer function
SO_4^{2-}	1.0	0.1	0.05	1.78	Reasonable, reduction
Na^+	104.1	2.39	1.20	17.1	Salinity, non-halophilic
K^+	22.0	0.86	0.43	4.3	
Cl^-	27.4	0.98	0.49	1.0	Salinity, counter ion
Mg^{2+}	1.0	0.02	0.01	5.21	
Fe^{3+}	0.050	0.003	0.0014	14.0	
Citric acid	0.100	0.019	0.0096	–	
Ca^{2+}	0.100	0.004	0.002		
Trace elements	$n_{Medium,4}$ (µmol/L)				
Mn^{2+}	5.1	0.28	0.14		
Zn^{2+}	12.5	0.81	0.41		
Cu^{2+}	2.50	0.16	0.08		
Co^{2+}	2.50	0.15	0.08		
MoO_4^{2-}	2.5	0.04	0.2		

Analysis of the medium shows that in the column "fraction from theoretical," the requirements for 2 g/L biomass with respect to yields and elemental balances are fulfilled. Ammonia and sulfate are present in a slightly higher concentration than required, but to a reasonable extent. Nitrogen will be the limiting element for final biomass concentration. Potassium, magnesium, and iron are present in unnecessarily high concentrations. Phosphate is highly overdosed as it serves two purposes, P-source and buffer. Sodium concentration is higher than required for biomass formation as well due to its function as carrier of salinity.

On this basis, as per calculation, for a medium of 50 g/L biomass, all salts necessary for biomass formation have to be increased by a factor of 25. On the other hand, salts that are not part of the biomass but are necessary for environmental conditions, e.g., sodium and chlorine for salinity, have to stay within given limits. Furthermore, our goal is to avoid any waste, especially of phosphate; therefore, it is necessary to increase the N content of the medium to get a molar N/P ratio of 10, according to the empirical formula of biomass composition. There are two degrees of freedom to go to higher final concentrations; firstly, to add larger volumes of the respective stock solutions to prepare the final medium and secondly, to increase the concentration of the respective salts in the stock solutions. The first option is limited by the fixed final volume while the second option is limited by the solubility of the salts in the stock solution and in the medium.

To get an overview of a possible route for arriving at an upgraded medium, we can adopt a thought experiment. Glucose is soluble in water up to 500 g/L. Choosing this value for the stock solution yields a factor of 2.5. The still missing factor of 10 can be reached by increasing the added volume in the dosing scheme up to 200 mL. Dispensing the 25-fold volume of the other stock solutions, except "salts," is possible as well, but fills up the 1 L final medium already up to 52.5%. A slight reduction of "magnesium" is possible, while the amount of "iron" could be strongly reduced. Phosphate has to be increased only by a factor of 2.5 with respect to stoichiometry. That can be handled by increasing the volume of stock solution "salts" by the corresponding factor, to get the volume to be added as 450 mL. Salinity still stays in the non-halophilic range <1 mol/L. However, this answer to the problem brings us close to the total volume restriction. The concentration of ammonium chloride has to be increased by an additional factor of 10 in the stock solution. However, this implies the risk of evaporation of ammonia into the gas phase of the bottle. To circumvent this problem, another N source like nitrate or urea is an option, or we could feed ammonia separately during cultivation to avoid too high a concentration in the bioreactor.

To conclude this part of medium design, a look at a reference medium for technical applications is advisable, guaranteeing high concentrations. A successfully tested medium is given in Tab. 3.5.

Solubility of a solute is defined as the analytical composition of a saturated solution, expressed in terms of the proportion of a designated solute in a designated solvent; in our case, the salts to be dissolved in water. Values for the solubility of

Tab. 3.5: A defined medium from literature for the production of recombinant proteins by *Bacillus*.

Defined medium (Zhang et al. 1996)	M (g/mol)	Final concentration (g/L)	Final concentration (mol/L)	Remarks
Glucose	180.16	112.50	6.24	
$(NH_4)_2SO_4$	132.14	16.50	0.12	
K_2HPO_4	174.18	1.50	0.009	
Monosodium glutamate	169.13	7.50	0.044	Precursor for proteins
$CaCl_2$	110.98	1.00	0.009	
		Final concentration (mg/L)	Final concentration (mmol/L)	
$MnSO_4 \, H_2O$	169.02	2.00	0.0118	
$CuSO_4 \, 5H_2O$	249.69	1.00	0.0040	
$ZnSO_4 \, 7H_2O$	287.53	10.00	0.0348	
$CoCl_2 \, 6H_2O$	237.93	0.80	0.0034	
$MgSO_4 \, 7H_2O$	246.48	0.578	0.235	
$FeSO_4 \, 7H_2O$	278.01	10.00	0.0360	

substances are listed in chemical handbooks or on dedicated websites, and are usually given for a defined temperature and pressure in g salt per 100 g water or in g/L, e.g., the solubility of NaCl is 360 g/L at 30 °C.

However, these values are valid only for pure salts. As we have mixtures of several salts, the different cations and anions can interact mutually. Calculation is done by applying the mass action law to the equilibrium between the solid and the dissolved phase:

$$Cat_{sCat}An_{sAn}(\text{solid}) \overset{H_2O}{\rightleftharpoons} s_{Cat} \cdot Cat^{sAn+}(\text{liquid}) + s_{An} \cdot An^{sCat-}(\text{liquid}) \qquad (3.13)$$

The exponents are the charges of the respective ions. Take note of the charge balance, $s_{Cat} \cdot s_{An+} - s_{An} \cdot s_{Cat-} = 0$. In further equations, the charges are omitted for simplicity.

Introducing the equilibrium constant $K_{diss,CatAn}$ leads to:

$$Cat_{sCat}An_{sAn}(\text{solid}) \cdot K_{diss,CatAn} = n_{Cat}^{sAn-} \cdot n_{An}^{sCat+} \ (\text{liquid}) \qquad (3.14)$$

The dissolved ions on the right side of the equation also include ions with origins from other salts in the medium that share the same ions. In sum, they shift the equilibrium to the left side of the equation, thus increasing the risk of precipitation. This is called the common iron effect. Checking the left side of the equation with a plausibility check tells us that the position of the equilibrium does not directly depend on the amount of solid salts in the medium. The term $Cat_{sCat}An_{sAn}$ rather means, a bit artificially, the number of moles in a liter of the undissolved salt. So, it is a material constant and can be combined with the dissociation constant $K_{diss,CatAn}$ to a new constant K_{SP}, called the solubility product constant:

$$K_{SP,CatAn} = n_{Cat}^{sAn-} \cdot n_{An}^{sCat+} (\text{liquid}) \qquad (3.15)$$

In principle, all possible combinations of ions after mixing the different salts have to be checked against their K_{SP} values. However, K_{SP} takes no account of pH, ion activity, and ionic strength, and should be employed with care. In addition to the ions provided by the medium, carbonate and hydrogen carbonate have to be considered as well, as these compounds are formed from CO_2 produced by respiration. Depending on the pH value of the medium, the concentration of hydroxide ions is an issue. The formation of sparingly soluble iron carbonate and hydroxide as well as the precipitation of calcium carbonate in alkaline solutions has to be mentioned.

Depending on the pH and the ionic composition of the solution, phosphates of iron, calcium, and other heavy metals with low solubility can precipitate. Critical is, for example, the formation of struvite, a sparingly soluble magnesium ammonium phosphate ($NH_4MgPO_4 \cdot 6H_2O$). Controlled by pH, temperature, and the presence of other ions in the solution, such as calcium, struvite precipitates when the concentrations of magnesium, ammonium, and phosphate ions exceed the solubility product. To prevent precipitation, chelating agents can be employed. They form soluble complexes with metal ions. Often used chelators are EDTA (ethylenediamine tetraacetate), citrate, tartrate acid, and gluconate. In our example, citrate is used to complex iron ions. The principle of complexation is also found as physiological performance in many microorganisms. Siderophores, for example, are excreted for the acquisition of iron, which is limiting in many environments due to its low solubility. Citrate – as used in our artificial medium – can also act as siderophore and is excreted by, e.g., *Aspergillus*.

The selection of buffers is an important issue. For biological buffers, significant factors among others are: good solubility, pH range, buffer concentration, sensitivity of pKa value to temperature and ionic strength, permeability through biological membranes, no interaction with other components, nontoxicity, and low costs. Besides the phosphate buffer system, as in the medium described above, Tris(tris(hydroxymethyl)-aminomethane), HEPES(N-(2-hydroxyethyl)-piperazine-N'-ethanesulfonic acid), and MOPS (3-(N-morpholino)-propanesulfonic acid) are examples of "biological buffers." This term has become popular and means buffers for biotechnological use.

Defined media, where all compounds are exactly known and calculated beforehand, are not always the best choice. For cost reasons and for supporting the cells with necessary vitamins, cofactors or additional carbon and nitrogen sources complex media are often employed in industrial fermentations. Furthermore, support for anabolism has to be provided by providing amino acids, co-factors, or vitamins. Essential or not, they are present in the cell in amounts high enough to obtain sufficient growth or production rates. The constituents are specific for different biological groups and may be even strain-specific. Selective and differential media are designed to support one organism and suppress others. One application is the cultivation of genetically engineered strains with antibiotic resistance. In addition, practical problems like foaming or sticking have to be checked. In addition, special ingredients can be necessary as chelating agents against precipitation of salts, protective agents against toxic or inhibiting compounds, or inducers for product formation. The choice of a medium is an important item in cost calculation, especially for low value products. Technical media, therefore, often have a different composition from lab media. Technical media, derived from residuals, e.g., from food processing, have to be checked against artificial media. Availability and handling issues are further points for consideration.

3.4 Complex and technical media

Like most living organisms, microorganisms also need more complex substances that they cannot synthesize themselves. These may include vitamins or certain amino acids (Tab. 3.6). In nature, these essential growth factors are found in degrading biomass, for example, or are provided by symbionts. In an axenic medium, they have to be provided artificially. In other cases, compounds are not strictly essential, but they can bridge slow metabolic steps or help as protective agents; a first glance is given in Fig. 3.4.

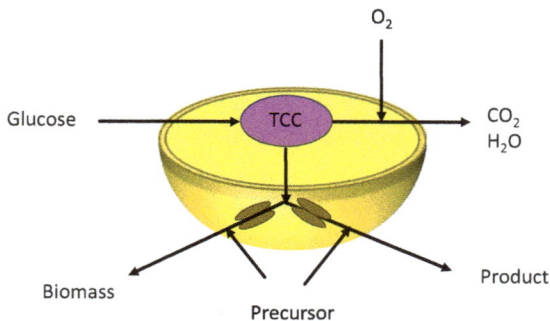

Fig. 3.4: The strongly simplified structure of a microbial metabolism for medium design; possible additives in the medium are for overcoming bottlenecks, marked with the brownish double ellipse.

The next idea to make the medium cheaper is to extract complex substances from nature or technical processes, e. g., from the food industry. These often contain the desired molecules. These so-called complex media are used to supply microorganisms with a broad spectrum of supporting molecules. The goal is to allow most microorgan-

Tab. 3.6: Some of the frequently required vitamins.

Vitamin	Amount (µg/g), estimated	Purpose	Remarks
Biotin	<1	Coenzyme for carboxylate reactions (CO_2-fixation, TCC)	
Vitamin B_{12}	0.02	Rearrangement reactions (e.g., glutamate mutase)	Can be produced only by some soil bacteria
Folic acid	<1	Coenzyme for carboxylate reactions	
Nicotinic acid (B_3)	<40	Precursor for NAD^+ and $NADP^+$	
Pantothenic acid (B_5)	<20	Precursor for coenzyme A	
Vitamin B_6	0.5	Precursor for pyridoxal phosphate and important coenzymes in transaminase and amino acid decarboxylation	
Riboflavin (B_2)	50	Precursor for FAD (coenzyme of flavoproteins)	
Thiamin (B_1)	2	Precursor of thiamine pyrophosphate (coenzyme of decarboxylases and transketolases)	
Vitamin K	1,000	Precursor of menaquinone (electron carrier in respiratory chain)	

isms to grow without specifically selected medium ingredients. Lysogeny broth (LB), a nutritionally rich medium, is primarily used for the growth of bacteria in the lab. LB medium formulations is used in molecular microbiology and has been an industry standard, especially for the cultivation of *Escherichia*, for decades. There are several common formulations of LB. Although they are different, they generally share a somewhat similar composition of ingredients, as given in Tab. 3.7.

Tab. 3.7: Composition of an LB medium.

Compound	Amount (g/L)	Purpose	Remarks
Yeast extract	5	Diverse organic compounds, vitamins (B), and trace elements	Auto-hydrolysate of yeast cells
Tryptone	10	Peptides, peptones, and source of amino acids	Casein, hydrolyzed by proteases trypsin and pepsin

Tab. 3.7 (continued)

Compound	Amount (g/L)	Purpose	Remarks
NaCl	5	Transport and osmotic balance	Varies, depending on the required salinity
Glucose	1	Easy to use energy and carbon source	Or another C source

This brute force method is easy to use in labs and brings most organisms to growth, but it is not defined well enough for quantitative physiological studies as uptake of specific substances cannot be followed by analytics. LB has its serious drawbacks for large-scale production. Substances such as yeast extract automatically, implying significant differences in composition between batches. It is obvious that the production of cell mass on other cell mass (yeast extract) and processed proteins will not be economically feasible. Other problems include lack of ability to be upgraded for high biomass concentrations, and technical problems like foaming. Nevertheless, in commercial production, yeast extract and tryptone are also occasionally used in smaller amounts for supply of essential amino acids and vitamins. In some bio-plants, yeast extract is stockpiled for a year to ensure against changes between consecutive production runs.

LB medium is a good choice but is also too complex on a technical scale. In such cases, simple, readily available materials are sought to produce so-called technical media. One of the most widespread technical media is sugar cane molasses (Fig. 3.5). It is a by-product of the sugar manufacturing process: once no more sugar can be crystallized from the raw crop, the residual product is molasses, still containing up to 20% of total sugar in the cane. The dark colored highly viscous liquid contains up to 400 g/kg sugars, so it is ideally suitable as a fermentation medium. Molasses, also called "blackstrap," contains a rich variety of trace elements and vitamins. The composition varies, depending upon the region of origin. An analysis of a commercially available molasses is shown in Tab. 3.8. The density is higher than the density of water, so values are given on mass per mass basis, rather than volume.

The carbon source is mainly sucrose. This is clear as molasses is a plant-derived product. Due to hydrolytic activities (sulfuric acid) during sugar production, sucrose is partly hydrolyzed to glucose and fructose, a mixture called invert sugar. This term evokes the change of direction of optical rotation during this process. Not all microorganisms can take up sucrose because of missing invertase activity to split sucrose. Yeast develops high activities during cultivation on molasses. Invertase for food application is actually produced by yeasts.

The high sugar concentration in the molasses makes it necessary to care for the osmotic pressure in the medium. Osmotolerant yeasts can grow in media with contents of 40–70% sugar. Therefore, it is possible that contaminants can come into the process from wild yeasts. On the other hand, the osmotic pressure of the solution is so high that our

Fig. 3.5: Molasses is a viscous and brownish liquid; pigments are from heat treatment during extraction of sugars.

Tab. 3.8: Composition of typical sugarcane molasses on a mass per mass basis.

Constituent	Unit	Value
Dry matter	% (g/100 g)	80
Total sugars	%	50
Of which, sucrose	%	67
Of which, inverted sugar	%	33
Nonsugar organic matter	%	10
N-containing nonsugar	%	4
Amino acids, peptides		
Measured as N·× 6,25		
Of which can be taken up by yeast	%	20
N-free nonsugar	%	6
Hemicellulose, organic acids		
Mineral elements	%	
Sodium		0.2
Potassium		3.5
Calcium		0.6
Magnesium		0.4
Chlorine		1.3
Sulfur		0.5
Phosphorus		0.1
Trace elements	mg/kg	
Selenium		0.02
Cobalt		0.5
Copper		9
Manganese		20
Zinc		10
Iron		200
Vitamins	µg/g	
Biotin		2.0
Pantothenic acid (vitamin B_5)		20.0
Inositol		4,000
Thiamine (vitamin B_1)		1.8

production strains suffer from suboptimal conditions. For batch processes, molasses has to be diluted to sugar concentrations of about 200 g/L. In Chapter 8, other process policies will be shown that use concentrations even as high as 500 g/L.

Even at the first glance, some minerals are not present in amounts sufficient to support balanced growth, up to high cell densities; see exercises. Therefore, additional nitrogen source ammonia is usually employed, but amino acids or proteins from other residuals can also be considered. This could be, for example, cheese whey or corn steep liquor (Tab. 3.9). Some small factories produce "organic yeast," based only on sustainable raw materials, where nitrogen source hydrolysates from proteins are an option. Trace elements like selenium are bonded to the organic matrix by the yeast cell, thus increasing their bioavailability for humans in deprived areas. Cobalt is available for use by animals, only after being bound, e.g., to vitamin B_{12} by yeasts or bacteria. Growth factors, including vitamins, are found in adequate quantities in numerous natural media. Nevertheless, they help to get the best growth rates and yields.

After harvesting the yeast biomass, the brown pigments have to be washed from the cells. Together with other residuals in the used-up medium, this creates a considerable wastewater problem, showing that bioprocesses are not automatically environmentally friendly in all aspects. For cultivation processes, which require a higher quality, high-test molasses is employed. The process follows the same pretreatment of sugar cane, but the steps of acidification, heating up, and sugar crystallization are skipped. This medium is clear, light brown syrup with a much better controlled composition than blackstrap molasses. Sugar content is up to 80%. The gentle processing makes addition of dry granular yeast necessary to care for invertase activity.

A complex medium with high nitrogen content and high nutritional value is corn steep liquor, a residual from the first washing and extraction step in the process of starch production from corn (Tab. 3.9). It is a brownish highly viscous fluid.

Tab. 3.9: Composition of corn steep liquor (expressed on a 100% dry matter basis).

Constituent	Unit % = [g/100 g]	Value
Dry matter	%	53
Proteins (N × 6,25)	%	45
Amino acids and peptides		
Fat	%	10
Ash	%	18
Mineral elements	%	
– Sodium		0.33
– Potassium		2.25
– Calcium		0.11
– Magnesium		0.66
– Phosphorus		3.12
Carbohydrates	%	37
– Lactic acid		20
– Phytic acid		10

The high lactic acid content caused by fermentative activity during production is typical. Phytic acid acts as phosphate storage in many plants. The formula $C_6H_{18}O_{24}P_6$ shows the high phosphorous content, thanks to six phosphate residues in the molecule. To guarantee availability, phytase activity has to be provided by the strains to be cultivated. On the other hand, corn steep liquor is rich in amino acids, including all possibly essential ones, making it a suitable medium or as adjunct to molasses for organisms that are auxotrophic in some amino acids.

With a similar function in media design, cheese whey can be employed. The 6% dry matter contains mainly lactose (4%), proteins (1%), and minerals (P, Ca, Na, and K). Of specific value are the high concentration of vitamins and cofactors, including vitamins A, B_{12}, and C, as well as choline and riboflavin. The focus of usage in fermentation lies in lactic acid bacteria and other anaerobic processes but also yeast biomass production (*Candida*). Lactose has to be hydrolyzed preceding cultivation in some cases, where the organisms do not show lactase activity.

Looking for other potential complex technical media, glycerol is a candidate. It is a highly abundant byproduct of biodiesel production. This was inspired using glycerol as a cheap carbon source for fermentation. Production of 1,3-propanediol (anaerobic) is of industrial importance, and succinic acid is promising and under development. Lignocellulose, in the form of straw or wood chips, is the most abundant raw material for possible use as renewable resource. However, prior to fermentation, hydrolytic pretreatment is necessary. Otherwise, the lignin network protects the cellulose against microbial attack. In fact, this is one of its biological functions. That is obvious when observing that wood decay takes months under natural conditions. White rot fungi degrade lignocellulose in nature, where in the first stage, cellulose (white) is left over for a while. This has evoked some scientific interest, as cellulose is a valuable material for other purposes. The idea is to use a medium without specific ions being cofactors for cellulases. This idea of influencing cell activity by targeted ion depletion in the medium, unfortunately, did not work out in this case but is an applied option for other processes. Enzymatic hydrolysis of straw with the lignin-degrading enzyme, laccase, derived from the mentioned rot fungi, is under investigation at the demonstration scale. Therefore, the usage of straw and wood as a fermentation medium meant to convert this cheap raw material into high value products or liquid fuels is left for the future.

3.5 Additional aspects – mutual interactions between cells and medium

Up to this point, we considered mainly medium composition under the condition of a constant elemental biomass composition and, therefore, the final biomass concentration. Nevertheless, the other way round is also true: media compounds also influence cell composition. This can be the accumulation of intracellular storage compounds or, especially for eukaryotes, changes in macromolecular cell composition, which has to be considered in media design.

As persons responsible for process development, we are interested not directly in elemental but in macromolecular composition of the biomass. In fact, the elemental composition is a consequence of the macromolecular composition. All organisms have a specific content of proteins, nucleic acids, carbohydrates, or lipids with which they adapt to a given environment. That holds, of course, also for a higher level of detail, meaning, e.g., the content of specific enzymes. The other way round is the idea to shift the macromolecular stoichiometry by a targeted deprivation or surplus of specific nutrients. A typical macromolecular composition is given in Tab. 3.10, with $e_{Mm/X}$ being the mass fraction of a molecular compound Mm in cell dry mass. Lipids, as a structural part of each cell, are listed here as well. Of course, there are differences with respect to the group of microorganisms and especially the growth rate.

Tab. 3.10: Typical macromolecular composition of microorganisms.

Species	Mass fractions $e_{Mm,X}$ (g/g) on dry mass basis					
	Protein	Nucleic acids (RNA/ DNA ≈ 7)	Carbohydrates	(Phospho)- Lipids	Storage compound	Others (ions, metabolites)
Virtuella generica	0.55	0.20	0.15	0.1	0.00	0.03

Now, we try to back calculate the macromolecular composition (Tab. 3.11) to the elemental composition. As a first attempt, representative molecules are chosen for which the empirical formula is known. These monomers and polymers could be defined further by data from public data bases or references.

Tab. 3.11: Elemental composition of macromolecules, based on representative mono- and polymers.

	Empirical formula	Molecular mass, M (g/mol)	Based on
Proteins	$C_{1489}H_{2403}N_{409}O_{453}S_{10}$	33603.42	Glucokinase *Bacillus subtilis*
Nucleic acids	$-C_{39}H_{49}O_{25}N_{15}P_4-$	1251.79	RNA chain of C,G,A,T/U
Carbohydrates	$-C_6H_{10}O_5-$	162.14	Glucose–H_2O
(Phospho)-lipids	$C_{42}H_{79}O_{10}P$	775.05	Dioleoyl-phosphatidy-glycerol
Others			

From this table, the elemental composition can be estimated, as shown in Tab. 3.12.

The calculated empirical formula of the virtual microorganisms *Virtuella generica* turns out to be quite close to the "measured" elemental formula of this species. This is only a crude estimate but justifies our view that the elemental composition is a projection of the macromolecular composition. For complex media, such calculations can

Tab 3.12: Comparison of the previously assumed elemental composition and the calculated elemental composition from the macromolecules.

Species	Mass fractions $e_{E,X}$ (g/g) and $e_{E,Mm}$ (g/g)					
	C	O	N	H	P	S
Virtuella generica	0.47	0.31	0.11	0.07	0.025	0.009
Proteins	0.532	0.216	0.170	0.072	0.000	0.010
Nucleic acids dAGCT	0.374	0.320	0.168	0.039	0.099	0.000
Carbohydrates	0.444	0.493	0.000	0.062	0.000	0.000
(Phospho)-lipids	0.651	0.206	0.000	0.103	0.040	0.000
Others						
Calculated *Virtuella generica*	0.499	0.277	0.127	0.067	0.024	0.005

give a first basis for possible yields of biomass from a given medium. This view can also be applied vice versa. Changing the elemental composition by changing the medium will possibly change the macromolecular composition and therefore the physiological state of the cells. This is especially true for species that can accumulate storage material. One example is the accumulation of bioplastic PHB by some bacteria during nitrogen starvation. The cells are then no longer in a position to synthetize proteins and nucleic acids. Catabolic reactions can nevertheless proceed, allowing the cells to produce ATP and accumulate carbon-rich compounds. Besides accumulation of storage material, prokaryotes are quite constant in their composition. A fixed ratio of lipids, proteins, and polysaccharides is necessary to build up the cell membrane and cell wall. The cell machinery will work properly only for an optimal protein-to-nucleic acid ratio. Eukaryotes like yeast (see running case study) can adapt to a changing supply of different nutrients with, e.g., changing protein content.

The actual concentrations of many compounds can influence growth velocity with respect to limitation or inhibition. In other cases, two or more compounds can positively interact in supporting growth, as we have sketched, for essential amino acids and growth factors. These aspects are discussed in the next chapter on kinetics. Optimization in media design is discussed under process strategies and in the case studies. The overall approach to design media is summarized in Fig. 3.6. It is important to understand that thinking always starts from the top and not from the bottom with the application of unaudited recipes.

Sustainability is an important factor in the public's acceptance of a bioprocess. Bioprocesses are not necessarily sustainable by themselves. Specifically, regarding the medium, at least the C source must be made from a renewable raw material. This is undoubtedly the case with glucose. This brings the bioprocess into a discussion of the competition between bioprocesses and food.

This also applies to residual materials such as molasses or glycerol, but is tolerated, in relation to the quantities used. Ideally, residual materials from other processes are

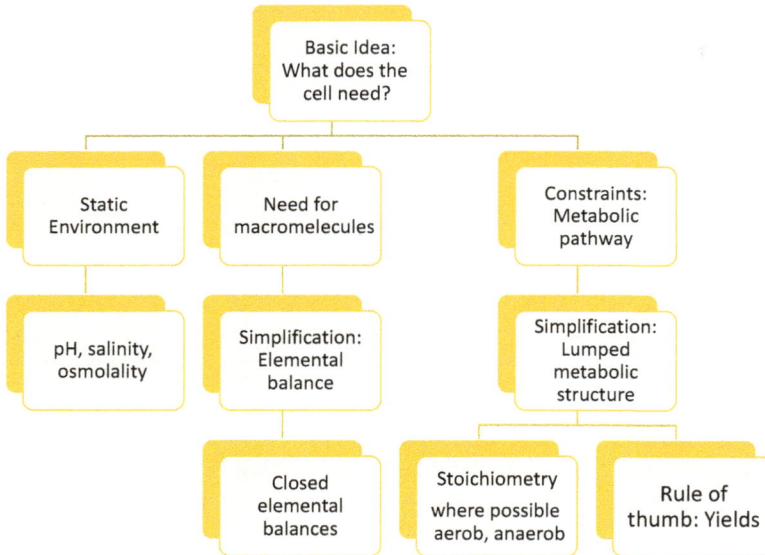

Fig. 3.6: The standard approach to medium design; the general idea to give the cell what it needs is broken down to different aspects of quantification.

used, for which no other utilization is possible. Slurry or straw for biogas plants is one example. Here, the residues are effectively disposed of and utilized for energy. Composting is the fermentation of solids such as plant residues, straw, or wood waste. Although it is a way of recycling, biotechnological recyclables are unfortunately not based on these materials, to this day. One reason for this is the very heterogeneous structure of lignin, for example, which makes biological attack very slow, which is what nature primarily intended. Chemical hydrolysis is possible but produces toxic substances such as certain pentoses. After all, straw is used to grow fungi and there are also attempts to produce enzymes or make them work, but these are slow processes that are also carried out outdoors in straw bales. The use of waste materials and the production of reusable products, avoiding the production of new wastes, is called "circular economy," where bioprocesses have the best options.

3.6 Media for microalgae – a clear matter of fact

Although microalgae may be more variable with respect to macromolecular composition, medium design follows the same principles as in the heterotrophic case, of course, without the organic C-source. Nevertheless, some aspects have to be considered, coming from the specific habitats of microalgae. Concerning different habitats of microalgae, one commonly distinguishes between freshwater and seawater algae.

Cyanobacteria, as the prokaryotic first photosynthetic organisms, have nutrient requirements differing from their eukaryotic descendants and can be found in freshwater and seawater or in terrestrial habitats. The group of freshwater microalgae is also comprised of species not only distributed in water bodies like lakes, but also species present in soil water, like the microalgae *C. reinhardtii*. Additionally, there are also microalgae species distributed in brackish water or on surfaces of plants, animals, ground, or rock. Hence, according to their different habitats, microalgae species can show very different nutritional requirements, especially concerning salinity and osmotic pressure. For industrial microalgae processes, the topic of contamination of reactor systems (either open systems like open ponds, raceway ponds, or also closed reactors) can be of importance. Here, highly selective growth conditions, like a high pH (10–11 for the cyanobacterium *Arthrospira platensis* for phycocyanin production) and/or a high salt concentration of sea water (for the green alga *Dunaliella salina* in β-carotene production) enable an axenic culture or also very little contamination for open systems in the larger, industrial scale. To start with, a look at the elemental composition (Tab. 3.13) is required.

Tab. 3.13: Elemental composition of microalgae and a bacterium for comparison (data are given as percentages).

	C	N	P	S	K	Mg	Ca	Fe
E. coli	47.0	11.0	2.5	0.9	0.012	0.08	1.27	0.002
C. reinhardtii	45.3	9.17	3.16	0.42	0.38	0.4	0.92	
Cyano								

With respect to the macro elements C, N, P, and S, the composition between the different species is similar. This is expected, as proteins and nucleic acids make up the main compounds. Storage and carbohydrates can change the level of these elements. Ions, not directly built into macromolecules, show some differences. Looking, e.g., to Mg^{++}, it is needed in higher quota for algae, as it is part of chlorophyll light harvesting system and is also a ligand to the CO_2-binding enzyme RuBisCo.

Now, we find out whether these differences affect commonly applied defined media listed in Tab. 3.14. Freshwater algae (e.g., *Chlorella* or *Chlamydomonas*) are often propagated in the TRIS-acetate-phosphate-buffer medium, abbreviated as TAP. In technical media, as described here, acetate as the organic C source is omitted. Seawater algae (e.g., *Emiliana*) are best cultivated in an enriched artificial sea water (EASW) medium. Cyanobacteria (spirulina, *Synechocystis*) need a special cocktail (BG for blue-green).

The first thing that stands out is the high content of Na^+ and Cl^-. The values correspond to their occurrence in seawater, causing the salinity and thus the osmolarity of the environment. Of course, the elements are not consumed in this concentration. It is

Tab. 3.14: Examples of commonly defined freshwater and seawater media for eukaryotic microalgae and cyanobacteria with ion concentrations (vitamins, organic buffers, and chelating agents are left out for simplicity).

	Freshwater, TAP medium		Artificial seawater medium, modified EASW		Cyanobacteria, BG11 medium	
Ion	(mmol/L)	(g/L)	(mmol/L)	(g/L)	(mmol/L)	(g/L)
NO_3^-	–	–	0.55	0.034	17.65	1.09
NH_4^+	7.0	0.13	–	–	–	–
K^+	1.64	0.06	8.77	0.342	0.351	0.014
Ca^{2+}	0.34	0.014	9.14	0.37	0.245	0.01
Cl^-	7.75	0.28	483	17	0.51	0.018
Mg^{2+}	0.41	0.01	47.2	1.15	0.304	0.007
SO_4^{2-}	0.51	0.049	25	2.4	0.305	0.029
HPO_4^{2-}	0.62	0.06	–	–	0.175	0.017
$H_2PO_4^-$	0.40	0.04	0.021	0.002	–	–
Na^+	0.3	0.007	416	9.6	23.4	0.54
F^-	–	–	0.066	0.001	–	–
SiO_3^{2-}	–	–	0.105	0.008	–	–
Br^-	–	–	0.725	0.058	–	–
Trace elements	(µmol/L)	(mg/L)	(µmol/L)	(mg/L)	(µmol/L)	(mg/L)
MoO_4^{2-}	–	–	0.006	9.8×10^{-4}	1.6	0.26
$Mo_7O_{24}^{6-}$	0.89	0.94	–	–	–	–
SeO_3^{2-}	–	–	0.001	1.27×10^{-4}	–	–
Zn^{2+}	77	5	0.254	0.017	0.77	0.05
Mn^{2+}	26	1.43	2.42	0.133	9.2	0.51
Fe^{3+}, Fe^{2+}	17.98	1.0	6.56	0.366	24.5	1.37
Co^{2+}	6.7	0.40	0.006	3.35×10^{-4}	0.267	0.016
Cu^{2+}	6.4	0.41	–	–	0.32	0.02
Sr^{2+}	–	–	82	7.19	–	–
Ni^{2+}	–	–	0.006	3.7×10^{-4}	–	–
BO_3^{3-}	180	11	372	21.88	46.3	2.72

therefore important to distinguish here between proportions that simulate the environment and proportions that are macro- or micronutrients. Also, K^+ and Ca^{2+} are highly overdosed, the latter to help as buffer and stabilize the alkalinity.

Note that the concentrations are calculated for much lower final biomass concentrations (e.g., 3 g/L) compared to heterotrophic cultivations (e.g., 50 g/L). So, absolute amounts in the medium must be set in relation to biomass produced or to other elements, preferably N. Nitrogen is mostly applied in the form of nitrate or ammonia, sulfur in the form of sulfate, and phosphorus as hydrogen phosphate or dihydrogen phosphate. As in other media, these also act as buffers in the concentrations given above. Concerning metal ions, Fe, followed by Mn, Zn, and Cu ions, are the most abundant ones in cells. These have to be also supplied by the medium in a suitable concentration. Also noteworthy is that some microalgae require vitamin supplementation in the medium, whereas others do not (here, EASW medium is complemented with thiamine, biotin, and vitamin B_{12}). In some cases, this may be attributed to their original habitat and potential symbioses with other microorganisms where there is a mutual exchange of essential metabolites. In lab applications, chelating agents and buffers are also applied.

While media compositions are often optimized by trial and error, it is worthwhile to consider trace elements in more detail as they are strain-specific and there is "a great deal of leeway," as somebody in the industry said. The basic difference between phototrophs and heterotrophs is, of course, photosynthesis. Chlorophylls contain Mg^{2+} as the central ion. Moreover, Mg^{2+} is an essential cofactor of the enzyme RuBisCo. Consequently, phototrophic cells and media contain a high concentration of magnesium. Current research disclosed some other examples: in contrast to *C. reinhardtii*, other freshwater algae like *Chlorella* and some marine algae are known to require boron. Soil, the original environment from which *Chlamydomonas reinhardtii* was isolated from, is a boron-poor substrate, which is assumed to be the reason for its lack of B requirement, compared to other algae species. Ionomic studies carefully investigating ion demand on the molecular level, revealed that selenium is required by "Chlamy" (as it is called in lab slang) for seleno-proteins. On the other hand, it has been shown that it inhibits the synthesis of chlorophyll. In general, the elimination of elements not required for algal growth also reduces the risk of contamination with unwanted organisms, which might require these substances. Strontium in EASW has to be given in even higher amounts for *Emiliania* for coccolith production.

The last point to be addressed is supply with carbon sources. It is also possible to grow several species of microalgae under heterotrophic or mixotrophic conditions for different reasons. Here, an organic C source is added to the medium, where usually acetate ("A" in TAP) or rarely a combination of glucose, acetate, glycerol, and sucrose is used. This may enhance growth, but organic carbon sources added to a medium might lead to an overgrowing by bacterial or fungal contamination. Even if heterotrophic production in closed, bioreactors are, for the time being, cheaper than phototrophic production – it is, in the sense, of applying microalgae to use CO_2 and sunlight. For small-scale lab applications, inorganic carbon is supplied by $NaHCO_3/Na_2CO_3$, serving additionally as carbonate buffer. While algae can take up CO_2, some can also use HCO_3^-. The optimal partial pressure of CO_2, $p_{CO2} = 0.5\%$. Note that this is about ten times higher than in air, a sign that photosynthesis develops in times when atmospheric CO_2 fraction is much higher. To overcome the diffusion barrier between gas bubbles and medium, CO_2 gas fractions for aeration have to be set between 5% and 10%. This makes application of off gas from gas combustion plants or heterotrophic fermentations possible.

3.7 Light – the little bit different energy source

Microalgae, as photoautotrophic microorganisms, need light as the primary energy source. Light is a form of electromagnetic radiation in a given spectral range and propagating in a defined direction. Light is physically measured as an energy flux, called light intensity. More specifically, for technical application and in the context of photo-biotechnology, the light energy flowing through a given area per time is called irradiance I_e (W/m^2). The index "e" recalls the energy flux. To get an idea of the primary irradiance from the sun and its spectral distribution, we look at measurements of sunlight, given in Fig. 3.7. In this graph, the spectral irradiance $I_{e,spec}(\lambda)$ (W/m^2 μm) is plotted over wavelength, meaning the contribution of light to the energy flux is divided into small slices (1 μm) of wavelength λ. The irradiance in a given wavelength range of interest is therefore given as:

$$I_{e,\text{Range}} = \int_{\lambda_{\min}}^{\lambda_{\max}} I_{e,\,\text{spec}}(\lambda) \cdot d\lambda \qquad (3.16)$$

The light energy E_A (MJ/m^2) in a spectral range falling on a given ground area in a given time interval, e.g., a day or a year, is calculated as

$$E_{A,\text{Range}} = \int_{t_1}^{t_2} I_{e,\text{Range}} \cdot dt \qquad (3.17)$$

The spectrum of light is classified into different wavelength ranges. The infrared range (IR) makes about 43% of the total solar irradiance (energy flux). Its role in photobioreactor operation is addressed later in this chapter. The ultraviolet range (UV) contributes about 7%. The different wavelength ranges of light have different physiological impacts. Only the photosynthetically active radiation (PAR) – from 400 to 700 nm – meaning, 50% of the solar energy, is used by microalgae for photosynthesis. Note that visible light is nearly of the same range. IR is not used. Biologists think that considering water as a strong IR filter, photosynthesis was "invented" by nature in water environment. The values are given, understanding a vertical position of the sun. However, the real irradiation depends on the time of day and the latitude of our position. Furthermore, we have to decide in which direction we have to measure light intensity. As an example, different measurements for Las Vegas are plotted in Fig. 3.8.

"Direct normal irradiance" is measured directly toward the sun. This may be important for our personal feeling or for photovoltaics, in cases when the modules can be turned toward the sun. "Global horizontal irradiance" is the radiation hitting a horizontal plane. This is indeed the measure we need to use to assess light impact on photobioreactors mounted on ground surface. Not all the photons find their way directly from the sun to a given spot under investigation, but are scattered by clouds or re-

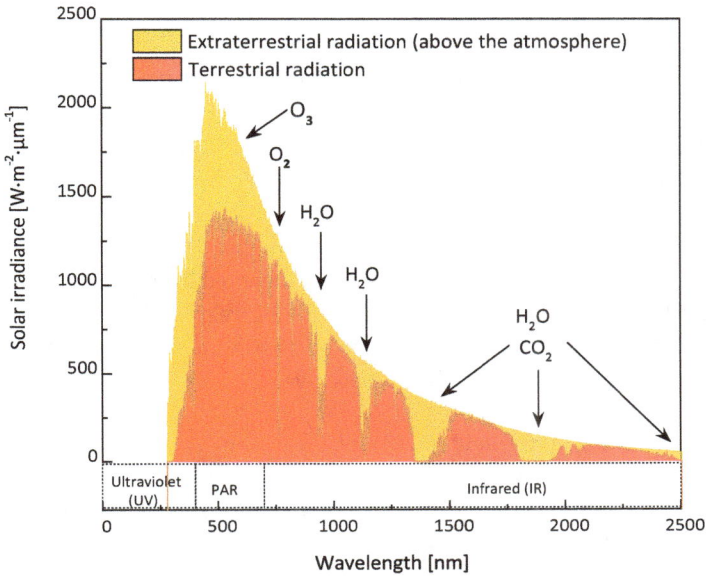

Fig. 3.7: Spectral irradiance of sunlight; absorption of light by the atmosphere, and especially the marked singularities are evoked, especially by water (clouds), oxygen, ozone, and carbon dioxide (greenhouse effect!).

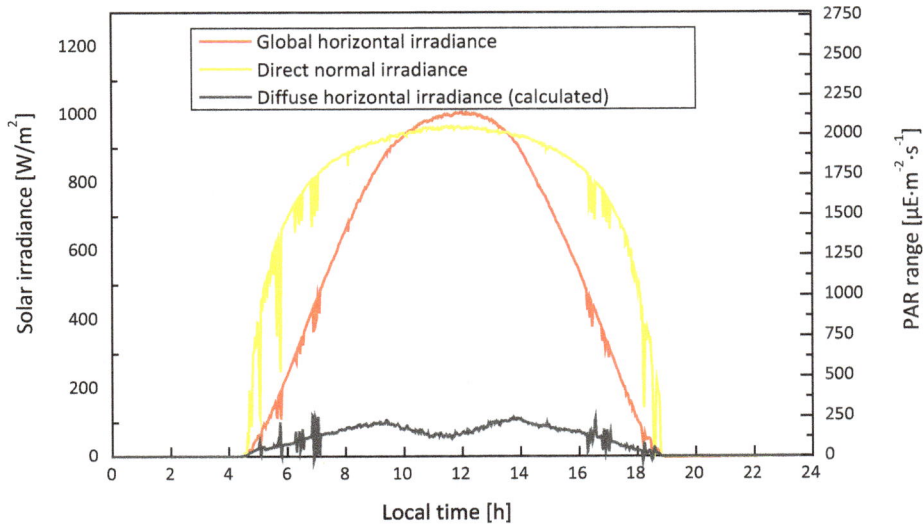

Fig. 3.8: Real data of irradiance in Las Vegas, 21-Jun-2011, a slightly cloudy day, with different ways to characterize the spatial aspects of irradiance.

flected from mountains, buildings, or trees. This fraction of radiation is summarized as "diffuse horizontal irradiance."

From the photosynthesis paragraph, we already now that the quantum efficiency of photosynthesis is wavelength-independent. This also justifies describing light over photon flux density (PFD), what is a useful and widely used variable in this context, and it is denoted here as $I_{h\nu}$ ($\mu E/m^2$ s). The index "$h\nu$" indicates that the variable means the number of photons impinging a given surface per s. The abbreviation "μE" is a tribute to Albert Einstein and means 1 micro-mole of photons. The photosynthetically active photon flux density is abbreviated as PPFD and denoted here as $I_{h\nu,PAR}$.

On a technical level, energetic considerations are in the foreground, so we need the relation between irradiance, on an energy basis and on a photon flux density basis. Each photon represents a given photon energy as given by Max Planck as

$$E_{h\nu}(f) = h \cdot f \text{ and } E_{h\nu}(\lambda) = h \cdot \frac{c}{\lambda} \tag{3.18}$$

$E_{h\nu}$ is the energy amount of a single photon (J); h is Planck's constant (6.626×10^{-34} J · s); c is the speed of light (2.998×10^8 m/s); λ is the wavelength of the photon (m); f is the frequency of the light wave (s^{-1}) (Hz). "Blue" photons, for example, have a higher energy content (400 nm, 4.97×10^{-19} J) than "red" photons (700 nm, 2.84×10^{-19} J). The mean value of photon energy in the PAR range (λ_{mean} = 550 nm, linear in λ!) is 3.61×10^{-19} J. The energy content of 1 E of a 550 nm photon is consequently $N_A \times 10^{-6} \cdot E_{h\nu}$ = 6.02×10^{23} /mol $\times 3.61 \times 10^{-19}$ J = 217.3 kJ/mol.

The needed relation now reads:

$$I_{e,Range} = \int_{\lambda_{min}}^{\lambda_{max}} I_{h\nu}(\lambda) \cdot E_{h\nu}(\lambda) \cdot d\lambda \tag{3.19}$$

Now, an overview about the situation at real locations should be taken (Tab. 3.15). This gives an idea how much sun energy can be harnessed.

Tab. 3.15: Energetic values of ground radiation.

Location	Latitude	Annual energy measured	Annual energy PAR	Power average PAR	Intensity annual average PAR	Max Intensity noon PAR
	(°N)	(MJ/m²)	(MJ/m²)	(W/m²)	($\mu E/m^2$ s)	($\mu E/m^2$ s)
Windhoek	−22.6	8,755	4,027	128	591	2,370
Las Vegas	36.1	7,190	3,320	105	488	2,180
Karlsruhe	49.0	3,231	1,486	47	218	1,940

We will need this table further down to calculate the possible microalgae production at different locations. It is somehow astonishing that only 2.5 times more sun energy is available at the tropical regions than in higher latitudes. The total annual solar energy impinging our planet is about 1.5×10^{18} kWh/a, while the world energy demand (2016) is 1.8×10^{14} kWh/a. Indeed, there is enough sun energy, nearly 10,000 times more than needed; the problem is that it is highly dispersed over large areas – one of the basic problems of microalgal biotechnology.

3.8 Questions and suggestions

1. Compare the list of vitamins for microorganisms (Tab. 3.8) with the list in the introduction (Fig. 1.3b).
2. Compare the sea water medium with a table of natural sea water.
3. Even the macronutrients for N- and P-supply are problematic with respect to sustainability. Why?
4. From which countries do molasses or glycerol come? Why is shipping costly?

Chapter 4
Kinetics – finding quantities for bioprocess reactions

One of the unique selling points of microbial reactions is the formation of biomass and complex high-value products from medium compounds. These reactions take place almost entirely inside the cells. For design and assessment of bioprocesses the velocity of the reactions and intrinsic stoichiometry has to be understood. A fundamental concern of this chapter is to simplify complex but basically known biological relationships and to transform them into manageable equations. Like the elemental balances in medium design, strict constraints like intracellular balances are helpful and often sufficient to set up macroscopically observable relations. Reaction engineering principles are applied to the metabolic network. As a consequence of simplifying, a priori unknown parameters must be defined and measured case by case. The following chapter starts with formulation of enzymatic reactions as the basic units of cellular metabolism. The second step is to find kinetic equations for the velocity of substrate uptake and growth. Finally, reasonable coefficients describing stoichiometric relations between the different pathways will be formulated.

4.1 Kinetics – the scaffold of reaction engineering

In physics and engineering, kinetics (Greek: "kinesis" = movement) is a term for description of the relationship between the motion of bodies and its causes, namely forces. By analogy, reaction kinetics is the study of rates (=velocity) of chemical reactions. The forces here are material concentrations. Kinetics is the scaffold for understanding and designing systems in reaction engineering. Kinetic parameters and other describing variables can be directly extracted from measured data.

Reaction rates R measure the number of reacting moles Δmol or masses Δm of a substance Comp in a given time interval Δt, e.g., the unit time. Related to volumetric concentrations they are given here as Q_{Comp} (mol/(L·h)) or as R_{Comp} (g/(L·h)). Growth and product formation turnover rates are an important measure for the efficiency or intensity of a bioprocess. The differences and the resulting rates are often calculated such that they become positive values. So, we speak of "product formation rate" and "substrate consumption rate," making both positive even though glucose concentration decreases. A second important measure is the amount of biomass or product being formed by a given amount of substrate used. This is measured as the biomass yield $Y_{X,S} = \Delta m_X/\Delta m_S$ or product yield $Y_{X,P} = \Delta m_P/\Delta m_S$. Yield is a considerable contributor to the cost structure of the production process. Here again, positive values are pur-

https://doi.org/10.1515/9783110773354-004

sued despite decrease or increase of the involved concentrations. Volumetric productivity P_V (g/L h) measures the amount of product that has been formed in a reactor with the working volume V_R for the duration of the fermentation process as P_V (g/L·h⁻) = $\Delta m / \Delta t / V_R$. For batch processes it is formally equal to the production rate, but interpretation is done in a sense of efficiently using the volume of the reactor.

A batch experiment based on the medium developed in the last chapter is shown in Fig. 4.1; characteristic numbers can be taken as a rough evaluation from the data sheet in Tab. 4.1.

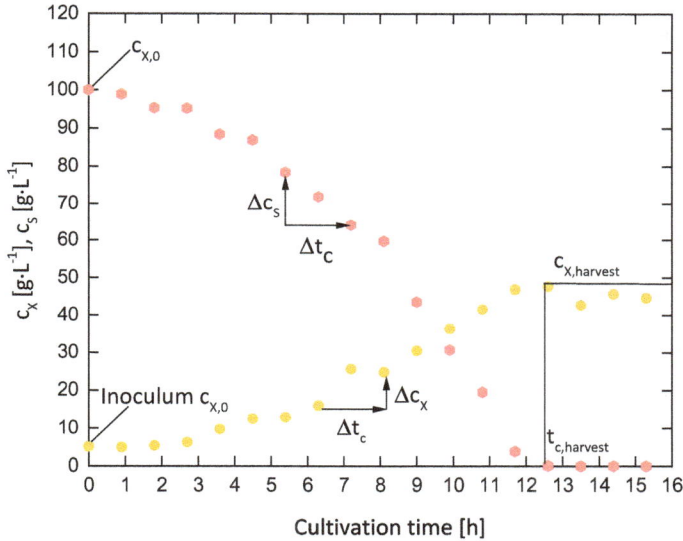

Fig. 4.1: Measurements of biomass and substrate concentration of a typical batch process. The differences are counted positively from the arrowhead backward.

The biomass formation rate and the substrate consumption rate are now:

$$R_X = \Delta c_X / \Delta t_C; \quad R_S = \Delta c_S / \Delta t_C; \tag{4.1}$$

Note that the term "rate" always means something that happens over time like a velocity. In this context, it is related to the volume in order to abstract from the concrete size of a reactor.

"Yields" is a term to describe the amount of something produced during a conversion process in relation to a related quantity from which it is produced. The "productivity" or more exactly the "volumetric productivity" measures the amount of something, e.g., the product in relation to the time and the volume used:

$$Y_{X,S} = \Delta c_X / \Delta c_S; \quad P_{V,X} = \Delta c_X / \Delta t_C; \tag{4.2}$$

Tab. 4.1: Calculated values for the experiment shown in Fig. 4.1.

	R_X (g/L/h)	R_S (g/L/h)	$Y_{X,S}$ (g/g)	$P_{V,X}$ (g/L/h)
$\Delta t_c = (5-4)$ h	13.0–10.5 = 2.5	86.5–81.0 = 5.5	2.5/5.5 = 0.455	13.0–10.5 = 2.5
$\Delta t_c = (10-9)$ h	36.0–29.7 = 6.3	43.6–29.2 = 14.4	6.3/14.4 = 0.438	36.0–29.7 = 6.3
$\Delta t_c = (12.5-0)$ h			42.8/100.0 = 0.428	47.8–5.0 = 42.8

Biomass formation rate, or in fact growth rate R_X, increases during the middle phase of cultivation and is obviously low at the beginning and even negative at the end. Substrate uptake rate R_S increases as well in the same interval. The yield coefficient $Y_{X,S}$ is quite constant showing only a small decrease. For the whole cultivation the overall yield from inoculum to harvest is $Y_{X,S} = 0.428$ g/g· lower than the expected 0.5 g/g, (see also Section 4.4). Volumetric productivity $P_{V,X}$ between the start of the fermentation and the harvesting time at the point at highest biomass concentration is 42.8 g/L h. The major contribution happens in the late middle phase of the process. Beyond these practical characteristic numbers, we have the ambition to describe the underlying reaction mechanisms and understand them as a function of process conditions. Some aspects are given in the following paragraphs laying the foundation for rational process design.

4.2 Enzymes as basic components – determining kinetics

Before starting with biological systems, we first refer to the mass action law, which is the fundamental approach in chemical reaction engineering.

Firstly, we consider a simple reaction equation:

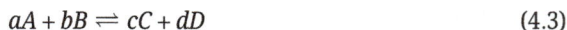

$$aA + bB \rightleftharpoons cC + dD \tag{4.3}$$

The equilibrium is found for a specific set of concentrations:

$$K_{\text{equ}} = \frac{c_C^c \cdot c_D^d}{c_A^a \cdot c_B^b} \tag{4.4}$$

Here K_{equ} is the equilibrium constant. The background is the idea that the reactions are due to collisions of reactant species involved. The frequency with which the molecules collide depends upon their concentrations or more precisely on their activity being dependent inter alia on temperature. The probability of a collision of two different reactants is given by the product of their respective concentrations. Whether a reaction takes place or not depends on the activation energy. This encourages us to formulate volumetric reactions rates R (mol/L h) as

$$R_{\text{forth}} = k_{AB,\,CD} \cdot c_A^a \cdot c_B^b \tag{4.5}$$

$$R_{\text{back}} = k_{CD,\,AB} \cdot c_C^c \cdot c_D^d \tag{4.6}$$

The parameters $k_{X,Y}$ are the reaction constants, e.g., from the compound X to Y. The unit depends on the order of the reaction. In cases where a single compound dissociates into two compounds, it has the dimension of a concentration. The equilibrium is reached when forward and back reactions are equal. The mass action law will be applied to enzymatic reactions in the next paragraph.

Enzymes catalyze almost all reactions in the cell. Isolated enzymes are also used in bioprocesses in vitro as effective catalysts. Substrate molecules bind to the enzyme at the so-called active site, where the conversion occurs. Thanks to their spatial structure enzymes exhibit a much higher substrate specificity compared to inorganic catalysts. In the past, the "lock-and-key model" was often used in this context. However, it has been shown that the spatial structure of the enzyme is flexible and adapts to the substrate. Today, this is referred to as spatial adaption "induced fit model." It represents the exact fit of the substrate molecule onto the active site excluding other molecules with a different structure. Finally, the products are released, and the enzyme is ready to bind to the next substrate molecule. Technical catalysts are often faster, but only at high temperatures, which cannot be used in bioprocesses. At ambient temperatures, the enzymes beat purely chemical catalysators. Biochemical reactions are proceeding far from their respective equilibrium. So, a backward reaction from the product to the substrate is usually but not always negligible. Figure 4.2 shows a representation of an enzymatic reaction process and will be used to set up the reaction equations following the mass action law.

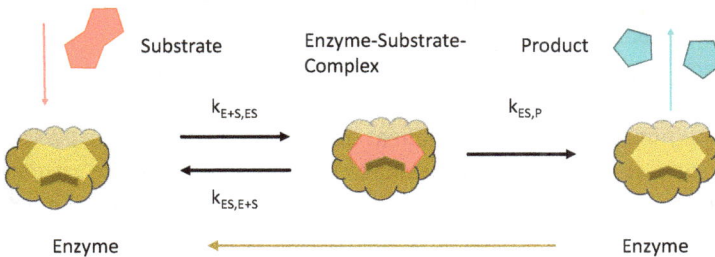

Fig. 4.2: Schematic representation of an enzymatic reaction process. The "pocket" of the enzymes, into which the substrate fits, is the binding site and shown here brighter than the enzyme itself. The small hook represents the reaction site. The arrows indicate direction of the respective reaction step.

For many enzymatic reactions it is assumed that the first step, the binding of the substrate S to the enzyme E, is reversible, while the reaction step and release of the product is irreversible leading to the reaction scheme:

$$E + S \xleftrightarrow{k_{AB,CD} k_{CD,AB}} ES \rightarrow P + E \tag{4.7}$$

Step-by-step the reaction rates are set up for each of the reaction steps involved:

$$R_{E+S,ES} = k_{E+S,ES} \cdot c_S \cdot c_E \tag{4.8}$$

for the forward reaction (second order) from enzyme and substrate to the enzyme-substrate complex,

$$R_{ES,E+S} = k_{ES,E+S} \cdot c_{ES} \tag{4.9}$$

for the back reaction (dissociation, first-order) of the complex, and

$$R_{ES,P} = k_{ES,P} \cdot c_{ES} \tag{4.10}$$

for the product formation itself.

Now we can set up material balance equations for the enzyme and the enzyme-substrate complex. The balance equations include all reactions in which the respective molecule is converted or formed. Contributions to the balance are counted positively along the counting arrows leading out of or into the balance boundaries. The arrows are chosen in a way that makes the numerical values positive. This gives a more intuitive view on the final results. R_{net} here means a possibly occurring accumulation. Starting with the free enzyme the balance reads:

$$R_{E,\text{net}} = -R_{E+S,ES} + R_{ES,E+S} + R_{ES,P} = -k_{E+S,ES} \cdot c_S \cdot c_E + k_{ES,E+S} \cdot c_{ES} - k_{ES,P} \cdot c_{ES} \tag{4.11}$$

For the enzyme-substrate complex we get accordingly:

$$R_{ES,\text{net}} = R_{E+S,ES} - Q_{ES,E+S} - Q_{ES,P} = k_{E+S,ES} \cdot c_S \cdot c_E - k_{ES,E+S} \cdot c_{ES} - k_{ES,P} \cdot c_{ES} \tag{4.12}$$

This equation is the same as the one before except for the sign. Both equations are therefore linearly dependent.

The single reaction steps are very fast, actually faster than we could observe during the bioprocess. This gives an argument for a further simplification step: the reduction of temporal resolution. In a virtual experiment the substrate concentration shall be kept constant by continuous feeding of new substrate. After some time, we assume that the system is in steady state meaning that the reactions take place, but the observable concentrations do not change. Under these conditions no accumulation of E and ES occurs leading to:

$$Q_{E,\text{net}} \approx 0 \quad \text{and} \quad R_{ES,\text{net}} \approx 0 \tag{4.13}$$

Furthermore, a reduction of the time horizon to reasonable periods can be envisaged, excluding effects like enzyme aging or adding fresh enzyme. The total amount of enzyme protein is therefore constant, and we note:

$$c_{E,\text{tot}} = c_E + c_{ES} \tag{4.14}$$

The set of equations (4.8)–(4.14) now contains all a priori knowledge from the mass action law and the additional simplifying assumptions. The concentrations of the two enzyme configurations are now obtained by solving the balance equations (4.13) and (4.14) for c_E and c_{ES} inserting the equations (4.11) or (4.7). Note that only one of the two equations is necessary:

$$c_E = \frac{c_{E,\text{tot}} \cdot (k_{ES,E+S} + k_{ES,P})}{k_{E+S,ES} \cdot c_S + k_{ES,E+S} + k_{ES,P}} \tag{4.15}$$

$$c_{ES} = \frac{k_{E+S,ES} \cdot c_S \cdot c_{E,\text{tot}}}{k_{E+S,ES} \cdot c_S + k_{ES,E+S} + k_{ES,P}} \tag{4.16}$$

This calculation can be found as the Computer Algebra sheet in the supplementary material. It allows us to deduce the required relationship between substrate turnover and substrate concentration.

R_P follows directly from eq. (4.8) by inserting eq. (4.14). As no metabolites are allowed to accumulate, the substrate turnover rate is directly accessible from R_P:

$$R_S \approx R_P \tag{4.17}$$

Finally

$$R_S = \frac{k_{ES,P} \cdot c_S \cdot c_{E,\text{tot}}}{\frac{k_{ES,E+S} + k_{ES,P}}{k_{E+S,ES}} + c_S} \tag{4.18}$$

This equation still leaves a certain degree of helplessness, as it is not clear from where we get the unknown kinetic parameters without elaborating fast analytical measures. However, we do not need the precise value of all parameters, as they are partially linearly dependent and are therefore not necessary to be known separately. A common approach in process engineering is to lump several parameters into new ones, which are relevant and comparatively easy to measure.

With

$$RE + S, ES_{SE,\text{tot}S,\max} \tag{4.19}$$

and

$$k_S = \frac{k_{ES,E+S} + k_{ES,P}}{k_{E+S,ES}} \tag{4.20}$$

we obtain the rate equations, which are given here on a g/L basis:

!

$$R_S = R_{S,max} \cdot \frac{c_S}{k_S + c_S} \qquad (4.21)$$

as a manageable form of simple enzyme kinetics. This popular form of enzyme kinetics is known as Michaelis-Menten kinetics. This is in honor of the two women, Leonor Michaelis and Maud Menten, who developed the underlying theory (in 1910).

$R_{S,max}$ (g/L·h) is the maximum substrate turnover rate and k_S (g/L) the limitation constant, also called the half-saturation concentration or Michaelis constant. The usual abbreviation K_M is replaced here by k_S to distinguish limitation constants for different substrates more clearly. A graphical representation is given in Fig. 4.3. The two a priori unknown constants can be deduced from the basic kinetic constants k or directly measured in macroscopic experiments.

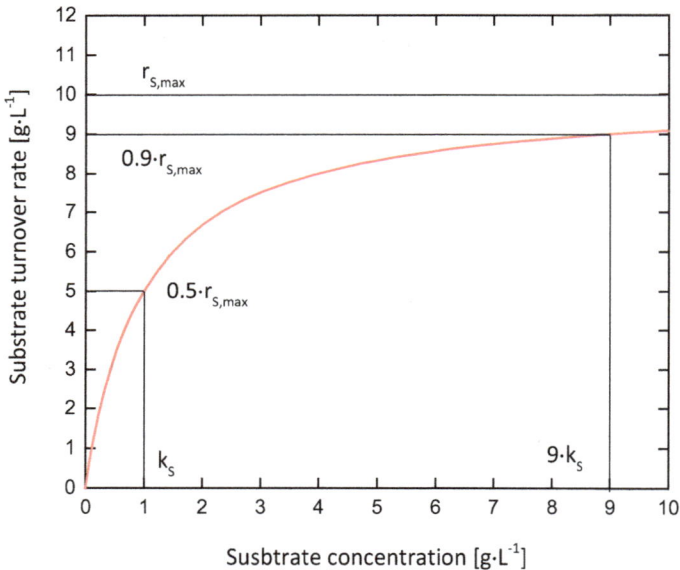

Fig. 4.3: Typical hyperbolic saturation curve of the Michaelis-Menten kinetics with $k_S = 1$ g/L [überall gleich] and $R_{S,max} = 10$ g/L. Two distinguished concentrations and the related substrate turnover rates are marked.

For low concentrations substrate turnover rate reflects the equilibrium of enzyme and substrate on the one hand and enzyme-substrate complex on the other. For the case of $c_S = k_S$ substrate turnover reaches the half maximum value as can easily be seen by substituting k_S in eq. (4.19). For high concentrations R_S asymptotically approaches $R_{S,max}$. However, even for ninefold concentrations of k_S only 90% of the maximum value is reached. Here the dissociation of ES to $E + P$ is rate-limiting (Tab. 4.2).

Figures 4.4 and 4.5 show the influence of the two formal parameters on the Michaelis-Menten curve.

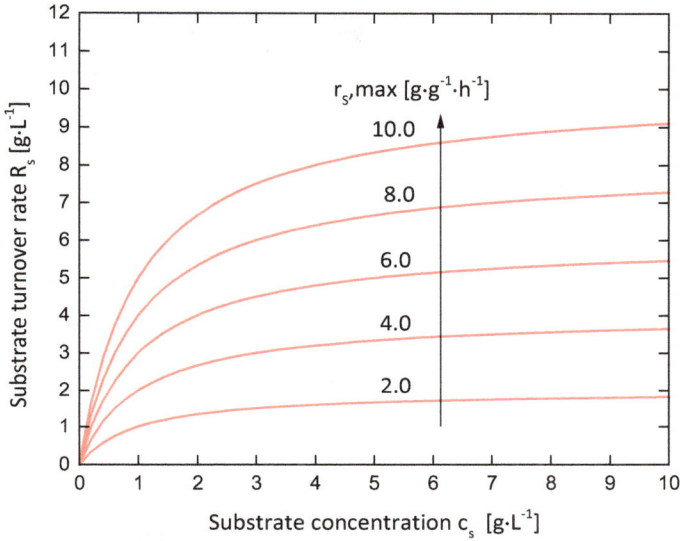

Fig. 4.4: Parameterized array of curves for different $R_{S,max}$ values.

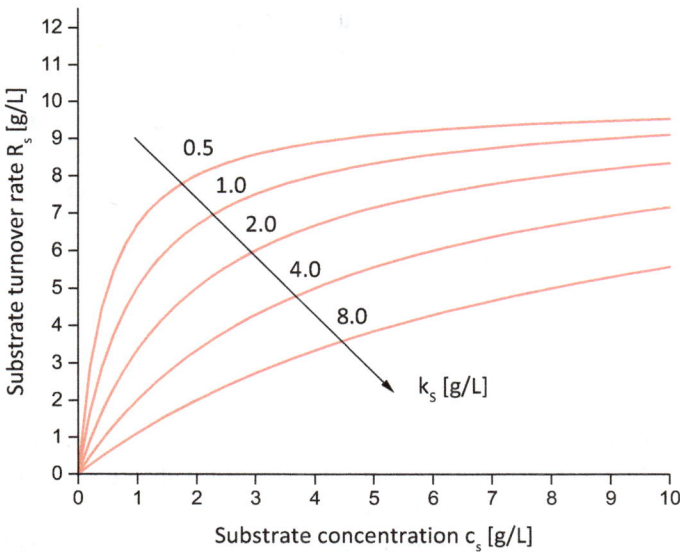

Fig. 4.5: Parameterized array of curves for different k_S values. The substrate turnover rate at a given substrate concentration depends nonlinearly on k_S.

To get an idea of the range of kinetic parameters some values are listed in Tab. 4.2.

Tab. 4.2: k_{Cat} turnover number.

Enzyme	Substrate	Product	Organism	k_{Cat}	k_S (mM)
Lactate dehydrogenase	L-Lactate + NAD$^+$	Pyruvate + NADH	*E. coli*	40	47,000 (lactate) 180 (NAD$^+$)
Lipase (EC 3.1.1.3)	Triacylglycerol	Fatty acid	*E. coli* BL21	1,500	20.4
Invertase	Saccharose	α-D-Glucose + β-D-Fructose	*Saccharomyces cerevisiae*	3.430	38
Aminoacylase (EC 3.5.1.14)	*N*-Acetyl-L-methionine	L-Methionine + acetate	*Spodoptera frugiperda* (Sf21)	134	2.72
Phytase	Phytate	Phosphate	*Aspergillus niger*	1,260	0.25
Penicillin amidase	Penicillin G	Penicillin-Rx	*E. coli*		0.003
Lactate dehydrogenase	L-Lactate + NAD$^+$	Pyruvate + NADH	*E. coli*	40	

Enzyme activity usually depends not only on substrate concentration but is also modulated by other molecules present in cell. Their physiological meaning is to adapt the activity of metabolic pathways to the present needs of growth conditions. In many cases the end product of a pathway is an inhibitor, which regulates the first metabolic step decreasing its activity avoid overproduction. In the following paragraph the formal impact of inhibitors is outlined but restricted to cases where the inhibitor binds to the enzyme reversibly.

In the case of competitive inhibition, the inhibitor molecule can bind to the enzyme, but the enzyme cannot convert it to a product. The situation, where the inhibitor molecule "competes" with the substrate molecule for the binding site is shown in Fig. 4.6.

As was done in the case of the Michaelis-Menten kinetics the formal deduction from the single reaction constants leads to:

$$R_S = R_{S,\max} \cdot \frac{c_S}{k_S \cdot \left(1 + \frac{c_I}{k_I}\right) + c_S} \tag{4.22}$$

The inhibitor only affects the apparent k_S value, so has a similar effect as shown in Fig. 4.5 for increasing k_S. Inhibition can be overcome by increasing substrate concentration. The inhibition constant is chosen not as factor to c_I but as a denominator to give it the unit k_I (g/L) of a concentration. This makes quantitative discussion more practical.

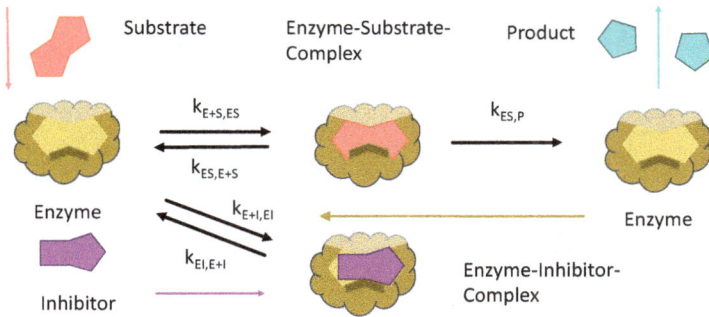

Fig. 4.6: Schematic representation of enzyme reaction mechanisms where inhibitor activity is involved, in this case as competitive inhibition.

There are other mechanisms of the action of the inhibitor. Unlike competitive inhibition the inhibitor may bind to sites other than the active sites, which are called allosteric sites. The inhibitor causes a conformational change in the enzyme affecting reaction constants. Here different cases are distinguished. In the case of noncompetitive inhibition, the substrate binds to the active site whether the inhibitor has already bound or not. The inhibiting effect is caused by preventing the enzyme from performing its catalytic action, so that only the dissociation back to enzyme and substrate is possible. This leads to an apparent reduction of the maximum reaction rate:

$$R_S = R_{S,\max} \cdot \frac{\frac{k_I}{k_I + c_I} \cdot c_S}{k_S + c_S} \tag{4.23}$$

Therefore, it has a similar effect on the maximum turnover rate as shown in Fig. 4.4 for decreasing $R_{S,\max}$. The reciprocal inhibition term could be applied to the denominator with the same justification. In the chosen representation the effect may be clearer.

In the case of uncompetitive inhibition, the inhibitor only binds to ES but prevents ESI from catalyzing product formation. This assumption leads to

$$R_S = R_{S,\max} \cdot \frac{c_S}{k_S + c_S \cdot \left(1 + \frac{c_I}{k_I}\right)} \tag{4.24}$$

This leads to a decrease in $R_{S,\max}$ but to an apparent decrease in k_S.

The three modifications (equations (4.22)–(4.24)) of the Michaelis-Menten equation (4.21) cover the formal possibilities of parameter modifications with respect to numerator and two summands of the denominator. In practice the enzymes do not always follow an either/or rule but could exhibit different mixtures of the deduced specific cases. A generalized scheme is shown in Fig. 4.7.

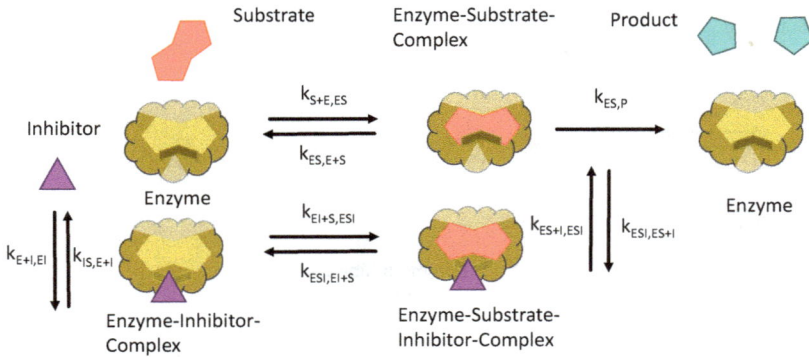

Fig. 4.7: Schematic representation of enzyme reaction mechanisms where a generalized inhibitor activity is involved being described as noncompetitive inhibition.

Inhibition is a mechanism useful for the cell to control, e.g., the flux along a metabolic pathway. In the case where the end-product is the inhibitor, it prevents its accumulation if it is not needed in smaller amounts. But the substrate itself can also be an inhibitor when present in higher concentration. The formal description reads:

$$R_S = R_{S,\max} \cdot \frac{c_S}{k_S + c_S + \frac{c_S^2}{k_{i,S}}} \tag{4.25}$$

Although it looks unnecessarily complicated the inhibition constant $k_{i,S}$ (g/L) is written as a denominator of the denominator to guarantee for the unit of a concentration and to give meaningful values. For concentrations going over the limiting range substrate turnover decreases again to $R_{S,\max}/2$ approximately at $c_S = k_{i,S}$ (see Fig. 4.8), where it is indicated for $k_{i,S} = 80$ g/L. The optimum value defined as $dR_S/dc_S = 0$ is reached for $c_{S,\mathrm{opt}} = \mathrm{sqrt}(k_S \cdot k_{i,S})$ at approximately $R_{S,\mathrm{opt}} = R_{S,\max}/(1+\mathrm{sqrt}(k \cdot k_{i,S})/k_{i,S})$.

The meaning of substrate inhibition seems to be a bit dubious at first glance. Nevertheless, in this way the cell can assign substrate, if present in excess, to another metabolic pathway or can manage large fluctuations. In enzymatic processes and in cultivations substrate inhibition can turn out to be a challenge for control.

Many enzymes need a coenzyme or a second substrate to work. In these cases, the situation becomes even more complex. In any case, no conclusions from measured curves to the underlying mechanisms should be drawn from a process engineering viewpoint. That is why the term "formal kinetics" is often used when kinetic relationships are applied to real processes.

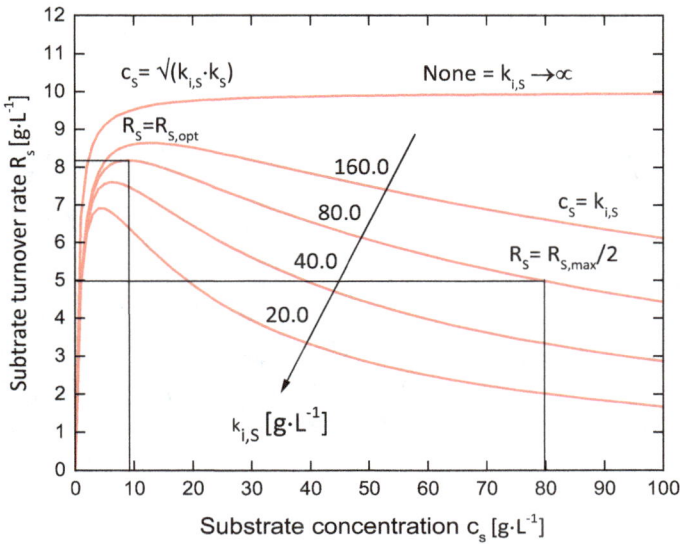

Fig. 4.8: Characteristic curves for substrate inhibition; the velocity curve rises to a maximum and then descends with increasing substrate concentration. Curves are parameterized with $k_{i,S}$.

4.3 The specific growth rate – describing growth by numbers

One of the first approaches (in the Occident) to finding a mathematical description of growth processes was given by the Italian mathematician Leonardo Fibonacci da Pisa (circa 1,200). He set up a sequence of numbers, namely:

$$(n_i) = (1, 1, 2, 3, 5, 8, 13, 21, \ldots) \tag{4.26}$$

where each subsequent number is the sum of the previous two. This sequence is called the Fibonacci sequence. The idea was that a pair of rabbits (=1) grow up over one generation and then give birth to another pair (=2). These grow up as well and become pregnant adding their offspring after one generation time to the other rabbits (=3) making a fast-increasing population n_i as the number of pairs grows. To the population belong the newborn individuals from the current generation, youngsters born in the last generation being pregnant in the meantime themselves, and adults – again pregnant – from the last generation. Note that among these there are some individuals from the penultimate generation and so forth to the first ancestors. So individual death is not considered in this sequence. The Fibonacci sequence behind this growth curve is a popular approach in science and art.

Before starting further calculations, the growth process can be visually followed by cultivation, where optical density reflects the biomass concentration as shown in Fig. 4.9. The pictures are taken in the sequence of the doubling time. During growth the culture looks more and more turbid. The maximum value for cell dry weight is approximately 40 g/L. Assuming 10% dry mass in the cell the residual being water, the cells would fill half of the reactor.

| 5g/L | 10g/L | 20g/L | 40g/L |

Fig. 4.9: Photographic pictures of a growing culture in a small lab bioreactor. The tubes inside the reactor just behind the glass wall are hardly visible even for moderate cell concentrations.

A corresponding graphical representation (Fig. 4.10) shows the strongly increasing cell density in a reactor.

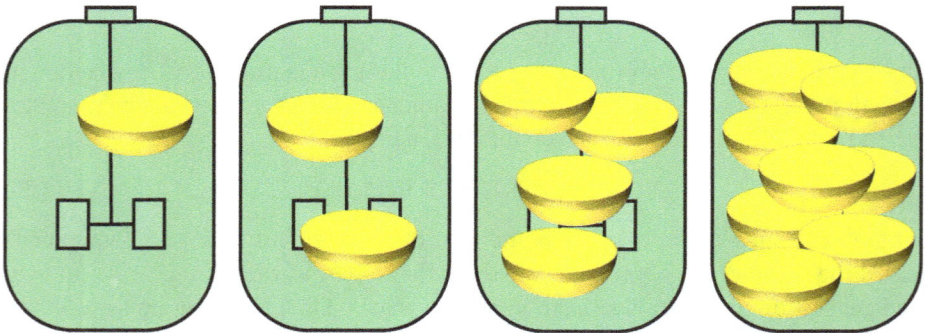

Fig. 4.10: Graphical model of the growth process; each "cell" represents 5 g/L biomass concentration.

To get an idea about how to describe the development of a growing culture it is helpful to perform a virtual growth experiment. We start with the assumption that a given initial number of cells, e.g., $n_{C,0} = 10^6$ cells/mL divide after a given doubling time t_d, e.g., $t_d = 1$ h to $n_{C,1} = 2 \cdot 10^6$ cells/mL. Here as well, no individual death is considered. This of course only holds if the cells do not react to changing medium composition or exhibit other physiological changes like aging. In the following calculation the sampling time t_S is chosen as a multiple of the doubling time t_d. The results concerning cell numbers $n_{Cells}(t_S)$ starting with the initial cell number $n_{Cells,0} = n_{Cells}(t_S = 0)$ of this "experiment" are listed in Tab. 4.3 in the first three columns.

Tab. 4.3: Data of a virtual growth experiment, where measurement of the cell number is calculated after multiple doubling times.

Sample number, n_s (–)	Sample time, t_S (h)	Number of cells, $n_{Cells}(t_S)$ (–)	New cells, Δn_{Cells} (–)	New cells per time, $\Delta n_{Cells}/\Delta t_S$ (h^{-1})	New cells per present cells, $\Delta n_{Cells}/n_{Cells}$ (–)
0	0	$n_{c,0}$			
1	$1 \cdot t_d$	$2 \cdot n_{c,0}$	$1 \cdot n_{c,0}$	$1 \cdot n_{c,0}$	1
2	$2 \cdot t_d$	$4 \cdot n_{c,0}$	$2 \cdot n_{c,0}$	$2 \cdot n_{c,0}$	1
3	$3 \cdot t_d$	$8 \cdot n_{c,0}$	$4 \cdot n_{c,0}$	$4 \cdot n_{c,0}$	1
4	$4 \cdot t_d$	$16 \cdot n_{c,0}$	$8 \cdot n_{c,0}$	$8 \cdot n_{c,0}$	1
. . .	. . .	. . .	. . .		1
10	$10 \cdot t_d$	$1{,}024 \cdot n_{c,0}$	$612 \cdot n_{c,0}$	$612 \cdot n_{c,0}$	1

The next three columns are evaluation results. Column 4 gives the number of newly grown cells Δn_{Cells} as the difference between the present number of cells $n_{Cells,s}$ at t_S and their number $n_{Cells,S-1}$ at the previous sampling time t_{S-1}. To understand growth as a dynamical process it makes sense to relate the increase in cell number to the observation interval $\Delta t_S = t_S - t_{S-1}$ as done in column 5. Not only the cell number, but also the number of new cells $\Delta n_{Cell,s} = n_{Cell,s} - n_{Cell,s-1}$ increases in each step. The more cells present in the sample, the higher is the increase in the next time interval. The relative increase $\Delta n_{cell,s}/n_{Cell,s-1}$ is therefore always constant (column 6). This last statement represented as proportionality reads:

$$\Delta n_{cell} \sim n_{Cell,s-1} \tag{4.27}$$

The numbers of cells form a geometric sequence where each term after the first is found by multiplying the previous one by a fixed, nonzero number called the common ratio. In our case this is 2, because cells double during the doubling time. As such it is not surprising that we constructed the sequence in this way. However, such geometric sequences appear quite often in describing natural or technical processes. Accumulation of these factors leads to

$$n_{cell,s} = n_0 \cdot 2^s \tag{4.28}$$

To understand growth as a continuous dynamic process it makes sense to formulate eq. (4.27) in differential intervals:

$$\frac{dn_{Cells(t)}}{dt} \sim n_{Cells}(t) \tag{4.29}$$

Not all cells will divide at the same time and they are not of equal size. In bioprocess engineering it is common to measure biomass as cell dry mass m_X:

$$\frac{dm_X(t)}{dt} \sim m_X(t) \tag{4.30}$$

Now it is reasonable to introduce a proportionality factor, traditionally named μ:

$$\frac{dm_X(t)}{dt} = \mu \cdot m_X(t) \tag{4.31}$$

The proportionality factor μ is called the "specific growth rate." The term "growth" represents the newly formed biomass, "rate" means "per time" and "specific" per biomass present. The unit is subsequently μ (g/(g·h)) or shorter (h^{-1}).

! Finally, the definition for the specific growth rate follows from the differential equation (4.31):

$$\mu = \frac{dm_X(t)}{m_X(t) \cdot dt} \tag{4.32}$$

The value of μ can be constant or can change during cultivation. In any case, it gives an insight on how the cells are doing. It also makes the process predictable.

With help of eq. (4.32) we can now calculate the course of cell dry weight during cultivation. This is differential equation (4.31) solved by separation of the variables:

$$\int_{m_{X,0}}^{m_{X,t}} \frac{dm_X(t)}{m_X(t)} = \int_{t_0}^{t} \mu \cdot dt \tag{4.33}$$

In mathematical textbooks the general primitive of a function of the form $f(x(t)) = dx(t)/x(t)$ is given as $F(x) = \ln(x(t))$. Applying this to our problem results in

$$[\ln m_X] = [\mu]_0^t \rightarrow \ln m_X(t) - \ln m_{X,0} = \mu(t - 0) \rightarrow \ln \frac{m_X(t)}{m_{X,0}} = \mu t \tag{4.34}$$

! Note that the integration of the right side is only possible for constant μ. Application of the exponential operator finally gives the required growth equation:

$$m_X(t) = m_{X,0} \cdot e^{\mu \cdot t} \tag{4.35}$$

This exponential function is the most fundamental equation to describe growth processes. It is also used in other fields in cases where the state of a system is proportional to its own changing rate. However, it is valid only for growth processes with constant μ. The Fibonacci sequence can be historically interpreted as a rough approximation of integration for the case of natural numbers.

The course of cell dry mass concentration of an exponentially growing culture is shown in Fig. 4.11. It is impressive that the curve increases faster and faster from the small value of 1–100 g/L after less than seven doubling times.

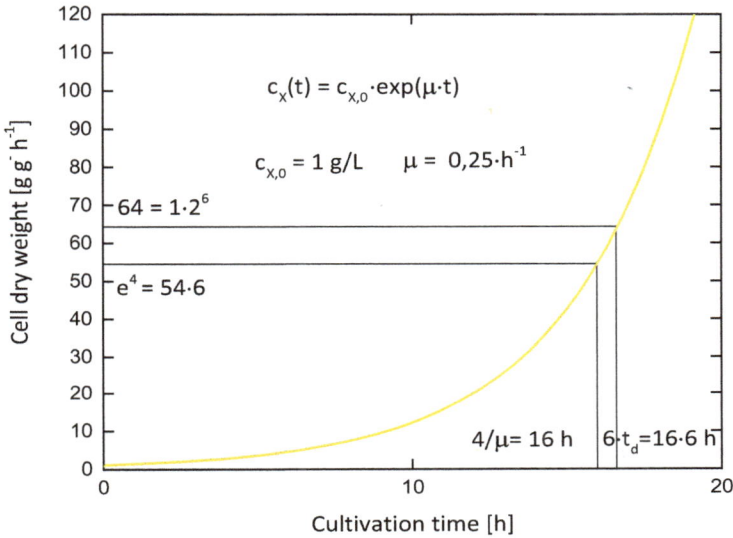

Fig. 4.11: Cell dry weight of an exponentially grown culture with the constant specific growth rate $\mu = 0.25\ \mathrm{h^{-1}}$.

Not all the different classes of microorganisms can grow so fast even under optimum conditions. In their natural habitat growth is limited by availability of nutrition or other environmental influences. Furthermore, the population is subject to cell death by many means. In bioreactors growth rates can be kept close to optimum for only a few generations. In many cases there are reasons to keep growth rates lower via an intentionally applied substrate limitation. The relation between substrate supply and growth is outlined in the next section; application examples are given in the next chapters (Fig. 4.12).

Typical spans of specific growth rates of bacteria range from 1.4 $\mathrm{h^{-1}}$ at maximum to 0.1 $\mathrm{h^{-1}}$ during production. Many eukaryotes like yeasts and filamentous fungi can reach rates of 0.5 $\mathrm{h^{-1}}$ but are often cultivated at lower rates. Microalgae and suspended plant cells are much lower in the range of 0.03 $\mathrm{h^{-1}} \approx 0.69\ \mathrm{day^{-1}}$, which corresponds to a doubling time of $t_d = 1$ day. Animal cell cultures often reach doubling times of only one week. The upper limits are only valid for optimum conditions and may vary strongly from species to species.

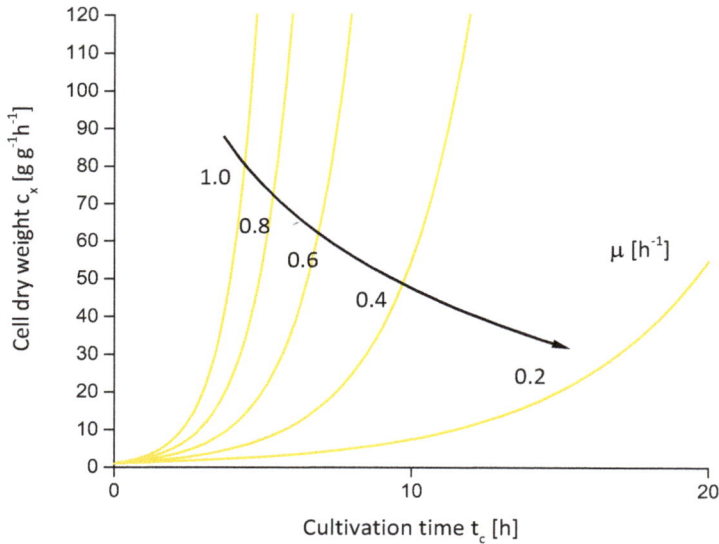

Fig. 4.12: Cell dry weight of exponentially grown cultures with different specific growth rates.

4.4 The yield coefficient – combining substrate turnover and growth

The question we are grappling with in this paragraph is how the cells "know" how fast and how long they can grow. Many environmental parameters like temperature, pH, or salinity of course play a role. Intracellular enzymatic steps also influence possible growth rates differently for different groups of organisms. Even in the case that all environmental parameters are at their respective optimum the cell is limited by intracellular steps, e.g., if an enzyme is at its maximum turnover rate. This gives reason to define a maximum specific growth rate μ_{max} (g/(g·h)). It is the task of metabolic modeling to find out the intracellular reasons for the value of μ_{max} for different classes of microorganisms. In the next paragraphs we focus on the role of substrate concentration.

From previous experience it is known that substrate concentration is one major factor that can be influenced by process engineering means. In the discussion of medium design (Chapter 3) we already used the relationship between initial substrate and final biomass concentration. From the experiment shown in Fig. 4.1 it becomes obvious that with progressively increasing biomass concentration substrate concentration also decreases faster, so we assume that growth and substrate uptake are coupled. In the time interval where c_S approaches zero, growth is no longer possible, implying that growth depends on the actual substrate concentration. Different experiments with different initial substrate concentration will result in different final biomass concentrations.

The first attempt to find an unspecific relation between growth and a limited environmental resource was formulated by Pierre-François Verhulst (1854). The background was the observation that growth as predicted by Fibonacci cannot go ad infinitum but has to stop somewhere. The imagination was a limited area offering enough space or food for a couple of animals but with a limited carrying capacity K of individuals. As this capacity becomes exhausted the animals must migrate or do not get so many offspring due to starvation. Food like grass could further grow but is divided by more and more animals. Finally, each individual gets so little food that only survival, but no propagation is possible. In addition to the linear growth model (4.29) Verhulst postulated that growth depends on the residual capacity $(K-N)/K$ (normalized from 0 to 1) of the habitat:

$$\frac{dN}{dt} = \mu_{max} \cdot N \cdot \frac{(K-N)}{K} \tag{4.36}$$

The solution of this differential equation with N_0 being the initial population is

$$N(t) = \frac{K \cdot N_0 \cdot e^{\mu_{max} \cdot t}}{K + N_0 \cdot (e^{\mu_{max} \cdot t} - 1)} \tag{4.37}$$

This so called "logistic growth curve" is a sigmoid function and a popular (as it has at least a closed solution) first trial to fit growth data. The formula also contains the idea of a constant yield. This means that the maximum possible number of individuals is linearly dependent on the "capacity" of an ecosystem. At this point, we can already anticipate that the yield $Y_{X,S}$ (eq. (4.2)) is constant within broad limits at least for a given substrate. For aerobic growth on glucose, it is often 0.5 g/g.

Nevertheless, the assumptions of the logistic growth curve do not really represent the situation in a bioreactor. In our batch, for example we see that growth ceases only very late, when glucose is nearly exhausted, so we have to look for more suitable assumptions.

The first one to describe growth as a function of substrate concentration was Jacques Monod in 1942 when he gave the famous Monod equation:

$$\mu = \mu_{max} \cdot \frac{c_S}{k_S + c_S} \tag{4.38}$$

Monod won the Nobel Prize (physiology or medicine) in 1965 for the operon theory, not for this equation.

Since then, the Monod equation has been employed to describe myriads of bioprocesses. The equation shows structural similarity to enzyme kinetics, so the assumption has been formulated that enzyme kinetics is an underlying mechanism. Substrate uptake as an enzymatic step in metabolism is the first candidate to be investigated as shown in Fig. 4.13. Understanding that the more biomass present the more substrate will be taken up, it is sensible to define the specific substrate uptake rate r_S (g/(g·h)) in analogy to the specific growth rate:

$$r_S = \frac{dc_S}{dt \cdot c_X} \tag{4.39}$$

as a measurable property of the cells. Following the attempt to understand the Monod equation in analogy to enzyme kinetics, r_S can be given as a function of the substrate concentration c_S:

$$r_S = r_{S,max} \cdot \frac{c_S}{k_S + c_S} \tag{4.40}$$

This is in accordance with the experiment in Fig. 4.1 that at lower substrate concentrations the substrate uptake rate is reduced. Substrate availability inside the cell then determines the possible growth rate. This concept is called "substrate limited growth."

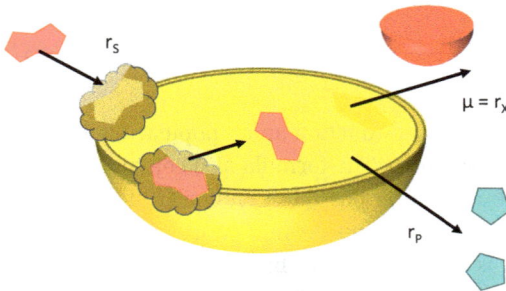

Fig. 4.13: The concept of substrate limitation; an enzymatic transport step (symbolized by the turning enzyme) carries substrate into the cell. This substrate flux determines the possible growth.

The substrate is then converted by hundreds of enzymes inside the cell. However, growth is regarded as a thermodynamic process where the cell could make "the best of it" under the restrictions of energy and material balances, stoichiometric relations, and other constraints. We have already discussed the elemental balance as such a constraint. Biology has provided different approaches to understand how much biomass can be built up by a given amount of substrate taken up. Of specific interest is the partitioning into anabolic pathways lumped as growth on the one hand and energy producing metabolic pathways on the other. A fundamental relationship determining the relation of growth and substrate uptake was found by S. John Pirt in 1965:

$$r_S = \frac{1}{y_{X,S}} \cdot r_X + r_m \tag{4.41}$$

To get a consistent system of terms the historic symbol μ is replaced by "r_X", where r is "rate" and "X" is biomass. The yield coefficient $y_{X,S}$ (g/g) is the same as $Y_{X,S}$ but here for differential dm_X and dm_S. To maintain a given specific growth rate the cell has to take up a proportional rate of substrate. An important point is that the yield coefficient is again constant for many growth conditions. This can be understood as an optimized energy handling and carbon annotation controlled by the cell. Metabolic energy is provided by a part of the substrate while the other part is used for growth under energy utilization. This balance is kept stable by the cells in order to avoid waste of energy and carbon. As long as the cell does not change this metabolic pattern this results in a constant yield. In addition, energy is partly used for nongrowth-associated purposes like movement, keeping the water balance, or maintaining membrane potential. This part of energy utilization is lumped into the "maintenance" term r_m (m in the original reference). Indeed, very often measurement data fit quite well to the linear relation between growth and substrate uptake. For clarity it is noted that $y_{X,S} = dr_X/dr_S$ is a kinetic parameter, while $Y_{X,S} = \Delta m_X/\Delta m_S$ is a measured process characteristic over a longer cultivation period. About the inner connection between substrate uptake rate and growth rate more background is given in the next chapter.

A first idea to describe biomass formation and substrate uptake simultaneously is by simply combining the Monod and the Pirt equations (4.33) and (4.37). A short inspection of this system of equations suggests that it is not a good idea. For $c_S = 0$ a positive substrate uptake rate is predicted, which is physically not possible. Turning the cause-effect chain we set the substrate uptake as a primary limiting step and reformulate the resulting growth rate following Pirt's linear relationship:

$$r_S = \frac{r_{S,max}}{k_S + c_S} \tag{4.42}$$

$$r_X = y_{X,S} \cdot r_S - r_m \tag{4.43}$$

This is a first simple physiological model of microbial growth and will be used as a standard approach throughout this book. According to the essence of kinetics these equations hold irrespective of the way c_S changes during the process.

Besides biomass formation and substrate consumption, product formation is of interest. As was done before we define a specific product formation rate:

$$r_P = \frac{dm_P}{dt \cdot m_X} \tag{4.44}$$

In addition, a formal relation between growth and product formation turnover is required. As the cell is highly stoichiometrically determined again a linear relation in analogy to Pirt's equation can be a good start:

$$r_P = y_{P,X} \cdot r_X - r_{m,P} \tag{4.45}$$

This equation was introduced to the field by Luedeking and Piret and bears their names. It is a flexible tool to fit different kinds of product formation patterns. The yield coefficient $y_{P,X} = dr_P/dr_X$ describes the ratio of product formed per biomass growth. But often product formation is coupled with catabolism, meaning substrate turnover. At least the carbon comes from there. For technical applications the ratio of product produced to substrate consumed $y_{P,S} = \Delta m_P/\Delta m_S$ is of interest. This is considered in the reformulation:

$$r_P = y_{P,S} \cdot r_S - r_{m,P} \tag{4.46}$$

Here product formation is coupled to substrate uptake where a small constant part of the substrate is reserved for other purposes. These two different formulations hold only for r_S and r_X in given thresholds, which is also the case for the Luedeking-Piret equation. As growth and substrate uptake are also nearly linearly dependent, the two variants can often not be distinguished from simple measurements. In the next chapter and some case studies the coupling of product formation to catabolism and anabolism (r_S, r_X) and the complexity of control patterns of the cells is analyzed in more detail.

Before trying out the kinetic equations for our standard batch process good guesses for reasonable values of the kinetic parameters are required. Table 4.4 gives some typical values.

Tab. 4.4: Typical specific growth rates of different microorganisms and substrates.

Organisms	Substrate	$r_{S,max}$ (g/(g·h))	$r_{X,max}$ (g/(g·h))	k_S (g/L)	$y_{X,S}$ (g/g)	Remarks
Saccharomyces cerevisiae	Glucose	2.52	0.44–0.60	0.180	0.485	Aerobic
	Ethanol		0.12–0.14		0.753	Anaerobic
	Glucose		0.53–0.63			
E. coli	Glucose		0.5–1.0	0.004	0.5	Aerobic
Aspergillus oryzae	Glucose			0.005		Aerobic
Klebsiella aerogenes	Glucose			0.009		Aerobic
	Glycerol			0.009		
Virtuella generica	Glucose		1.0	1.0	0.5	Aerobic
Virtuella generica	Glucose		1.0	1.0	0.05	Anaerobic

Some trends can be observed from Tab. 4.4. Aerobic microorganisms can reach a yield of 0.5 g/g growing on glucose. Sometimes it is a bit lower, but never higher. The yield also depends on the substrate. With a higher degree of reduction (ethanol) higher yields are obtained. Anaerobic growth on glucose results in a low yield of about 0.03–0.05 g/g. Here, the product is always formed in a high yield product per substrate of e.g., 0.5 g/g. As substrate uptake is often higher for anaerobic growth, similar growth rates as for aerobic growth are obtained.

4.5 Kinetics can also work linearly – basic kinetics of processes with alternative redox and energy supply

Reference has already been made to electro-biotechnology in Chapter 3. The mode of action is explained in more detail in Fig. 4.14. The process energy is provided by electrical current from a sustainable source. The electrons can then transfer to oxidized coenzymes and thus reduce them. The coenzymes then diffuse to suitable enzymes where the desired reaction takes place. The discharged coenzymes diffuse back to the electrode in this cycle. In other variants, synthetic mediator molecules can also be used to transport the electrons to the enzymes or microorganisms. In principle, enzymes or entire living microorganisms can also be immobilized directly on the electrode. One problem is the low current density on the electrode surface. Porous conductive materials are used to increase the surface area. However, only a voltage of about 600 mV can be built up in order to avoid electrolysis. In any case, there is a linear relationship between current and product formation rate:

$$R_{\mathrm{Prod}} = y_{e,\mathrm{Prod}} \cdot I_{\mathrm{elec}} \qquad (4.47)$$

The yield coefficient $y_{e,\mathrm{Prod}}$ is an integer in molar terms, namely the number of electrons transferred per product molecule.

The great advantage of the process is that an electron is only transported if a reaction occurs. Processes can therefore be set up that use 100% of the substrate used. This is referred to as 100% carbon efficiency. The energy efficiency is also over 90%. For these reasons, active research is being conducted into processes for binding CO_2, e.g., to hydrogen. The aim is development of "power-to-fuel" processes that supply liquid fuels in particular. These will not be completely replaced (aircraft, trucks, etc.). This synthesis process is actually comparable to the second stage of photosynthesis. In this respect, the production of plastic materials is also conceivable. To date, the direct transfer of electrons into a living cell has not been properly understood and is still being researched. The fuel cell process works in exactly the opposite way as far as energy flows are concerned. The "bio-fuel cell" contains living microorganisms and can not only simply eliminate organic substances in wastewater, but also convert their chemical energy into electrical energy. This process is already being used in practice.

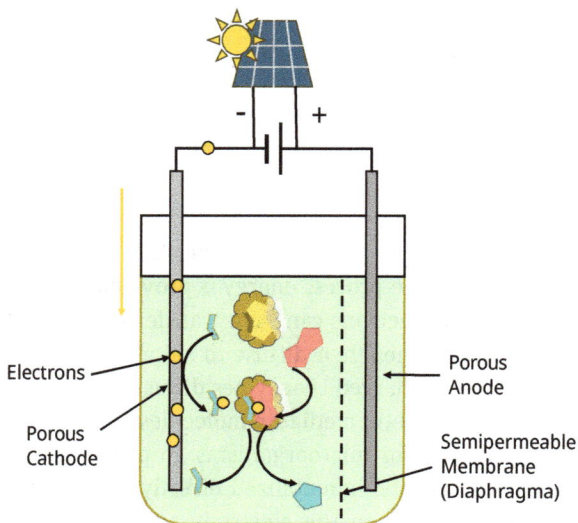

Fig. 4.14: Scheme of an electro-biotechnological process; the electrons, preferably from sustainable source, enter the reactor from the left side, are transferred to a mediator or a co-factor and then to a product.

! The cable bacterium *Electronema* is "Microbe of the Year 2024," also the year of publication of this book. The cells can form thin current-conductive filaments of thousands of individuals. How do the bacteria conduct electricity and how far is there still to go before they can be used as bio-cables?

The growth of microalgae is an example for a transport / reaction system with linear kinetics as well. The substrate "light" is not taken up by an enzymatic process, but by passive linear absorption. That happens in the light harvesting complexes. In case of eukaryotic algae they are located in the chloroplasts. The first reaction step is water splitting in a protein center called PSII. For this reaction a minimum of four photons is necessary to convert $2 H_2O$ into $1 O_2$ and $4 H^+ + 4e^-$ independently of their respective energy. Excess quantum energy is lost via heat or fluorescence. The energy of four additional photons and the generated proton gradient is then used in another protein complex called PS1 to gain 3ATP and 2 NADPH/H^+. That is just enough for fixation of one CO_2 catalyzed by the enzyme RuBisCo. The produced glucose is then further metabolized to biomass with a similar yield as in the heterotrophic case. For now, it is important to see, that a constant flux of photons leads to a constant increase of biomass, a linear kinetics. More details from engineering view can be found in Section 6.7 or in the microalgae book by Griehl/Posten.

4.6 The batch process – the simplest form of a bioprocess strategy

Now all information can be assembled to get a simple model to describe a batch process. Remembering the physiological model:

$$r_S = \frac{r_{S,\max}}{k_S + c_S} \tag{4.48a}$$

$$r_X = y_{X,S} \cdot r_S - r_m \tag{4.48b}$$

$$r_P = y_{P,S} \cdot r_S - r_{P,m} \tag{4.48c}$$

the concentrations c_X, c_S, and c_P can be calculated to have a complete set of equations. This should be done by setting up the material balances for biomass and substrate. For the batch process the balance equations are in the form of simple differential equations of accumulation and reaction term:

$$\frac{dc_X}{dt} = \mu \cdot c_X \tag{4.49a}$$

$$\frac{dc_S}{dt} = -r_S \cdot c_X \tag{4.49b}$$

$$\frac{dc_P}{dt} = r_P \cdot c_X \tag{4.49c}$$

These reactor equations together with the physiological equations form a complete model to represent batch cultivation. Note the minus sign for substrate to get a decreasing substrate concentration for a positive turnover rate. Even for more complicated cases the biological sub-model and the reactor sub-model should always be strictly separated during building up and for writing them down. Unlike in setting formal kinetics, e.g., only for biomass, these models cannot be solved explicitly but must be numerically integrated. A simulation is shown in Fig. 4.15.

Batch processes are usually divided into several phases. During the lag phase the cells adapt to the environmental conditions inside the reactor, which may be different from the parameters in the pre-culture. The specific growth rate increases only after some time, e.g., one hour, to the value predicted by kinetics. This can happen, e.g., in cases where the medium in the pre-culture contains supplementary compounds that are not present in the technical medium or vice versa. Adaptation is not represented in the model equations. As the lag phase is unproductive it should be kept as short as possible, e.g., by adjusting the conditions in the shaking flasks or the cultivation in the stage before.

During the "exponential phase," μ is close to but not equal to $\mu_{\max}$ for several hours as substrate concentration is far in the saturation range of Michaelis-Menten kinetics. Do not use the outdated term "logarithmic phase." This is actually the state of

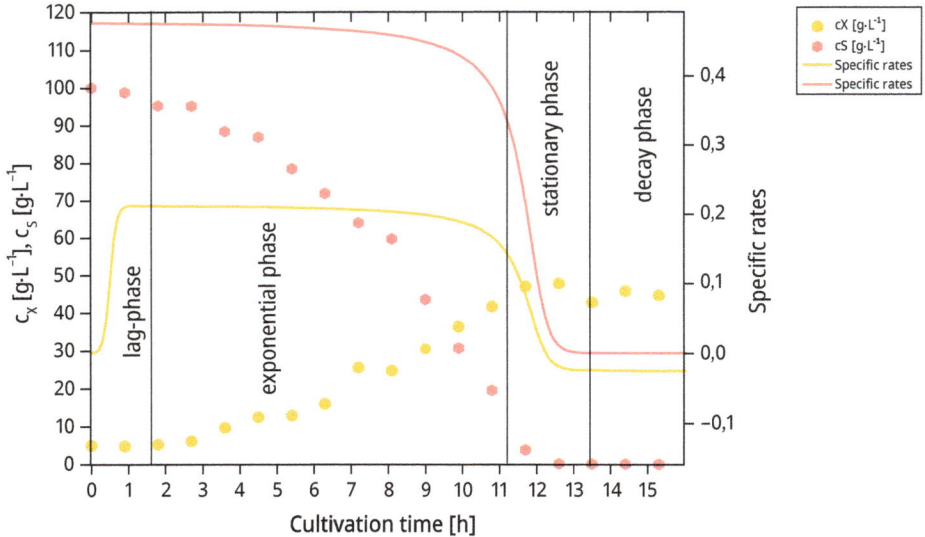

Fig. 4.15: Simulation of batch cultivation with the parameter set for the model organism *Virtuella generica* (Tab. 4.4) showing the course of the concentrations and the specific turnover rates.

the reactor we want to achieve as long as possible. This is no longer the case in the "stationary phase," where substrate concentration falls to the order of magnitude of the limitation constant k_S and becomes therefore the decisive value for growth. In many cultivations accumulation of an (possibly unknown) inhibiting product could also contribute to the stationary phase. After the substrate is completely exhausted a decrease of cell dry mass is observed, which is also visible by the negative specific growth rate. That does not mean that the cells really perish in the "decay phase." In fact, they fuel their maintenance demand by respiration on intracellular material like storage compounds, a process called "endogenous respiration." The optimum harvesting point is shortly before entering the decay phase.

To evaluate data from discrete measurements more exactly than using intervals as in Tab. 4.1 some assumptions are necessary. The simplest one is that μ is the remaining constant between two measurement points. This neglects time changing effects of limitation and maintenance. Nevertheless, the question remains as to when the biomass for the denominator should be determined. The exponential growth formula can then be used for evaluation of μ:

$$c_X(t_2) = c_X(t_1) \cdot e^{\mu \cdot (t_2 - t_1)} \rightarrow \mu = \frac{1}{t_2 - t_1} \cdot \ln\left(\frac{c_X(t_2)}{c_X(t_1)}\right) \tag{4.50}$$

or in a simpler notation:

$$\mu = \frac{1}{\Delta t} \cdot \ln\left(\frac{c_{X2}}{c_{X1}}\right) \tag{4.51}$$

Now we assume that $y_{X,S}$ and r_S are constants as well:

$$y_{X,S} = \frac{\Delta c_X}{\Delta c_S}; \; r_S = \frac{\mu}{y_{X,S}} \tag{4.52}$$

Application results are listed in Tab. 4.5, The limitation constant cannot be determined because the process runs through the limitation phase very fast. To have an idea of the maintenance, plot μ versus r_S and read off the μ-axis.

Tab. 4.5: Calculated differential values from the measurements in Fig. 4.1.

T (h)	c_X (g/L)	c_S (g/L)	μ(h^{-1})	$y_{X,S}$ (g/g)	r_S (h^{-1})
0.0	5.0	100.0	−0.0454	−0.1667	0.2721
0.9	4.8	98.8	0.1332	0.3552	0.3749
2.7	6.1	95.1	0.2774	0.4066	0.6823
5.4	12.9	78.4	0.2393	0.4729	0.5060
6.3	16.0	71.9	0.2419	0.7139	0.3388
8.1	24.7	59.6	0.2330	0.3589	0.6492
9.0	30.5	43.6	0.1950	0.4554	0.4283
9.9	36.4	30.7	0.1495	0.4674	0.3198
10.8	41.6	19.5	0.1343	0.3406	0.3944
11.7	46.9	3.8	0.0183	0.2103	0.0870
12.6	47.7	0.1	-0.1232	-4.65E+01	2.65E-03
13.5	42.7	1.6E-03	0.0248	1.22E+03	2.02E-05
15.3	44.6	4.3E-07			

It becomes obvious that the solution of the differential equation is more accurate than using differences. For very low substrate concentrations unrealistic values for r_S are observed. This is due to respiration on intracellular storage compounds making the simplified assumption $r_S = \mu/y_{X,S}$ wrong for negative μ, as then of course no substrate is produced, but used for maintenance from intracellular storages. Extreme care should be taken when applying formal equations without knowing the background and constraints.

To determine the parameter $y_{X,S}$ and μe for the whole physiological model we plot μ versus r_S (Fig. 4.16). We see that our basic physiological model holds. The slope of the straight line is the yield coefficient and the intercept at the μ-axis the maintenance. The limitation constant k_S cannot be determined from batch processes as the limited phase is very short.

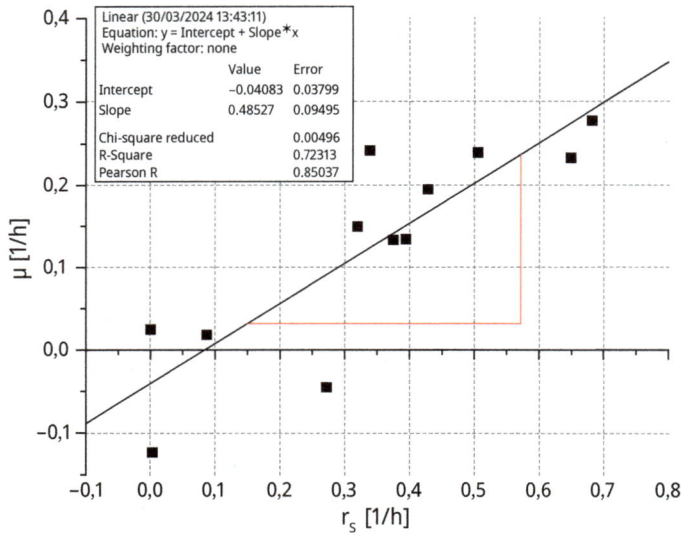

Fig. 4.16: Reconstruction of the specific substrate rate and the specific growth rate from measured data.

! To plot a growth curve on logarithmic paper and make a linear regression gives in principle a value of μ. Nevertheless, it is not up to date in the computer age. The same holds for the Lineweaver-Burk diagram, a transformation of enzyme kinetics. Calculating the reciprocal Michaelis-Menten kinetics a linear plot is obtained, which allows us to read the kinetic parameters. Measurement errors are also transformed leading to limited accuracy. This gives a first idea about an appropriate model selection, but later computer evaluation is more reliable.

To avoid inaccuracies by calculating specific rates as difference quotients, a formal deduction is recommended.

? Solving the balance equation for substrate at an interval Δt and substituting the exponential growth equation leads to

$$\frac{d}{dt}c_S(t) = -r_S \cdot c_X(t) = -\frac{r_S \cdot c_X(t_1) \cdot e^{\mu \cdot t}}{e^{\mu \cdot t_1}} \tag{4.53}$$

$$c_S(t) = -\frac{r_S \cdot m_X(t_1) \cdot e^{\mu \cdot (t-t_1)}}{\mu} + c_S(t_1) + \frac{r_S \cdot c_X(t_1)}{\mu} \tag{4.54}$$

Solving for r_S and rearranging with $t = t_2$ leads to

$$r_S = -\frac{\mu \cdot (c_S(t) - c_S(t_1))}{c_X(t_1) \cdot \left(e^{\mu \cdot (t-t_1)} - 1\right)} = -\frac{\mu \cdot \Delta c_S}{c_X(t_1) \cdot (e^{\mu \cdot \Delta t} - 1)} = -\mu \cdot \frac{\Delta c_S}{-\Delta c_X} = \mu \cdot \frac{1}{y_{X,S}} \tag{4.55}$$

This justifies the initial assumptions to take $y_{X,S}$ directly from the data and calculate r_S from μ and $y_{X,S}$.

4.7 Questions and suggestions

1) In 2010 the specific growth rate of mankind was 1.2% per year. After how many years is the number of people expected to have doubled?

2) Some bacteria have a specific substrate uptake rate of 10 g/g h. If a human had the same metabolic speed, how many sacks of sugar would they need per day? Discuss the result compared to the approximated real energy need of a human. How many orders of magnitude would it be more or less?

3) Recombinant proteins can be produced in a batch culture. The specific growth rate is $\mu_+ = 0.2\ h^{-1}$. In the inoculum 1% of the cells have lost their plasmid and can now grow faster with $\mu_- = 0.3\ h^{-1}$ because they do not spend metabolic energy on product formation. What is the percentage of cells without the plasmid after $t_{End} = 12\ h$?

4) Plot an exponential function, a logistic function, and a quadratic function in one diagram and try to change parameters so that they become similar. Is it always easy to decide what kinetics it is?

Chapter 5
Bioreactors – designing a home for the bioreaction

Bioreactors show a huge diversity depending on the product, the microorganisms employed, and the production environment. A bioreactor cannot be simply bought off the peg. This chapter starts by considering what the different purposes are. This is also the starting point for a biochemical engineer to design new reactor concepts with new features. In order not to reinvent the wheel a look at commonly applied reactors, taking on board the terms and concepts to describe their performance is helpful. For this starting point the stirred tank reactor is chosen. From there the view goes to different reactor types specialized for bioprocesses. This includes, in the last paragraph, a walk through several orders of magnitude of dimension to assess current designs.

5.1 Not only a vessel – for what a bioreactor is needed and what it can be

On hearing the word "bioreactor" many people think immediately of a stainless steel vessel or a small glass reactor in the lab. Here we want to avoid this shortcut thinking and try to specify the requirements applying to the reactor. This includes how the reactor supports the goals of the process and the technical constraints. In this book there is no space to set up a complete "user requirement specification" (see supplementary material) but is important to realize the basic concepts. First and foremost, the bioprocess should take place in a defined space, where the biocatalysts, microorganisms, or enzymes do their job without being dispersed into the environment. This it is something like an exoskeleton to stress a picture from biology. Secondly, to enable the microorganisms to grow and carry out material conversion, suitable conditions with respect to temperature, pH, or medium concentrations have to be maintained. This implies, thirdly, transport of water and chemical compounds into and out of the reactor. With these very general statements a general definition can be given:

> "A bioreactor is an enclosure in which in the presence and involvement of a biocatalyst material, biochemical conversions take place."

This definition does not imply that the reaction space is closed in a strict sense. Depending on the specific task, different exchange mechanisms between reactor and environment are at work but must be defined and controlled. On the other hand, the definition implies that the reactor is a manufactured device. These last two points make up the difference, e.g., to the biofilm on a stone or to our intestine, where material exchange also happens, and microorganisms feel happy. In terms of process engi-

https://doi.org/10.1515/9783110773354-005

neering a bioreactor is a technical transport/reaction system. Starting from an understanding of bioprocesses we are now collecting things to be transported.

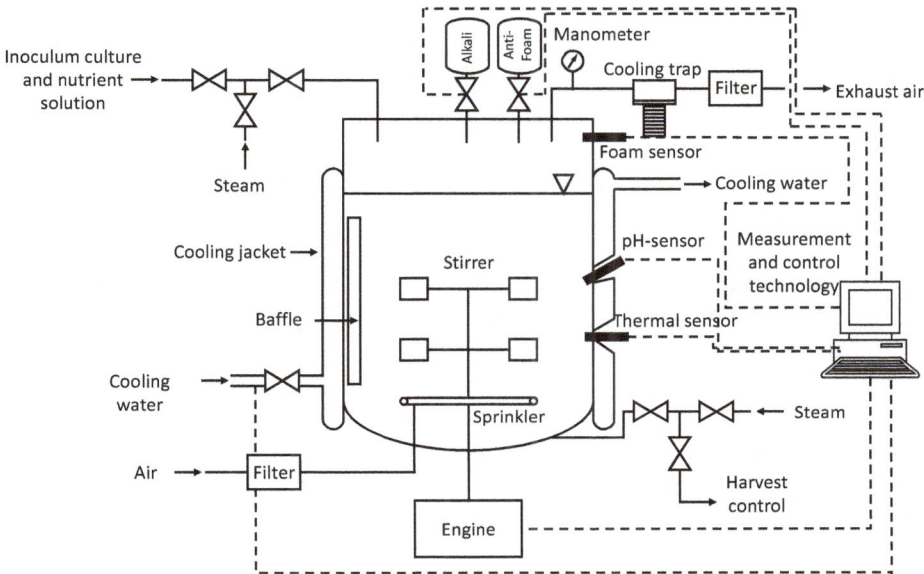

Fig. 5.1: Scheme of a typical bioreactor with peripheral devices. Follow all transport lines into and out of the reactor while reading the next paragraphs.

The medium is to be pumped from a medium flask or container through pipes into the reactor. The included substrates have to be further transported via the next convective step all through the reactor space to the cells. This usually involves mixing. The last step is diffusion to and enzymatic transport by the cells. Other liquids to be dosed are the inoculating material, and titration and antifoam agents. Following the process further we come to sampling and harvesting. Similarly, gaseous substrates enter the reactor driven by pressure, and are then macroscopically spread by bubbles and further by mixing and diffusion. These processes require peripheral equipment outside the reactor like pumps, valves and pipes, and specific structures inside, like stirrers. Figure 5.1 gives an overview. More details are given below and in the chapter on fermentation (Chapter 11). Energy also must be provided and be removed. Mechanical energy is needed to drive most of the transport processes. It is transported via the stirrer shaft and dissipated by the impeller. In fact, this is a critical point in reactor design and needs special attention. All bioprocesses are in principle exothermic, so we have the issue of transporting heat out of the reactor. All material and energy fluxes have to be somehow observed and controlled, so transport of information is an indispensable issue. Different sensors make process variables accessible, while actuators control fluxes according to the needs of the microorganisms and according to our idea of the process. More about measurement techniques is given in Chapter 10. Unfortunately, gathering information is sometimes neglected making process development a game of trial and error.

The needs of microorganisms are at the center of the considerations. But outside the reactor humans work with their own requirements, so we must respect these as well. "Safety first" is the motto of labor and environmental protection. No microorganisms should leave the reactor unintentionally, e.g., via the off gas. Neither should any microorganisms enter the reactor to keep an axenic culture, which is important for product safety and economic risk minimization. Of course, the whole equipment is subject to governmental regulations. Specific features of bioreactors that address these issues are summarized under the terms "aseptic" or "hygienic design," for example heat (steam) sterilization before starting the cultivation. Small reactors must be brought to an autoclave, while large reactors have their in-built devices that allow "sterilization in place" (SIP). Cleaning is also an issue, a typical housewife and househusband problem. Nobody likes to scrub a fermenter, as it is sometimes called disrespectfully. Large reactors would need to be entered by laborers using mountaineering equipment. In modern reactors appropriate installations make the reactor capable of cleaning in place (CIP). Manual operation is also not state of the art, but automation based on the sensors and actuators mentioned above is.

Economic efficiency is the goal a company looks for. This has impact on the material, number of sensors or (CAPEX, OPEX), and energy savings. Bioreactors are not only used for production but also for scientific reasons. Highly equipped bioreactors have advantages over shaking flasks, which are in principle bioreactors as well. This concerns the possibility to measure and control environmental variables, allowing for frequent sampling or delivering of more reproducible results. Typical biomass concentrations are tenfold higher due to better oxygen transfer. Here the costs are not balanced against a chemical compound as a valuable product but against amount of data and quality of gathered information. This justifies the costs of better equipment.

5.2 Revisiting the stirred tank reactor – the most important issues to learn from chemical engineering

! Stirred tank reactors are the most widely used reactors in chemical and biological processing due to advantages such as simplicity of construction, low capital and operating costs, and operational experience. A stirred tank reactor is a cylindrical vessel in which the uniform mixing of the reactants is achieved by means of an integrated stirrer or agitator. Since a bioreactor is the place where the biochemical reaction takes place, bioreactors are the heart of the bioprocesses.

Standard geometry of bioreactors may vary depending on the process and organism requirements. The reactor shape is mostly cylindrical with a flat bottom, but curved, dished, and conical bottoms are also used. Curved bottoms support the mixing of the medium especially at the bottom and prevents sedimentation at the edges. However, a standard design with a flat bottom is preferable due to the operational costs. The bioreactor construction material is typically made of glass if small scale bioreactors are needed. For industrial applications and pilot scale implementations, stainless steel installations are used as construction material. High quality (Cr-Ni, V4A) is necessary to prevent corrosion induced by high salt contents and nickel discharge. Electrolytic polishing of the surface to a finish roughness of 0.8 μm is preferred to prevent micro-organisms attaching to small grooves. The ratio of height (H_R) to diameter (D_R), re-ferred to as aspect ratio of a stirred tank bioreactor is usually chosen between 2–3 depending on the number and arrangement of impellers and the reactor application. A higher aspect ratio enables better aeration, heat transfer, and mixing as it is used for microbial applications. Lower aspect ratios are generally needed when animal cell cultures are used since reduced shear stress, mixing, and aeration are essential. There are three basic reasons for a vertical shaped bioreactor: The contact time between bubbles and liquid is longer and therefore oxygen can be better treated in the biore-actor. The reactor can be better cooled down due to the higher specific area (area/vol-ume). Better lighting area is achieved for phototrophic organisms

Adequate mixing and aeration are important tasks for bioreactors. Agitation for the purpose of mixing is obtained by the power-driven agitators and the baffles. Baf-fles are used to increase turbulence and prevent rotating and swirling of the fluid. The number of baffles ranges between four and six depending on the diameter of the bioreactor. Dimensions of baffles and impellers are a fraction of the reactor diameter and are depicted for a typical case in Fig. 5.2.

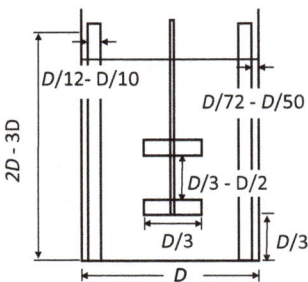

Fig. 5.2: Geometry of a typical stirred tank reactor; most of the dimensions are related to the reactor diameter.

The use of multiple impellers improves mixing and effective gas-liquid mass transfer. The common working volume of a bioreactor is 70–80% of the total volume of the bioreactor. The remaining volume, i.e., the headspace, is a restraint system for foam and depends on its formation. Furthermore, increasing aeration rate brings more bubbles into the system and the liquid surface needs space to rise. The same holds for the liquid rising at the reactor wall due to the central vortex. Gas is introduced into the bioreactor via a perforated sparger, which is located under the bottom impeller, with a moderately smaller diameter compared with the impeller diameter. Besides the minor influence of the sparger, impeller type plays an important role in terms of gas dispersion. A wide variety of impeller sizes and shapes allows us to create various flow patterns inside the bioreactor. Impeller type is basically chosen depending on the viscosity of the fluid and the sensitivity to shear stress. Three different impeller types are commonly applied in stirred tank bioreactors. These impellers are: three-blade propeller, six-blade Rushton turbine, and Intermig (Fig. 5.3).

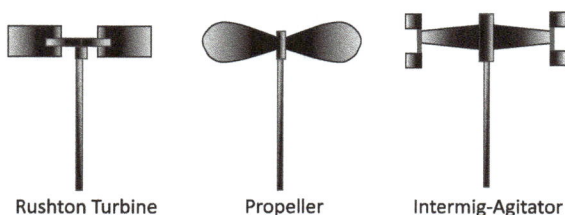

Rushton Turbine   Propeller   Intermig-Agitator

Fig. 5.3: The three most important impeller types for bioreactors.

For flat-bladed turbines the most common generic term is "Rushton turbine" after J.H. Rushton. Their blades are flat and set vertically along an agitation shaft, which produces a unidirectional radial flow; compare in Fig. 5.4. The standard design features six vertical blades, although four and eight are also common. Rushton type impellers are commonly used in fermentations of cell lines that are not considered shear-sensitive, including yeasts, bacteria, and some fungi. Propeller type agitators resemble the design of marine propellers, in that case to generate propulsion for ships. Consequently, propellers transfer water mainly in the axial direction. Together with bending of the flow at the bottom of the reactor a low-shear axial mixing is achieved. The industrially developed "Intermig" agitator is a predominantly axial-pumping impeller that comprises several stages arranged on top of one another and each staggered by 90° angles. A typical layout has two stages and two flow directing elements. The decisive difference between the Intermig and other mixing systems is the arrangement of the two blades: the inner blade and the two-stage outer blade are oriented in the same radial plane but with opposing blade angles. This generates a primary axial pumping flow in opposing directions in the inner and outer sections, therefore also supporting axial mixing.

Rotation of stirrers in a stirred tank bioreactor is enabled by application of electrical power. From the agitator mechanical energy flow goes further into the suspension. Here the specific (volumetric) power input drives hydrodynamic behavior with respect to mixing. This includes homogenization, suspending, and dispersing for heat, gas, and

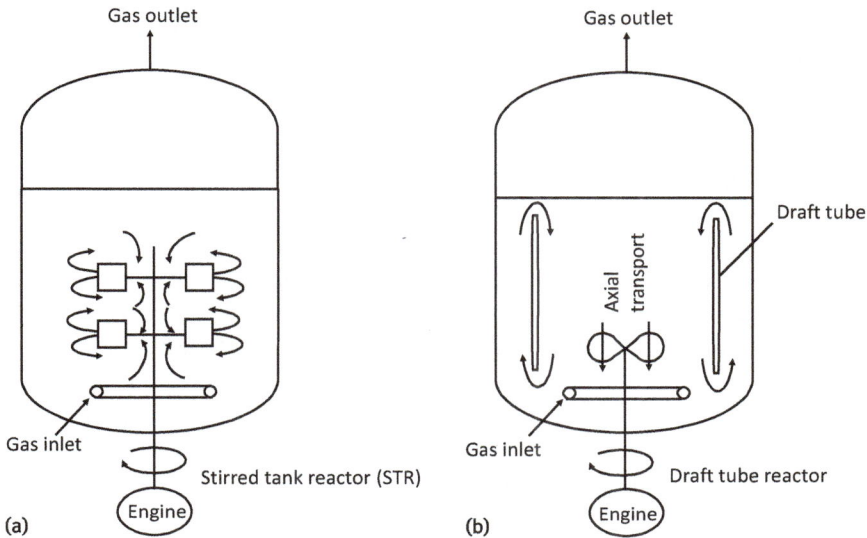

Fig. 5.4: The basic flow patterns of the Rushton turbine and the propeller being different in primarily generating radial or axial flow respectively.

material transfer, the most important processes for operation and scale-up (Tab. 5.1). Firstly, the power input $P_{R,imp}$ (W) has to be characterized and guaranteed.

Tab. 5.1: Typical power inputs for different tasks in bioprocess engineering.

Task	Homogenization		Suspending		Dispersing	
Example	Distribute heat jacket vs. volume	Avoid high local medium concentrations	Stop cells from sedimentation	Keep oil drops from floating	Crushing air bubbles	Tearing flakes
Typical power Input (W/kg)	0.1	0.2	0.5	1.0	2.0	5.0

The most important tasks can therefore be sorted according to increasing power input. If the power input is high, the less time-consuming tasks are completed at the same time. Even if you do not develop a reactor but operate it or order it from a company, you need to know such values and have an idea of how this is implemented in terms of equipment. The following explanations are to be understood in this sense.

! "God made the laws of nature, the devil made hydrodynamics."
. . . freely adapted from Werner Heisenberg.
. . . but don't worry, we make user friendly approximations.

The power input depends on the rotational speed of the stirrer, geometry of the stirrer, and fluid characteristics. To get as close to physics as possible, we first observe a flat plate of surface area A_{Plate} moving through à fluid, i.e., the medium, with a speed of v_{Plate} in normal direction. In this context, "normal" means in direction perpendicular to the surface. The plate experiences a resistance, called the "drag" force acting opposite to its relative motion. The medium in front of the plate is pushed backwards, upwards, or downwards, creating increased pressure in front of the plate accelerating the fluid. Vortices form behind the plate, which leads to negative pressure. The pressure difference multiplied by the plate surface is the force. Frictional forces can also occur, but these are low in the plate example. Experience has shown that drag increases quadratically with speed. This brings us to the drag equation:

$$F_{Drag} = \frac{1}{2} \cdot c_D \cdot \rho_{liqu} \cdot A_{Plate} \cdot v_{Plate}^2 \tag{5.1}$$

In this equation ρ is the density of the fluid and c_D (–) the drag coefficient. This is a purely empirical factor that can at best be calculated using numerical flow simulation. In the plate case it is about 1.1. Essentially streamlined ships, for example, have a value of 0.1, but we explicitly want to transfer the mechanical power into the medium as effectively as possible. The power to be transmitted is calculated by multiplying the drag force by the speed:

$$P_{Drag} = F_{Drag} \cdot v_{Plate} = \frac{1}{2} \cdot c_D \cdot \rho_{liqu} \cdot A_{Plate} \cdot v_{Plate}^3 \tag{5.2}$$

Now comes the transfer to the case of the impeller with flat plates as agitator blades, the Rushton turbine. Basically, it can be understood as a product of torque, being the driving force of the blades multiplied with the impeller diameter (m), and the angular velocity measured as rotational speed n (s^{-1}), in practice often given in (min^{-1}). This results in the following equation:

$$P_{R,imp} \sim \rho_{liqu} \cdot n_{imp}^3 \cdot d_{imp}^5 \tag{5.3}$$

In bioprocess engineering the range is 1,000 (kg/m^3) $< \rho_{liqu} <$ 1,200 (kg/ m^3). You may be wondering where we left the area A. But it gets even better. To relate the power input to the volume, you have to divide by the volume. At this point, we assume a standard stirring vessel in which the vessel diameter and stirrer diameter always have a fixed ratio. This results in

$$P_{R,imp}/V_R = \text{Ne} \cdot \rho_{liqu} \cdot n_{imp}^3 \cdot d_{imp}^2 \qquad (5.4)$$

The dimensionless parameter Ne is called the power number (also called Newton number relating the resistance force to inertial force). However, these simplifications now make us dependent on knowing the "constant" Ne as a function of the geometry and other circumstances.

The power number depends on the fluid flow pattern of the stirred tank bioreactor. The stirrer makes its own environment and changes turbulence. However, this in turn changes the power input. The turbulence can be described by the Reynolds number Re_{imp} (–), indicating whether the flow in the stirred tank is in the laminar, transitional, or turbulent regime. In general, the Reynolds number (Osborne Reynolds, 1842–1912) is the ratio of inertial forces to viscous forces within a fluid. Hence, the power characteristic ($N_P(\text{Re}_{imp})$) is of practical interest. The Reynolds number is listed in textbooks for different given application situations. For a rotating impeller it can be given as

$$\text{Re}_{imp} = \frac{\rho_{liqu} \cdot n_{imp} \cdot D_{imp}^2}{\eta_{liqu}} \qquad (5.5)$$

where η_{liqu} is the dynamic viscosity of the medium. The typical range is from 1 mPa · s for clear water to 1 Pa · s for highly viscous media or products; higher values may occur, e.g., during polysaccharide production. For such cases special agitators have been invented.

For Re_{imp} between 1 and 100 (the laminar regime) the Newton number Ne is inversely proportional to the Reynolds number and decreases from about 100 to about 1. For 100 < Re < 1,000 turbulence is located in the transition area. For Re > 1,000 for the impeller induced turbulence, actually the turbulent regime, power number is independent of Re_{imp}. This is the normal operational case, with some examples given Tab. 5.2.

Tab. 5.2: Typical power numbers at the turbulent regime for some commonly used stirrer arrangements.

Stirrer	Rushton, 6 blades	Inclined blade, 6 blades	Intermig 07, 3-stage	Propeller, 3 wings
Ne (–)	5–6	1.5–2	0.9	0.5–07

A low Ne number is not automatically bad, as other criteria like axial mixing may be better.

The presence of aeration-induced bubbles results in a reduction in the power requirement and transfer since the gas bubbles decrease the density of the fluid and affect the hydrodynamic behavior of the fluid around the impeller. The effect of aeration, which is of course the usual case, upon power consumption is expressed as

$$PR_{R,gas} = 0.72 \cdot \left(\frac{P_R^2 \cdot N_{imp} \cdot D_{imp}^3}{f^{0.56}} \right)^{0.45} \tag{5.6}$$

$P_{R,gas}$ is the gassed power, P_R is the unaerated power, and f_{Gas} is the volumetric gas flow rate. For the above given correlations several variations are given in the literature considering many only qualitatively known influences like reactor geometry. In real application the manufacturers of bioreactors give data sheets with real values. Some typical values are given in Tab. 5.3.

Tab. 5.3: Typical values for power transfer and hydrodynamic parameters.

Case	V_R (L)	N_{imp} (min^{-1})	Re_{imp} (–)	N_P (–)	P_R (W/m^3)	t_c/t_m (–)
1	0.1	1,000		10	0.7	
2	0.1	2,000		5	3.0	
3	10	300			2.5	
4	10	500			5.0	
5	2,000	50				
6	2,000	200				

1) Minireactor, one stirrer, low speed; 2) minireactor, one impeller, high speed; 3) lab reactor, one stirrer, low speed; 4) lab reactor, three impellers, high speed; 5) medium size production, low stirrer speed, blade; and 6) medium size production, high stirrer speed.

> ? Find other values for different reactors and impellers from supplier manuals, check, and amend the above table. Do not seek formulas; even better for data from measured graphs.

Average energy dissipation rate per unit mass ($\bar{\varepsilon}$ (W/kg)) is defined as

$$\bar{\varepsilon} = \frac{P_R}{V_R \cdot \rho_{liqu}} \tag{5.7}$$

where V_R is the liquid volume in the bioreactor and ρ_{liqu} is the liquid density. It represents the power input per mass of medium.

Using eqs. (5.5) and (5.11), eq. (5.12) is obtained:

$$\bar{\varepsilon} = \frac{N_p \cdot N_{imp}^3 \cdot D_i^5}{V_R} \tag{5.8}$$

After the mechanical energy is transferred from the stirrer to the medium it has to be dissipated throughout the reactor. The local energy dissipation rate $\varepsilon(x)$ over average energy dissipation rate ($\bar{\varepsilon}$) represents the dimensionless unit $\varepsilon / \bar{\varepsilon}$. Knowledge of the overall and the local distribution of energy dissipation is necessary for a successful scale up, bubble breakup or coalescence, estimation of shear stress, and many other process applications. Now a closer look at the flow patterns will be visualized for understanding the basic action and impact of the agitators. This includes the flow profile, the turbulence expansion, and the shear rate, which can be understood as density of the isolines of the flow profile.

A significantly high local energy dissipation occurs at the Rushton turbine especially at the blade tips. This causes stress for sensitive microorganisms. During scale-up, the velocity of the stirrer tips increases due to higher circumferential speed, so lower stirrer speed is necessary. This can be a limiting step. Therefore, the choice of appropriate impeller is vastly important especially when shear sensitive microorganisms are used (Fig. 5.5).

Turbulence expansion [m^2/s] and flow profile (arrow & flow lines)

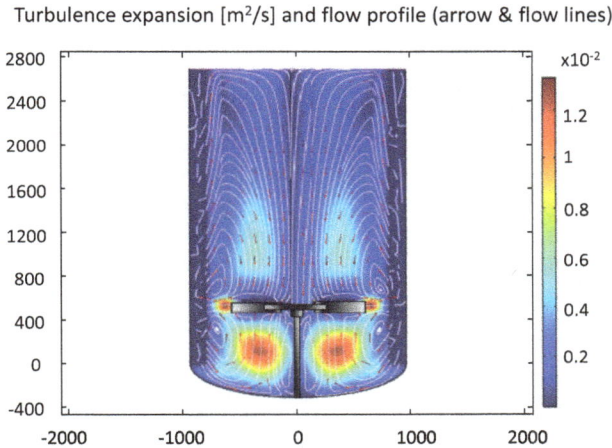

Fig. 5.5: CFD simulation of flow profile and turbulence expansion of a reactor with one Rushton turbine. The unit m^2/s becomes kinetic energy gradient per kg fluid.

Lack of an axial component leads to a significantly low medium movement in the upper part of the reactor. This must be compensated by several agitated layers. Axial flow impellers create a top to bottom motion as Fig. 5.6 demonstrates.

The local shear stress is much lower than for the six-blade impeller. Only one level is necessary as fluid flow is directed first to the bottom and then bent to the top reaching the whole reactor. The Intermig type impeller (Fig. 5.7) compared with the Rushton turbine, transfers more uniform energy to the fluid and accordingly demands less power consumption in order to obtain the same degree of mixing and mass transfer coefficient accompanied by lower shear stress.

Efficient mixing is necessary to care for an ideal homogenization of all compounds. This is to prevent sedimentation of the cells, transportation of gas from the medium to the bubbles and vice versa. Not unimportantly, heat has to be brought to the reactor wall by convection. Fresh medium and titration agents are usually dosed locally and have to be suspended; otherwise, local gradients would occur changing cell physiology.

Turbulence expansion [m²/s] and flow profile (arrow & flow lines)

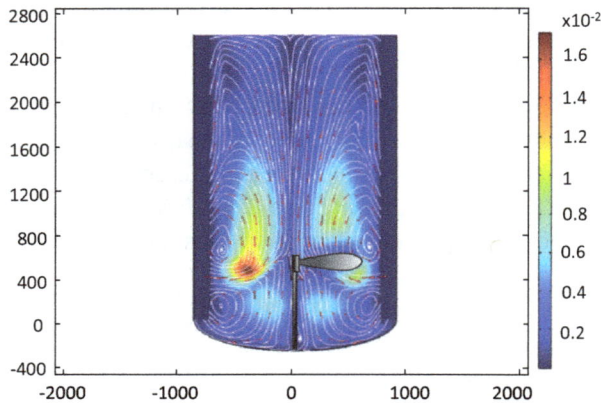

Fig. 5.6: CFD result for an axially acting impeller.

Turbulence expansion [m²/s] and flow profile (arrow & flow lines)

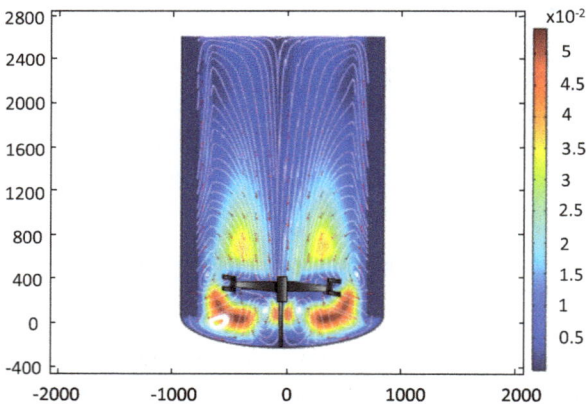

Fig. 5.7: CFD simulation of the impact of an "Intermig" agitator.

Mixing efficiency of stirred tank reactors are evaluated by mixing time. Short mixing times are necessary for a proper bioreactor design. Mixing time, t_M, denotes the time required to achieve a certain degree of homogeneity which is a measure of the quality of a mixture. The mixing time measurements are performed by injecting a tracer into the stirred vessel. Tracers mainly include acidic or alkaline solutions, salt solutions, heated solutions, and colored solutions. Tracer concentration is measured at fixed points by various methods such as sensor method, discoloration method, and optical and colorimetric methods.

The mixing time depends on liquid properties, the stirrer type and size, the specific power input, and bioreactor geometry. At the required degree of homogeneity (M), tracer concentration reaches within ±5% range of final concentration (c_f) after fluctuation. Circulation time (t_C) is defined as the time for the fluid in the stirred tank vessel to circulate one loop. The degree of homogeneity M is then defined as

$$M = 1 - \frac{\Delta c}{\Delta c_0} = \frac{c_t - c_i}{c_f - c_i} \tag{5.9}$$

where c_f is the final/equilibrium tracer concentration, c_t is the tracer concentration at time point t, and c_i is the initial tracer concentration. Δc_0 is the concentration difference at the beginning of mixing and Δc is the concentration difference at respective time t. The temporal course of the concentration fluctuations can be approximated by an exponential approach as follows:

$$\frac{\Delta c}{\Delta c_0} = k_M \cdot \exp\left(-\frac{t_M}{t_C}\right) \tag{5.10}$$

where k_M is a constant, t_M is the mixing time, and t_C is the circulation time. Mixing time can also be expressed in dimensionless form in a turbulent flow regime:

$$N_i \cdot t_M = N_M = \text{const.} \tag{5.11}$$

where N_i is the rotational speed of the stirrer and N_M is the mixing number. Mixing number reveals the number of stirrer rotations required for a specific degree of homogenization. The course between mixing number and impeller Reynolds number is called mixing time characteristics and can be created for various stirrer types. In a laminar flow regime, N_M decreases with increasing impeller Reynolds number whereas in a turbulent flow regime, when $\text{Re} > 10^4$, mixing number N_M is constant, having no relation with impeller Reynolds number.

The last point to discuss is heat transfer. Bioreactions are generally exothermic. We remember that in aerobic bioreactions half of the substrate is completely oxidized in respiration. A heat balance can be set up as

$$\Delta c_{Gluc} \cdot H_{C,Gluc} = Y_{X,Gluc} \Delta c_X \cdot H_{C,X} + Q_{heat} \tag{5.12}$$

Therefore, for the volumetric heat flux q_{heat}:

$$q_{heat} = r_{Gluc} \cdot c_X \cdot (H_{C,Gluc} - y_{X,Gluc} \cdot H_{C,X}) \tag{5.13}$$

For an example we set $r_{Gluc} = 1$ g/(g·h), $c_X = 50$ g/L, $y_{X,S} = 0.5$ g/g, $H_{C,Gluc} = H_{C,X} = 16$ MJ/kg we obtain $q_{heat} = q_{heat} = 400$ kJ/(L·h). That means that the medium would heat up by 100 °C ($c_{P,H2O} = 4$ kJ/(kg·K). Indeed, high temperatures can occur in haystacks and even self-ignition, induced by microbial activity. Silos filled with organic powders have also been reported to burn in cases where humidity has led to microbial contamination. Discharge of heat in bioreactors goes by convective transport to the reactor wall and then through it by thermal conduction into the cooling liquid. With the help of Fourier's law this process can be described:

$$q_{heat} = k_{heat} \cdot A_{wall} \cdot \frac{T_{medium} - T_{cool}}{D_{wall}} \tag{5.14}$$

where A_{wall} is inner surface of the reactor contributing to heat transfer, and k_{heat} is the material's heat conductivity (W/mK). In practice, the reactor wall is not a plane but has different inside and outside surface areas. Together with surface roughness this means that d_{wall} cannot be exactly determined. A laminar medium film complicates the problem further by making heat transfer dependent on medium turbulence. The final heat transfer from the medium into the wall material is described by the overall heat transfer coefficient. The material is not known by the operator and heat radiation may contribute. So, a practical value is provided by the manufacturer, the thermal transmittance $U = k_{heat}/d_{wall}$ (W/m$^2 \cdot$ K). This lumping of influences and replacing exact values by lumped experimentally obtained values is a common approach in chemical engineering.

Cooling jackets are preferred in pilot and medium size production reactors as they do not expose additional surface area (biofilm formation), do not reduce active working volume, and facilitate cleaning. In large reactors heat exchangers are sometimes not avoidable. This also holds for small glass lab reactors, where a "cooling finger" can be positioned. Glass exhibits only 1% of the heat conductivity. This has also impact on lab work. Cooling down a shaking flask in an ice bath may happen only with 1 °C/s, which is too slow for some purposes.

Finally, we come back to mechanical power transfer, where another specific issue of bioreactors must be envisaged belonging to the topic of hygienic design. To prevent leakage between shaft and vessel a mechanical seal is built in as a shaft bearing; details are made clear in Fig. 5.8. This type of seal restrains the medium via two rings, from which two surfaces are in direct contact with each other. One ring turns with the shaft and is called the "rotary ring"; the stationary ring is fixed at the reactor vessel. A spring element presses the two rings against each other thus forming the elements of a face seal. Medium forms a lubricating film between the surfaces when the shaft rotates. The pressure in this film balances the seating force. Next to this primary seal an additional secondary seal formed by O-rings (toric joints) must be foreseen. Only one mechanical seal would allow microorganisms to escape to the outside or to enter the reactor via the lubricating film. The double mechanical seal with one at the product side and one on the outside allows for application of sealing liquid or steam for deactivation of cells approaching the room between the two seals.

To avoid all these problems, magnetic coupling between the outer engine and an inner shaft has also been developed. For small scale this principle is already known in the form of a magnetic stirrer using a magnetic stir bar. Applications can also be found for smaller bioreactors. For large dimensions magnetic coupling has not been widely established.

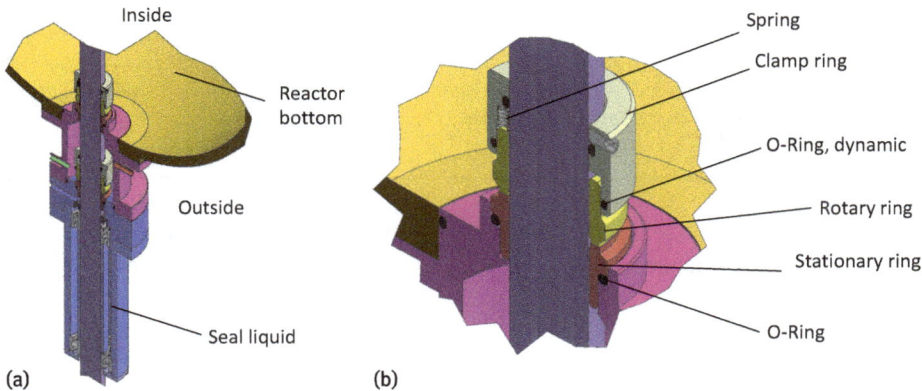

Fig. 5.8: CAD drawing of a double-acting seal: (a) overview and (b) details.

The STR can be universally applied and is in the lower and middle scale among the most common fermentation vessel. Some disadvantages have to be mentioned as well. Firstly, energy dissipation rate is very high at the tips of the stirrer blades. The second point is that axial mixing is worse than radial mixing. These points are addressed by the draft tube reactor (Fig. 5.4), where the fluid is forced in the axial direction by a propeller, then bent by the dished ground plate (tori spherical dished end), and finally reversed to the top. The occurring vortex can also suck foam back into the fluid. This reactor is especially of advantage for sensitive cells at lower oxygen transfer rates.

5.3 Pneumatically driven bioreactors – a soft way of energy transfer

But who said that stirrers are the only means to make something move? We can think more generally of ways to transfer mechanical energy into the medium and care for adequate flow patterns. Forces of the technical scale can be easily applied either by pneumatic or hydrodynamic pressure. Flow control is gained by static installations adjusted to the energy transfer and the specific needs (Tab. 5.4).

Tab. 5.4: Possible means of mechanical energy supply and appropriate flow pattern control.

Energy transfer/flow control	Stirrer (mechanical)	Gas bubbles (pneumatic)	Pump (hydraulic)
Baffles	Stirred tank reactor	–	–
Draft tube	Draft tube reactor	Air lift reactor	Jet reactor
None	Cup of coffee	Bubble column	Jet reactor

> **!** Bubble column bioreactors are pneumatic reactors in which aeration and agitation are enabled by gas bubbling without using any mechanical stirring device. Therefore, bubble column bioreactors in general have no moving parts. The main internal structure is the gas sparger, which is usually shaped as a perforated plate. The preferred area of application is slow but high-volume cultivation, such as penicillin production with *Aspergillus*.

In the simplest case bubble columns are tubes standing upright. A typical aspect ratio $H_R/D_R = 10$ or even higher allows for a long residence time of the bubbles. While the columns can be up to 100-m high, the working volume can be 300 m^3 or even up to 3,000 m^3. Occasionally when tall columns are preferred, perforated horizontal plates are used to break up the coalesced bubbles (Fig. 5.9).

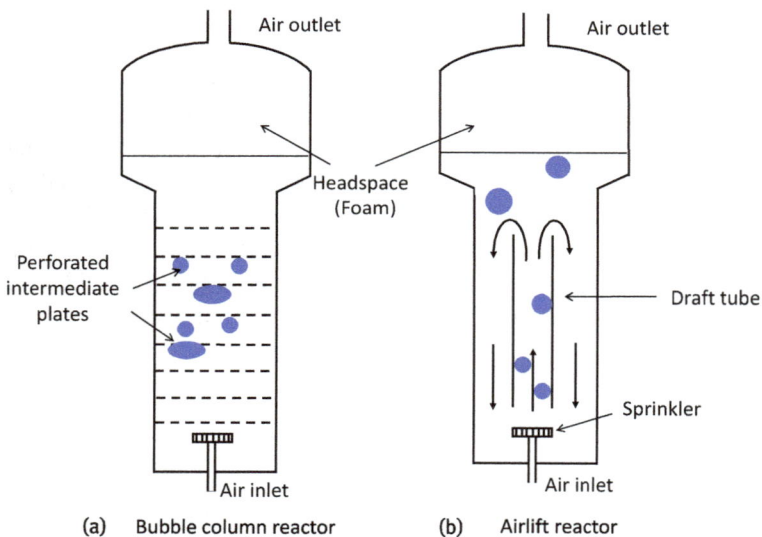

Fig. 5.9: Scheme of the two basic designs of pneumatic columns reactors: a) bubble column; b) airlift reactor. Bubble coalescence occurs on the way up; perforated plates lead to dispersion again. Bubbles size will also increase due to decreasing hydrostatic pressure.

Bubble column performance is influenced by the hydrodynamic behavior of the bubbles. Sparger design, medium viscosity, column diameter, and especially gas flow rate have an important influence on the development of different flow regimes. As the introduction of gas at the bottom is the only energy input specific attention is required. In contrast to the STR gas flow F_G is not normalized to the volume but is based on the cross-sectional area A_R. This approach delivers a superficial gas velocity U_G (m/s):

$$U_G = \frac{F_G}{A_R} \tag{5.15}$$

The idea behind it is that rising bubbles evoke similar effects along the height of the longitudinal axis of the reactor with respect to energy and mass transfer. At low superficial gas velocities (≤ 0.05 m/s) bubbles have uniform shape and rise through the column without interacting with other bubbles and thus a homogeneous flow pattern occurs. With the increase of gas flow rate, e.g., up to 1 m/s, a heterogeneous flow regime occurs due to the unstable and turbulent motion of the bubbles. This effect leads in particular to bubble dispersion and coalescence. Here large bubbles emerge. This can even go so far that the bubble diameter reaches the diameter of the reactor. The rising bubble leaves no space for the backflow of the fluid. This phenomenon, known as slug flow, is sometimes intended in small tube systems, as the bubbles press liquid layers through the tubes. In larger reactors, it is dangerous. Cases are reported where slug flow even led to destruction of the reactor. In the homogeneous flow regime gas holdup ε_G is proportional to U_G, which is physically intuitive assuming constant and same rising velocity of all bubbles. In the heterogeneous regime ε_G shows a disproportionately low increase with U_G.

Since no mechanical stirring device is applied, bubble column bioreactors require less energy or, to say it the other way round, do not allow for high energy and mass transfer compared to the STR, comprising an agitation system. The volumetric power input exerted to the fluid in the bubble column depends on the superficial gas velocity and can be determined as follows:

$$\frac{P_R}{V_R} = \rho_L \cdot g \cdot U_G \tag{5.16}$$

where P_R (W) is the power supplied for aeration, V_R (m³) is the liquid (working) volume, ρ_L (kg·m⁻³) is the liquid density, g (m/s²) is the gravitational acceleration constant, and U_G (m/s) is the superficial gas velocity. The rationale behind this is that the bubbles must overcome the hydrostatic pressure at the bottom of the reactor. The locally distributed values are shown in Fig. 5.10 as simulation results.

Low capital cost, gentle environment for sensitive microorganisms, easy cleaning of the vessel, and lack of mechanical stirring device are the main advantages of bubble column bioreactors, while low energy transfer and accordingly mass transfer are limitations. Applications in industry that fit this profile include cultivation of filamentous fungi for citric acid and penicillin production. Later we will see that low growth rates, e.g., 0.1 h⁻¹ for *Aspergillus*, are accompanied by a low oxygen consumption rate. Beside the advantages, disadvantages such as foaming and the formation of coalescence-induced gas bubbles decreases the bioreactor performance. The high pressure at the bottom may not affect the microorganisms directly but influence the CO_2/HCO_3^- balance and therefore pH value. Low axial mixing time is another problem, which is tackled by a reactor design described in the next paragraph.

Speed profile of the liquid phase [m/s] Shear rate [log (1/s)] Volume fraction of the gas phase [%]

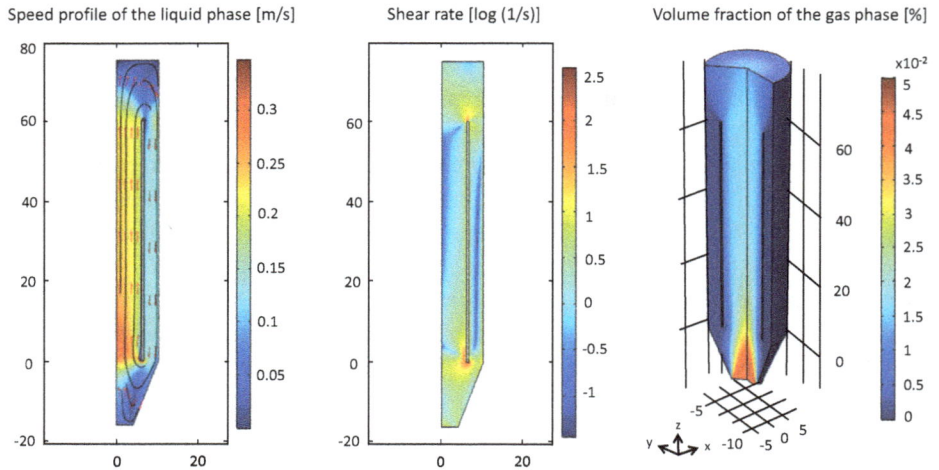

Fig. 5.10: CFD simulations of some hydrodynamic parameters of the bubble column.

Airlift bioreactors are likewise pneumatic tower-shaped reactors, in which agitation is enabled by gas sparging. The design of the airlift bioreactors includes a draft tube, which is situated in the center of the reactor. The draft tube creates two distinct zones in the reactor where only one of these zones is sparged using a gas. The density difference between the aerated zone, the riser, and the not aerated zone, the downcomer, results in a pressure difference between the two columns and subsequently to a liquid circulation allowing an adequate mixing. In the riser gas-liquid upflow emerges while in the downcomer a fluid downflow exists (Fig. 5.9b). This principle was already invented 230 years ago for pumping suspensions.

In general, airlift bioreactors have two basic configurations. These are internal loop airlift bioreactors and external or outer loop airlift bioreactors. In the internal loop airlift bioreactors, the bubble column is separated by an internal baffle so that a riser and a downcomer occur, whereas in the external or outer airlift bioreactors, riser and downcomer are separate tubes connected by short horizontal sections at the top and the bottom. External loop airlifts have a better mixing efficiency and a faster liquid circulation since the density difference between fluids in the riser and downcomer is greater as a result of the extended distance.

Due to the lack of a mechanical agitator system, construction cost and energy consumption accordingly reduce, less maintenance is needed, and an easier sterilization is possible compared to the bioreactors that contain agitator shafts. Moreover, despite the advantages of airlift bioreactors, the need for relatively higher pressures, greater air throughput, and the inefficient foam breaking should be considered depending on the process requirements. Owing to the controlled liquid flow and equalized shear forces, airlift bioreactors have high efficiencies opening new possibilities in aerobic bioprocessing technology. Therefore, airlift bioreactors are used in many applications such as the growth of shear-sensitive cells, the production of single cell proteins, and immobilized enzyme reactions.

The volumetric power input exerted to the fluid is defined by the degree of turbulence and can be determined as follows:

$$\frac{P_R}{V_R} = \frac{\rho_L \cdot g \cdot U_G}{1 + \frac{A_d}{A_r}} \tag{5.17}$$

where A_d (m^2) is the cross-sectional area of the downcomer and A_r (m^2) is the cross-sectional area of the riser.

The third means of energy transfer, namely hydrodynamic pumping, is used mainly in the wastewater field to increase residence time in the reactor while maintaining good mixing. Apart from that the use in biotechnological processes is limited. The basic schemes can be taken from Fig. 5.11.

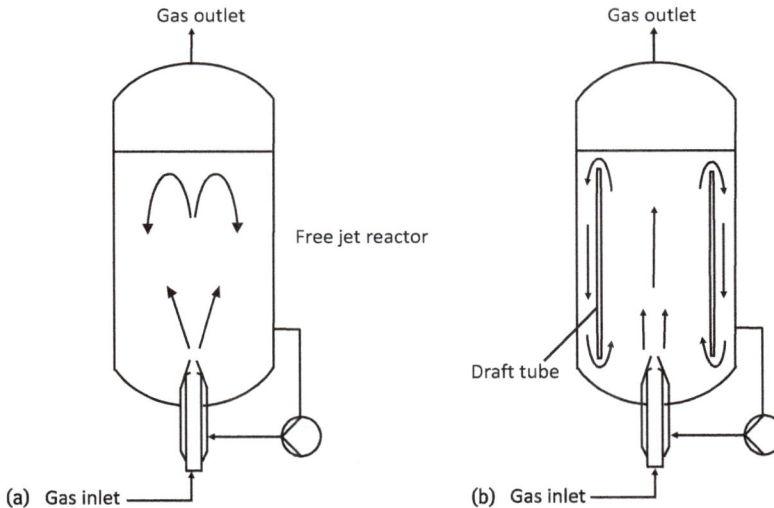

Fig. 5.11: Jet reactors with (a) free jet and (b) without draft tube.

5.4 Other types of bioreactors – translating demands into design

Standardization is a general demand in bioreactor design to keep things simple and cheap. Nevertheless, specific designs for specific applications can exhibit significant advantages to justify tailored designs. Gradual improvements with respect to aspect ratio, specific stirrers (e.g., to cope with high viscosity), or operational parameters mark one direction. Another thing is to introduce additional degrees of freedom. Until now medium, cells, and dissolved gases have been treated as ideally mixed. Bubbles are an exception as they experience additional buoyancy forces. This is not optimal as the residence time cannot be exactly adjusted to fulfill the task of oxygen supply or mixing simultaneously. Decoupling of solid phase (cells), medium (inclusive dissolved gases), and gas phase with respect to transport means and patterns is worthy of consideration. Gas could also be applied by tubes or membranes, and medium and/or cells may be really solid giving room for application of gravitational or structural forces. These could be fibers or granules as substrate and immobilized cell pellets or biofilms for the biomass. Table 5.5 is an attempt to structure the different possibilities.

Tab. 5.5: Possible ways shaping transport processes of gases, medium, and biomass.

Gas or substrate supply/substrate, biomass	Surface	Bubbles	Membrane	Liquid film
Dissolved	Shaking flask	Submers (regular case)	Membrane reactor	
Solid	Emers reactor	Biomass pellets	Biofilm	Trickle bed

Now a few reactor types will be picked out and sketched.

Before agitated bioreactors came into use, biomass, e.g., filamentous fungi propagated by swimming on the surface of the medium. In this way the mycelium could take substrate from below while oxygen uptake was enabled by the air above the medium surface. Citric acid as a product was excreted into the medium. This surface cultivation is called an emers cultivation in contrast to the submerse cultivation with the biomass being suspended in the medium that is normal nowadays. Production is possible in simple flat stackable vats without any auxiliary equipment. Sometimes this approach is still employed for spore germination. Some plants also need direct contact with air for propagation. The degree of process intensification is of course very low.

Biomass can also form a more solid phase as pellets (filamentous fungi) or cell flocs (wastewater). These aggregates are large enough to be treated as a separate phase independent from medium hydrodynamics. This can be utilized not only for cell harvesting but also for retaining the cells in the bioreactor by gravity or structural means like sieves (compare Chapter 9). The most widespread application is the upflow anaerobic sludge blanket reactor (UASB). Wastewater enters the reactor from the bottom via a jet or nozzle and flows upward. The upward fluid velocity in the reactor is kept low enough to allow the biomass pellets to sediment against the flow with the same velocity. Gas bubbles from anaerobic metabolism (biogas) care for additional turbulence. The gas is separated at the top or at several layers by deflectors (baffles) and a gas collector (caps, domes). In the case of low upstream velocities axial gradients of substrate concentration may emerge. This effect may even be intentionally supported. The effect is that easily metabolized medium compounds are used up at the bottom, while the microorganisms at the top are forced to take up the more resistant compounds. Such a gradient also occurs inside the pellets leading to different bacterial consortia along the gradient. Thus, the message is that the continuous stirred tank reactor (CSTR) comprising ideal mixing is not always the best choice. A perforated plate can keep them back in case higher medium flows are possible reducing axial gradients. In this case partial medium feedback can be envisaged via the nozzle, an application of the jet reactor principle.

Autonomously arising flocs can be regarded as biofilms, where the cells attach to each other either passively or actively. Biofilms in the narrow sense grow on the surface of solid substrates, here meaning an inert matrix. In such a biofilm the cell density and the activity can be much higher than in suspension culture, especially in anaerobic processes. The substrate can be formed as small (cm) parts called carriers, e.g., from plastic with high surface area. A popular shape is a spoked wheel. Such overgrown carriers can sediment much faster allowing higher flow rates. The moving bed biofilm reactor (MBBR) improves reliability, simplifies operation, and requires less space than traditional wastewater treatment systems. In addition, slowly growing and sensitive cells like mammalian cell lines can be cultivated using this principle. The cells do not form a biofilm as such but settle in pores of a porous carrier. In a trickle bed reactor, the carriers do not flow freely but form a packed bed. The medium does not flow through the bed upwards but is trickled from above. The gas phase is now the continuous phase allowing for good and controllable gas transfer via diffusion through the extensive water film on the carriers. This principle can therefore also be applied to aerobic processes. Note that a composter is also a bioreactor with biofilms growing on the solid substrate and kept wet in a trickle bed. To make use of cellulosic wastes, trickling hay bales with enzyme solutions has been tried out on a large scale. A great future will be awarded to biofilm reactors. Current research is in the direction of completely controlled biofilms. One means is the application of substrates either by a fluid medium or by diffusion through a membrane that is the substrate. Spontaneous biofilm formation is replaced by molecular linkers attached on one side to the substrate, and on the other to a specific epitope of the cells. In this way only strains of interest are part of the biofilm, while contaminants are washed out. Zero growth conditions are maintained without washout and carbon is mainly assigned to an extracellular product.

Besides mass transport issues, other very important aspects of reactor design include the interconnected aspects of scale, economics, product, and microbial physiology. In general, larger reactors are cheaper than small ones calculated on a volume basis. Only considering the expenses for stainless steel reactors the surface-to-volume ratio leads to cost proportional to $V_R^{2/3}$. Measurement equipment is necessary only once. Unfortunately, stirred tank reactors have their limitations of scale in the range of a few thousands m^3. For high-value products like pharmaceutical proteins, quite expensive reactors are needed to fulfill safety and quality issues. This generates high volume-specific costs, but these are justified by the high market price, which is calculated also based on downstream operations and finally on payback of development costs like clinical studies. For cheaper bulk products like citric acids or penicillin very high reactor volumes are necessary. This makes employment of bubble columns sensible, even if compromises in oxygen transfer rate are unavoidable. This is made easier insofar as fungal cultivations need lower oxygen transfer due to lower growth rates but benefit from lower shear rates. The bottom of the value chain is represented by wastewater treatment and anaerobic digestion. Here nobody speaks about stainless steel or axenic operation. Reactors are made from metal sheets or concrete. Peripheral devices are kept to a minimum.

Now we should spend some time considering how to design, calculate, and decide dimensions for such large plants. Scale-up follows the usual rules: first describe limiting parameters and then calculate their development during scale. In bioprocesses this is not quite easy. While calculation of things like energy transfer and flow patterns may be possible using common correlations or CFD simulations, cell reactions in response to the environment are not known a priori. The situation is abstracted in Fig. 5.12.

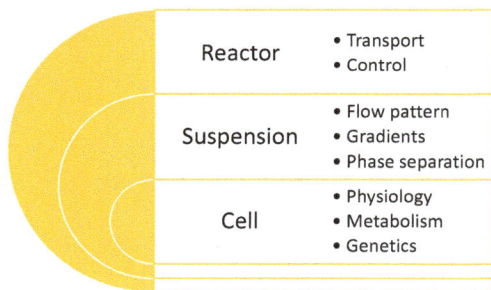

Fig. 5.12: Situation bioreactor calculation on three different layers.

On the outer level the production goals are defined, e.g., high productivity or high product concentration. Direct access to the actual transport rates of substrates, gases, or energy flows is possible by valves and pumps. On the medium level, design flow patterns can be specified by internal installations like baffles and mixers. Calculations using the transport level parameters are done by dynamic balance and reaction equations. However, some parameters like turbulence or diffusion can only be roughly estimated. On the cell level we have to face the dominant inaccuracy with respect to kinetic parameters. Kinetics is the interface between the cell and the medium level. Consequently, the design process must go the other way around: first determine these parameters, then calculate and define optimum conditions in the bio-suspension. Here limitations like mass transport by diffusion (e.g., bubbles) can be rate-limiting. Finally, the necessary external transport parameters have to be fixed. No additional limitations should occur on this level but be avoided by design. So, we need small bioreactors to measure layout parameters for big ones. This approach is called "scale-down." In the following examples we will follow the read thread from very small to very big.

The highest specific costs result if scientific data are the product itself. Strain characterization or medium design are typical examples of scientific goals to be achieved in bioreactors. Scanning the possible parameter space can lead to a combinatorial explosion. To cope with this situation, very small reactors on the µL scale can be designed allowing for a high degree of parallelization. In small scales, diffusion contributes positively to homogenization. An example is shown in Fig. 5.13.

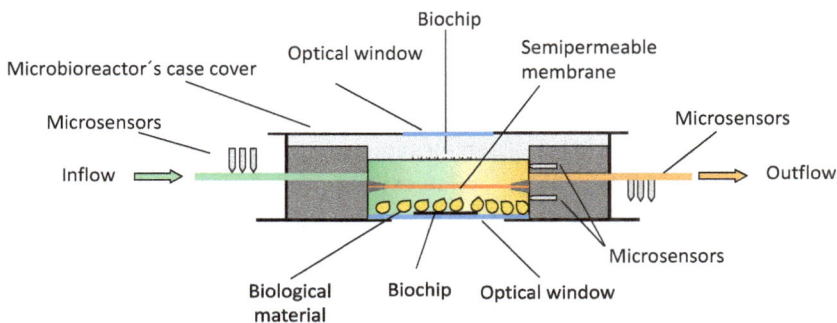

Fig. 5.13: Microreactor for investigation of cell growth (©CytoBioTech).

The cells under investigation are kept in a small chamber and are separated from the upper part by a membrane. Fresh medium is supplied in this upper part, while compounds for supplying the cells can diffuse through the membrane. To fulfill the desired goal of data acquisition sensors-on-a-chip and microsensors are employed. Microscopy as an information source is not out of fashion but foreseen here by optical windows. Note that neither bubbling nor mechanical mixing is necessary. To increase experimental throughput microtiter plates (µL-scale) are employed. Fast screening of strain libraries is the preferred task to be accomplished. In the best case, gas exchange is possible but no medium exchange. Growth can be observed by an optical reader giving data for biomass, pH, pO_2, or fluorescence. Nevertheless, chemical analysis is possible only at the end of a batch run. So, the results cannot really be extrapolated to anticipated process conditions. The last step towards small (pL) scale is the adaptation of microfluidic devices making single cell analysis possible. This opens a new dimension in process development. In bioprocesses we observe only medium growth characteristics of all cells. In reality, a heterogeneity in physiological parameters has to be considered depending on partly unknown reasons. This holds even for monoclonal cultures (Fig. 5.14).

Cell-to-cell-heterogeneity is an omnipresent phenomenon in bioprocesses. Microfluidic single-cell bioreactors offer an elegant way to analyze the dynamics of micro-

(a)   (b)   (c)

Micro-fluidic single cell platform   Single cell bioreactor   Bacteria colony

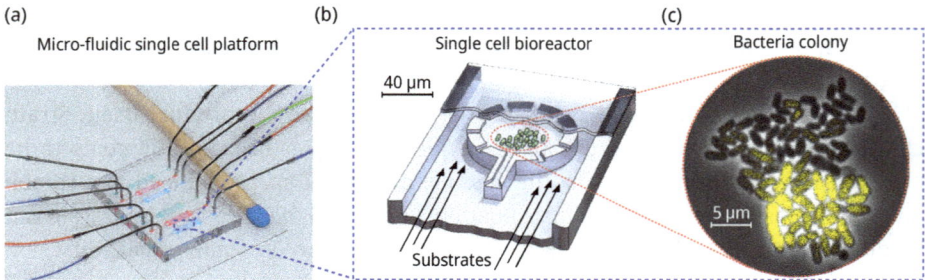

40 µm

Substrates

5 µm

Fig. 5.14: (a) Sketch of a platform to work with and to observe single cells; (b) schematic drawing of the cultivation chamber; the arrows indicate differently controllable substrate channels; (c) monoclonal bacterial cells with recombinant GFP; although they originate from a single cell in the chamber, they behave differently with respect to GFP activation (© Alexander Grünberger).

bial populations at the single-cell level. One example is to measure changes during the cell cycle. Synchronized cultivations are an expensive alternative and are delivering only qualitative data. Other areas of interest are changing cell morphology, cell-to-cell interactions, or fast reactions to stepwise changes in medium composition, mimicking a badly mixed reactor.

Back at the mL scale we find the shaking flasks as a stepping stone in process development. The specific power input (P/V [kW/m³]) depends mainly on shaking frequency. Typical values are around 1 kW/m³ but can be increased up to 3 kW/m³ for shaking frequency of, e.g., 300/min. This is actually in the order of magnitude of stirred tank reactors. Small filling volumes and baffles can increase the number further. Large flasks (e.g., 2,000 mL) can reach even higher numbers up to 6 kW/m³. With respect to oxygen transfer the sealing cap could be a bottleneck, but headspace aeration with small tubes is also an option. Measurements of p_H and p_{O2} are possible with internal optical sensor pads. Color changes are read out via light beams. The shaking flask is really a suitable tool for multiple parallel experiments toward process development.

A better option for observation of cultivation are parallel mini reactors (about 200-mL scale) allowing for control of the most important parameters and even for fed-batch processes. One commercially available reactor system is shown in Fig. 5.15. This is a compromise between parallelization and process similarity. There are sets on the market with four or more (six-pack) of such minireactors sharing the same peripheral devices. Several batches with different parameters are possible in direct comparison. In addition, sterilization, a high degree of automation, and data acquisition are possible.

Not only for James Bond but also for sensible cells and high value products "shaken not stirred" is mandatory. Animal cells do not have a rigid cell wall and cannot stand strong agitation. Here reactors have been developed that realize noninvasive mixing by smoothly rocking the whole reactor. One example is shown in Fig. 5.16. The plastic bag (cell bag) contains the cell suspension, while the support (rocker) carries out slow rocking (rolling or pitching) movements. This induces wave-like movements of the fluid,

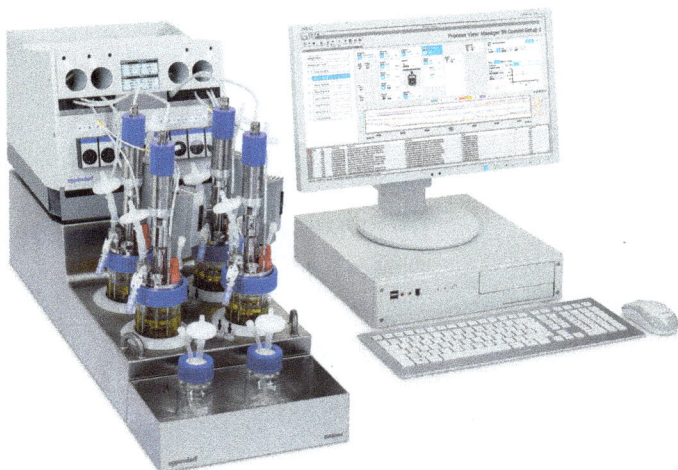

Fig. 5.15: Dasgip multi-parallel mini reactor installation with process and control periphery included (©Eppendorf). Visible are the magnetic stirrer through the headspace, off-gas cooling and membrane vent filters.

which gave rise to the generic name as propagated by a company of "wave bioreactor." Oxygen transfer happens via the headspace. The working volume from less than 1 L up to several 100 L is even good for production of small charges of high value products like antibodies.

In the last picture the reactor is only a plastic bag. This is meant to be disposed of after use. Sterilization is done by radiation during the manufacturing process making thermal sterilization superfluous, a big advantage. Plastic bags are of course cheaper than stainless steel reactors. Changing the bag after each cultivation allows for higher flexibility producing small charges for the preclinical and clinical market. Current discussion is around whether this really pays back. The main parameters discussed are duration of a batch and the added value of the product. The application of single-use bioreactors (SUB) in biopharmaceutical research and production has increased enormously in recent years. They are employed for production of biopharmaceuticals like vaccines and biosimilars in all primary process steps, in particular for small- and medium-volume scales (Fig. 5.17). SUBs are primarily used in processes in which protein based biotherapeutics from mammalian cell cultures are the target product. Additionally, they can be used for cultivation of plant cell cultures, microorganisms, and microalgae, as well as for special products in the food and cosmetics sector. A large variety of single-use bioreactors and single-use mixing systems with volumes of up to 2,000 L is presently on the market. These systems differ in terms of mixing, type of power input, and gassing strategy. An outer support container is engineered and fabricated to fully support each an SUB fermentation bag and allow easy access for operation. It contains the mixing drive, possibly a silicone electric resistive heating

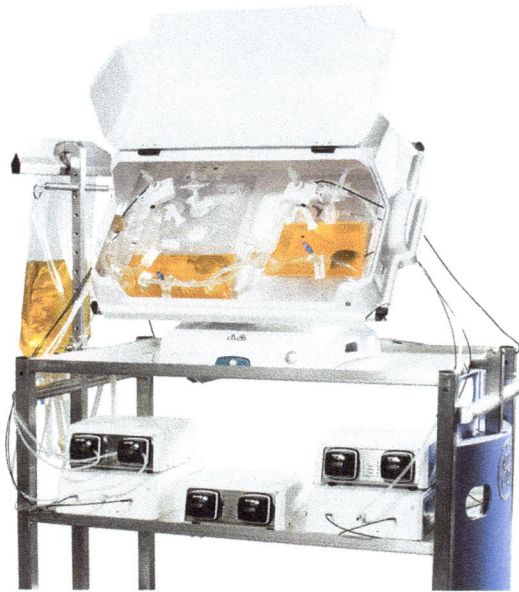

Fig. 5.16: Wave rector for animal cell culture with several disposable plastic bags (©GE healthcare).

blanket or water jacket, and optional controllers for mixing. With respect to plastic waste there is an ongoing discussion about sustainability of disposable reactors.

Having already discussed a few things about standard reactors we now make a jump to really big vessels but employed for low value products. The "biotower reactor" for aerobic wastewater treatment carries this direction to extremes. Figure 5.18 shows a picture and the internal structure of this reactor. The reactor should not be mixed up with an anaerobic digester. It is basically a bubble column, but with immediately visible differences. First, the reactor is not tall as expected for bubble columns. The air does not leave the sparger upwards but downwards against the wastewater stream being, of course, the medium. This measure leads to a first gas dispersion. A shield leads the bubbles to the outer regions of the reactor where they can freely rise. The bubble column effect appears now in the opposite direction compared to the usual geometry of bubble columns. Such a "tower-biology" needs more pumping energy and is more expensive than an activated sludge tank while the main advantage is space saving. In fact, the reactor integrates a whole wastewater plant, especially as it comprises also sludge separation. This is done in the ring-shaped structure at the top. Sludge can sediment (flocculation occurs naturally) and can be removed or recycled while the cleared water is discharged. The biotower reactor is therefore an example for integration on the process level.

Back to our roots: What did our ancestors do to produce ethanol without stainless steel or disposable plastic bags? Beer and wine were usually produced in amphorae

Fig. 5.17: Complete ready-to-use unit using disposable bags, here in the left steel vessel. Further downstream processing like harvesting and extraction units can be designed as single use equipment in the same rack (© Sartorius).

(a)

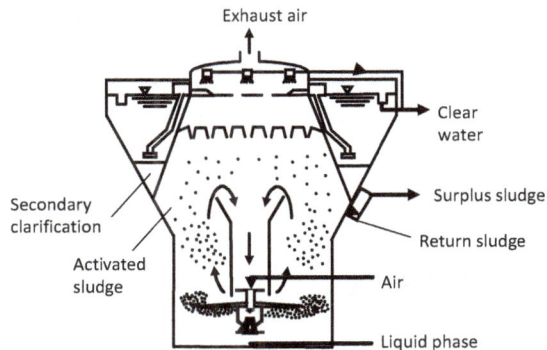

(b)

Fig. 5.18: Appearance and sketch of the internal structure of the "biotower" reactor (©infraserve).

or inconspicuous ceramic pots. Sometimes archeologists cannot even decide whether it is a jar or an urn without investigating the content by modern analytics. Maybe, bioprocess engineers could sometimes contribute to this integration of sciences. An especially nice ancient fermentation vessel is shown in Fig. 5.19.

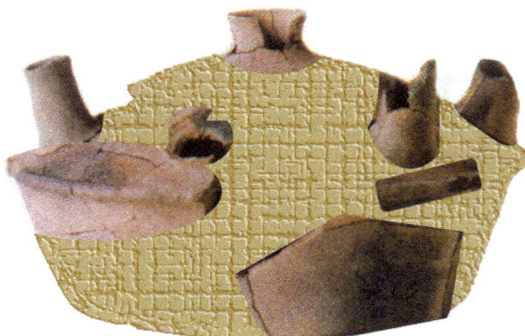

Fig. 5.19: Virtual reconstruction of an ancient fermentation jar; the small "chimneys" are probably designed as outlet for CO_2 keeping the foam back in the vessel.

5.5 Gas transfer – supplying microorganisms with gaseous compounds

Besides dissolved organic substrates and minerals, gaseous compounds also must be supplied to the microorganisms as educts. With respect to biochemical engineering gas supply requires special consideration and technical implementation. Low solubility requires constant feeding. Gases are also endpoints of biochemical reaction chains like respiration and must be removed from the reactor. The transport of dissolved gases into the medium is a common reason for limitations in process productivity. The final strived product can also be a gas. This makes in situ product removal easy and can potentially reduce downstream costs. All these facts are reason enough to have a closer look at the biological, physical, and technical aspects of gas turnover in bioreactors.

Oxygen (O_2) is needed by most aerobic microorganisms for respiration and almost only for this purpose. The cells take it up from the medium similarly to most gaseous compounds by diffusion. Usually, it is supplied to the reactor as part of normal air, but also pure O_2 can be employed in the case of a high oxygen demand or in order to reduce bubble volume. As pure oxygen is much more expensive than air, the decision to apply it depends of course also on the product value.

Carbon dioxide (CO_2) is a product of respiration and has to be removed from the reactor. To some degree carbon dioxide is fixed in anaplerotic sequences, e.g., for closing the tricarbonic acid cycle. There are some hints that strong stripping leads to extension of the lag phase. Phototrophic organisms need CO_2 for photosynthesis and in this case, oxygen must be removed. Further on, CO_2 has side effects on the medium as it leads to acidification.

Nitrogen as N_2 is usually not involved in technical bioprocesses, so together with the noble gases it is referred to as inert gas. Being part of normal air, it is fed into the reactor and contributes to gas exchange and mixing. Some anaerobic bacteria and some phototrophs can reduce nitrogen to ammonia. As about 1% of the worldwide energy is used to convert atmospheric N_2 to NH_3 in the Haber-Bosch process it is a worthwhile goal to find a feasible bioprocess to perform this nitrogen reduction. Inert gases are also used to strip traces of oxygen from the medium for cultivation of strictly anaerobic organisms.

Methane (CH_4) is produced in biogas plants along with CO_2 by methanogenic microorganisms. Energy for mixing and product separation is provided by the bubbles inclusively, an absolute "must" in bioenergy production. Methane can also be a substrate for special gas fermentations.

Hydrogen (H_2) is also involved in biogas formation as an intermediate compound. How to stop the process producing hydrogen instead of methane at that point is an open question in ongoing research. Producing hydrogen by water splitting employing microalgae is another option for the future. H_2 as substrate is an option as well, e. g., to synthesize liquid fuels from hydrogen and carbon dioxide.

Some volatile organic compounds could possibly also act as substrates or are products like volatile flavors. Ethanol and acetaldehyde are products in alcoholic fermentation and can be subject to desired or undesired stripping out. The bulk chemical isoprene is discussed as a gaseous product from bioprocesses avoiding energy consumption for cell separation and disruption.

5.5.1 Transport processes – the journey of gases through the reactor

Gas transfer to the reactor and into the medium can be understood as a sequence of different transport steps as shown in Fig. 5.20. Gas enters the reactor with the flow $F_{Gas,in}$ (L/h), where values are given for normal conditions. Related to the working volume of the reactor volume the volumetric gas flow $f_{Gas,in} = F_{Gas,in}/V_R$ (L/(L·h)) is obtained. In technical environments this is often given as gas volume per reactor volume per minute (vvm). The content of a particular gas G in the gas mixture is given as molar fraction x_G (mol/mol). As an example for aerobic cultivation, air enters the reactor via a gas pipe ending in a sparger. This is in simple cases a tubular ring with pores to form and release gas bubbles. The bubbles rise driven by their buoyancy. Due to the turbulences evoked by the stirrer the bubbles are dispersed in the reactor. Oxygen diffuses through the bubble/medium interface and is more or less homogeneously distributed by convective transport. The volumetric rate with which oxygen enters the fluid phase is called the oxygen transfer rate (OTR (g/(L·h))). The cells are in direct contact with the medium and take up the oxygen by diffusion through the cell membrane. The amount of oxygen consumption by biomass is calculated as oxygen uptake rate OUR (g/L·h). Carbon dioxide leaves the reactor in the opposite direction out of the cells also by diffusion with the carbon dioxide production rate (CPR), through the medium and into the bubbles with the carbon dioxide transfer rate (CTR). Finally the bubbles burst at the surface of the medium and release the gas into the headspace, from where it is set free into the environment with the flow rate $F_{Gas,out}$.

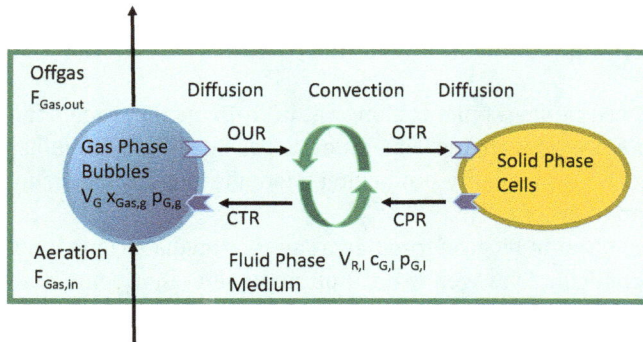

Fig. 5.20: Exchange of oxygen and carbon dioxide between bubbles, medium, and cells. The oxygen transfer rate (OTR) from the gas phase to the liquid phase is driven by the concentration difference, while the oxygen uptake rate (OUR) is controlled by the cells.

The oxygen balance in the medium as the basic equation finally reads:

$$\frac{dc_{O_2}}{dt} = OTR - OUR \tag{5.18}$$

It balances gas transport and metabolic gas turnover. In the following paragraphs a closer look at the different transport steps is outlined.

To understand the route of oxygen out of the bubble into the fluid we have to revisit two basic physical laws. The first one is Henry's law (eq. (5.19)), which gives a linear relation between the partial pressure of a gas in the gas phase and the concentration in the fluid phase. A prerequisite is that both phases are in direct contact and equilibrium conditions are adjusted (Fig. 5.21).

The equilibrium partial pressure of a gas G in the gas phase is given as

$$p_{G,Gas} = x_{G,Gas} \cdot P_{Gas} \tag{5.19}$$

leads to the saturation concentration $c^*_{G,Fluid}$ in the liquid:

$$c^*_{G,Liquid} = H_{G,Liquid} \cdot p_{G,Gas} \tag{5.20}$$

In this linear relationship, the Henry coefficient $H_{G,Fluid}$ is a proportionality factor. Some important Henry's coefficients are listed in Tab. 5.6. Note that Henry coefficients are dependent on temperature or salt concentrations in the case of a medium, so they can be looked up in chemical engineering handbooks or have to be measured. Before-

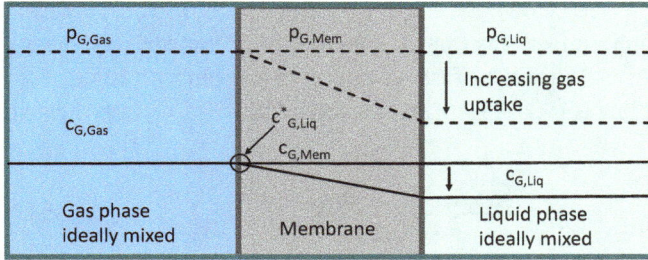

Fig. 5.21: Course of partial pressure and concentration in the gas phase and liquid phase separated by a membrane. Partial pressure for equilibrium conditions is the same in the gas and the liquid phase but drops with increasing uptake by the cells. Concentration in the membrane can be different due to a different solubility.

hand check the definition of H as it differs between the respective books with respect, e.g., to g and mol or bar and kPa.

Tab. 5.6: Henry's coefficients for different gases and solvents at 298 K (≈ 25 °C).

Henry's coefficient	Value (g/L/bar)	Major dependencies	Comments
$H_{O2,H2O}$	41×10^{-3}	T, salts, surfactants	Often limiting
$H_{O2,H2O}$ 35 °C	34×10^{-3}		Decreasing with increasing temperature
$H_{CO2,H2O}$	1.48	pH	Plus dissociates carbonate species
$H_{CH4,H2O}$	22×10^{-3}		Biogas
$H_{H2,H2O}$	1.54×10^{-3}		Biogas, extremely low
$H_{O2,Decane}$	570×10^{-3}		Hydrophobic solvent, similarity to cell membranes
$H_{NH3,H2O}$	490	pH, T strongly decreasing	Undesired stripping during sterilization, highest solubility for gas in water

As an example, we can calculate the oxygen concentration in water for normal air as gas phase at 25 °C:

$$c_{O_2,H_2O} = 41 \times 10^3 \frac{g}{L\,bar} \cdot 0.21\,bar = 8.61 \frac{mg}{L} \tag{5.21}$$

This is a very low value in comparison to other dissolved nutrients leading to multiple adverse consequences. Firstly, the amount of oxygen stored in the medium is so low that it would be immediately consumed by the microorganisms unless it is not constantly supplied by aeration. Secondly, the driving force, being the concentration difference over the water film (see below), is very low leading to a slow transport from the bubbles into the medium. In fact, this step can be limiting in industrial fermentations. Thirdly, uptake by the cells has to be fast even at these low concentrations.

The second important law is Fick's law of diffusion. The flux $F_{Diff,Gas}$ (g/s) through a membrane is given as

$$F_{Diff,Gas} = \frac{k_{Diff,Gas} \cdot \Delta c_{Gas} \cdot A_{Bub}}{d_{Membrane}} \qquad (5.22)$$

This equation is motivated by considering that a higher flux is possible for larger membrane areas and lower for increasing membrane thickness. The diffusion coefficient $k_{Diff,Gas}$ is a material property. The driving force of this mass transport is the concentration difference over the membrane. For a membrane separating a gas phase and a liquid phase $\Delta c_{Gas} = c^*_{G,Gas} - c_{G,Liq}$, assuming that the concentration on the membrane surface at the gas-phase side is in equilibrium with the gas phase. That may not be entirely true, as additional boundary layers exist on both sides of the membrane. Quasi-stationary conditions follow a linear progression of the concentration inside the membrane.

For application of Henry's and Fick's laws to gas transport from bubbles to the medium we imagine a virtual liquid boundary layer (film) of thickness d_{Film} between the bubble surface and the free fluid as first achieved by Nernst (1904). Behind this "film theory" stands the observation that close to the bubbles the water layer is laminar, and molecules are somehow ordered. Oxygen molecules can pass this boundary layer only by molecular diffusion. Gas transfer is therefore diffusion limited. For oxygen, that reads:

$$F_{Diff,O_2} = \frac{k_{Diff,O_2, H_2O} \cdot \left(c^*_{O_2,H_2O} - c_{O_2,H_2O} \right) \cdot A_{Bub}}{d_{Film}} \qquad (5.23)$$

This equation supports understanding of the gas transport but is not simply applicable to real situations inside the bioreactor. While the diffusion coefficient k_{Diff} is tabulated, the thickness of the film d_{Film} is hardly measured and is not exactly defined. This gives reason to introduce the gas transfer coefficient k_L (m/s):

$$k_L = \frac{k_{Diff}}{d_{Film}} \qquad (5.24)$$

Its value can be measured with macroscopic means, e.g., for defined plane surfaces. The total bubble volume related to the fluid volume $\varepsilon = V_{Bub}/V_{Liquid}$ is the gas holdup. However, in a bioreactor the total bubble surface area A_{Bub} is usually not known. The next step towards a measurable number is to relate the bubble volume to the fluid volume giving the volumetric exchange area as $a_{Bub} = A_{Bub}/V_R$. Finally, a_{Bub} and k_L are multiplied to give the volumetric gas transfer coefficient $k_L a$ ($1 \cdot s^{-1}$), or the "$k_L a$

value" for short. This is the most important coefficient for bioreactors designed for aerobic processes.

This procedure of combining nonmeasurable physical parameters to get macroscopically measurable characteristic coefficients is a common approach in process engineering (compare with derivation of enzyme kinetics). The oxygen transfer rate can now denoted in a compact form as

$$\text{OTR} = k_L a \cdot \left(c_{O_2}^* - c_{O_2} \right) \tag{5.25}$$

Typical values for the abovementioned parameters are given in Tab. 5.7.

Tab. 5.7: Typical parameters values describing gas transfer.

Variable	Abbreviation and unit	Values for oxygen	Comments
Diffusion coefficient	$k_{\text{Diff,O2}}$ (cm²/s)	2.42×10^{-5} (25 °C)	CO_2 2.1; N_2 2.0;
Film thickness	d_{Film} (m)	0.5×10^{-3}–1×10^{-4}	Increasing with velocity and turbulence of phases
Gas transfer coefficient	$k_{L,\text{O2}}$ (m/s)	8×10^{-4}–3×10^{-4} 1.5×10^{-4}–8×10^{-4}	For d_{Bub}>2 mm for d_{Bub}<2 mm, increasing size
Bubble ascent velocity	v_{Bub} (m/s)	0.2–0.3 0.02–0.3	For d_{Bub}>2 mm for d_{Bub} <2 mm
Volumetric surface exchange area	$a_{\text{Bub}}{}^a$ (m2/m³)	1,200–120	For d_{Bub} = 1–10 mm, ε_{Gas} = 20%
Gas hold up	ε_{Gas} (Vol.%)	10–40	$a_{\text{Bub}} = 6 \cdot \varepsilon_{\text{Gas}}/d_{\text{Bub}}$ for spheres
Volumetric gas transfer coefficient	$k_L a$ (1·s⁻¹)	0.04–0.4	Depends on surface active compounds
Volumetric oxygen transfer rate	OTR (g/L·h)	<5	

Besides this "stagnant film" model more complex models can be employed considering a partial exchange of water in the film ("surface renewal" model) or assuming another boundary layer inside the gas phase ("two-film" model). For precise investigation of the molecular mechanisms involved in mass transfer the diffusion coefficient is replaced by the Schmidt number Sc = v/k_{diff} describing the relation between kinematic viscosity and molecular diffusivity. This helps to predict values for different gases, fluids, relative velocities, and temperatures. Nevertheless, for practical applications equations are condensed as described above and k_L and/or $k_L a$ have to be experimentally determined. A rough correlation between $k_L a$ and aeration rate/agitation is given as follows:

$$k_L a = A \cdot \left(\frac{P_R}{V_R}\right)^a \cdot U_G^b \tag{5.26}$$

where A, a, and b are empirical parameters. As other correlations also exist, great caution is necessary. The best way is to ask the supplier if a new reactor shall be purchased or to rely on measurements.

For bubble columns it is mainly determined from bubble diameter d_B and the gas holdup ε_R. The k_L value is nearly constant ignoring changes in film thickness due to different rising velocities:

$$k_L a = \frac{k_L}{d_B} \cdot d_B \cdot \frac{6g}{d_B} = 6 \cdot \frac{k_L}{d_B} \cdot \varepsilon_G \tag{5.27}$$

This holds at last for the laminar regime (compare with the microalgae in Chapter 8), where the bubble diameter can be adjusted by the sparger. A commonly applied formula for the general case of bubble columns is

$$k_L a = 0.467 \cdot U_G^{0.82} \tag{5.28}$$

The background is that bubble diameter no longer depends on the sparger but on turbulences and can therefore no longer be measured. A first impression of how different aeration and agitation rates look in reality is depicted in Fig. 5.22.

How much oxygen do cells need to grow? The oxygen demand is quantified by the oxygen yield coefficient $Y_{X,O2} = \Delta m_X / \Delta m_S$ similar to $Y_{X,S}$ for the substrate. A typical value is $Y_{X,O2} \approx 1$ g/g for aerobic growth on glucose. This value can be understood making the rough assumption that for $Y_{X,S} = 0.5$ g/g half of the glucose is used to build up biomass and the other half is oxidized in a molar ratio of 6 mol O_2/1 mol glucose corresponding to 192/180 g/g ≈ 1 g/g. On the level of the specific turnover rates, we obtain:

$$OUR = r_{O_2} \cdot c_X \tag{5.29}$$

As glucose uptake, growth rate, and oxygen uptake are stoichiometrically coupled, r_{O2} depends on the limitation conditions. The specific oxygen uptake rate itself is assumed to be kinetically limited at partial pressures below 10% or even 5%.

Carbon dioxide production is described by analogy as

$$CPR = r_{CO_2} \cdot c_X \tag{5.30}$$

Assuming that oxygen is almost purely used in respiration and carbon dioxide mainly produced there, it makes sense to look at the stoichiometric relation between carbon dioxide production and oxygen consumption, the respiratory quotient (RQ):

$$RQ = \frac{CPR_{mol}}{OUR_{mol}} \tag{5.31}$$

Fig. 5.22: Pictures for different aeration rates (vvm) and agitation speeds (rpm).

The respiratory quotient is a lumped but nevertheless useful measure to follow changes in the metabolic pattern of fermentations. RQ of 1 mol/mol indicates aerobic growth on glucose. Formation of side products like ethanol in the case of yeasts leads to higher values as CO_2 is formed in the anaerobic pathway. Growth on substrates with higher or lower degrees of reduction than glucose leads consequently to changes in the RQ value as well. So, this parameter is a useful measure to follow changes in the metabolic pattern of fermentations.

5.5.2 Measuring gas transfer

Determination of OTR and CTR is possible via balances over in-gas and off-gas. In the following equations the ideal gas law is employed. The definition of F_{Gas} refers to standard conditions $T_{Gas} = 273.15$ K, $P_{Gas} = 1.013 \times 10^5$ N/m^2 and the universal gas constant $R = 8.314$ J/mol· K. The molar volume $V_m = R \cdot T_{Gas}/P_{Gas} = 22.4$ L/mol. In the first step we determine the volumetric molar gas flow to the inlet:

$$F_{Gas,mol,in} = \frac{f_{Gas,in} \cdot P_{Gas}}{R \cdot T_{Gas}} \tag{5.32}$$

and to the outlet accordingly.

$$F_{Gas,mol,out} = \frac{f_{Gas,out} \cdot P_{Gas}}{R \cdot T_{Gas}} \tag{5.33}$$

Note that the volume flow into and out of the reactor is not necessarily the same. Why?

In the second step the volumetric mass flows for the different gas compounds, here O_2, ensues as

$$f_{O_2,mass,in} = f_{O_2,mol,in} \cdot M_{O2} \cdot x_{O_2,in} \tag{5.34}$$

and for the other gas species and the exhaust gas, respectively.

Usually, we know the gas composition of air beforehand $x_{O2} = 0.2095$, $x_{CO2} = 0.0004$, $x_{N2} = 0.7901$ (inclusive other inert gases), but that must be checked depending on the gas origin like bottles with artificial air or the space from which the air is pumped. The composition of the exhaust gas is obtained from measurements. In any case the total molar balance holds for the exhaust gas:

$$x_{O_2} + x_{CO_2} + x_{inert} = 1 \tag{5.35}$$

The inert gases (mainly N_2) leave the reactor, as the name already suggests, unchanged with respect to moles and mass:

$$f_{N_2,in} = f_{N_2,out} \tag{5.36}$$

We can now derive in the third step the formula for the volumetric transfer rates:

$$OTR = f_{O_2,mass,in} - f_{O_2,mass,out} = \frac{f_{O_2,mol,in} \cdot M_{O_2} \cdot x_{O_2,in} \cdot (1 - x_{CO_2,out}) - x_{O_2,out} \cdot (1 - x_{CO_2,in})}{1 - x_{O_2,out} - x_{CO_2,out}} \tag{5.37}$$

$$CTR = f_{CO_2,mass,out} - f_{CO_2,mass,in}$$
$$= \frac{f_{CO_2,mol,in} \cdot M_{CO_2} \cdot x_{CO_2,in} \cdot (1 - x_{CO_2,out}) - x_{O_2,out} \cdot (1 - x_{CO_2,in})}{1 - x_{O_2,out} - x_{CO_2,out}} \tag{5.38}$$

Note that OTR is defined with a reverse sign than CTR to get positive values for the normal operational conditions. In cases where more CO_2 is being produced than O_2 is being consumed (RQ > 1) the exhaust gas flow is greater than the inlet flow.

Now we make use of the measured OTR value to determine the $k_L a$ value from eq. (5.39) as follows:

$$k_L a = \frac{OTR}{c_{O_2}^* - c_{O_2}} \tag{5.39}$$

Off-gas measurements are not provided in all processes, particularly lab reactors, are, as the off-gas analyzer is expensive. An estimation can be obtained by the so-called dynamic "$k_L a$ measurement" or "gassing in–gassing out" method. The approach can

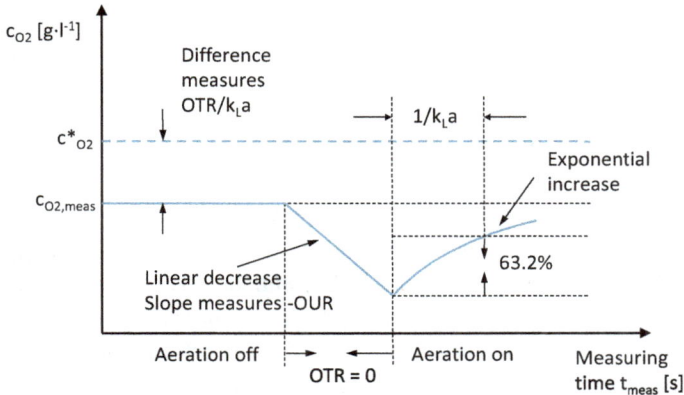

Fig. 5.23: Course of oxygen concentration during an experiment to determine OTR, $k_L a$, and c^*_{O2} during an ongoing cultivation.

be carried out in a running fermentation starting from steady-state conditions (OTR = OUR) as shown in Fig. 5.23. The saturation concentration $c^*_{O2,out}$ has to be determined for the gas-phase composition in the reactor, practically the exhaust gas. In fact, this value is not known without off-gas measurement. To start the measurement, aeration is stopped (OTR = 0). A linear decrease of oxygen concentration according to $dc_{O2}/dt = OUR$ can be observed and determined from the measurements. Switching on aeration brings the system again to steady state, where c_{O2} follows an exponential curve (5.40) with the formal solution of the differential equation:

$$c_{O_2}(t) = \left(c^*_{O_2} - \frac{OUR}{k_L a} \right) \cdot \left(1 - e^{-k_L a \cdot t} \right) \tag{5.40}$$

We know already the oxygen concentration for the steady-state condition, which allows us to substitute $OUR/k_L a$ by $(c^*_{O2} - c_{O2,meas})$ and make further simplification by cancelling the unknown term $k_L a \cdot c^*_{O2,ex}$ leading to

$$c_{O_2}(t) = c_{O_2,meas} \cdot \left(1 - e^{-k_L a \cdot t} \right) \tag{5.41}$$

It is not necessary to wait until equilibrium is obtained. For $t_{air,on} = 1/k_L a$ the curve reaches a value of 63.2% of its way from the minimum to the already measured steady state $(1 - e^{-1} = 0.632)$. Another way is to plot the values on a logarithmic scale and read out the slope. This dynamic measurement is not very precise as, e.g., if the sensor is not fast enough to resolve the curve precisely or in cases where a considerable head space aeration occurs. Furthermore, c^*_{O2} changes during the oxygen increase phase of the experiment. These effects could be considered by logging the full data set and estimating the parameters in a computer program including time constants of the side influences. Small steps of the operational pressure can also be applied, which overcomes the inaccuracies caused by residual bubbles after switching off aeration (dynamic pressure method DPM).

5.5.3 Alternative means of gas supply

Bubbling is the most commonly used means of oxygen supply for aerobic microorganisms. Aeration by gassing can easily be applied and besides they contribute to mixing. However, there are some concerns. Some sensitive cells like animal cells can be harmed by surface tension of the bubbles. Floatation leads to attachment of cells to especially small bubbles. Rising and bursting at the surface can even destroy cells. From an engineering view, increasing gas transfer is at the cost of higher gas holdup and shear stress, which is not really welcome. For special cases other forms of gas supply should be found.

One obvious solution is surface aeration like in shaking flasks, where oxygen diffuses from the head space into the liquid. Shaking, besides providing mixing of the culture, increases the surface area for better mass transfer. This approach depends on a large surface-to-volume ratio, which is the case in shaking flasks due to the relatively thin liquid layer. Furthermore, cell concentration and specific growth rate and therefore volumetric oxygen demand are usually lower compared to bioreactors.

Employment of membranes is a means to separate the gas and liquid phase and to give additional degrees of freedom for geometric design of larger bioreactors. As an example, a membrane reactor on the pilot scale is depicted in Fig. 5.24.

Such membrane reactors are commonly employed, e.g., in animal cell culture.

In living nature bubbling is an absolute exception. Thus, it is worth looking at how plants and animals solve the transport problem of bringing oxygen or carbon dioxide from air or water to the cells. Leaves use stomata to provide an unimpeded means of air transport close to the cells. Chloroplasts are often arranged close to the cell wall to minimize the diffusion path to a few μm. In fact, CO_2 transport is often a limiting factor for growth. C4 plants like maize can bind CO_2 during the night to make profit from additional hours for gas transport. A technical solution in greenhouses is CO_2 gassing. Some plants also manage air transport over longer distances. The most remarkable examples are reed and alder. They can grow with their roots underwater or can at least survive long periods of flooding. Oxygen supply to the roots is possible by so called aerenchym and internal aeration channels. This feature is technically used in artificial wetlands for wastewater purification. Animals with a high oxygen demand are facing comparable problems. Most of the vertebrates need lungs or gills to provide a large surface area (200 m^2 in humans) for diffusion, the first step of oxygen uptake. As in bioreactors, convective transport is the next step. Comparing solubility of oxygen in water and blood (up to 70-fold higher in blood) reveals that nature invented a great solution for this problem by binding O_2 to hemoglobin. The last step from the blood to the tissue again is diffusion. This requires a minimization of the diffusion length by a network of blood capillaries.

Fig. 5.24: Picture of a membrane reactor for plant and animal cell culture. The membrane tubes are wrapped around a pivotable upper and lower frame. Thus, the membranes themselves form the agitator (© B. Frahm, Bayer Ag).

Hardly anyone in the public knows that many of the products we consume, such as pharmaceuticals, food, and plastics, are the result of a stirring process. While the stirrer usually carries out its work invisibly, Vera Meyer has created a monument for them here. The growth of the mycelium is based on the stability of the stirrer (Fig. 5.25).

Fig. 5.25: The stirrer rediscovered as "The Prince's Flower" by the scientist and artist Vera Meyer (vita in chap. 12.8); the "flower" is a mushroom (© Vera Meyer, Image credit Martin Weinhold).

5.6 Questions and suggestions

1) Usually, it is considered that the diffusion resistance of oxygen from the medium to the cells is negligible compared to oxygen diffusion out of the bubbles. What is the volumetric surface area of cells assuming $c_X = 10$ g/L (dry weight), cells are spheres of $d_{cell} = 2$ μm diameter, and the water content of the cells is 90%? Compare it with a typical bubble surface.

2) In a cultivation with *E. coli* we measure $c_{O2} = 5$ mg/L, $c_X = 10$ g/L and $r_X = 0.5$ g/g · h. Suddenly aeration stops. Calculate the time t_{crit} until the dissolved oxygen is exhausted. This happens similarly after taking a sample in an unaerated tube.

3) Make a list of the physical parameters determining OTR and discuss qualitatively how they can be influenced by technical control parameters. What are the limitations of manipulating OTR?

4) Better mixing also means higher shear forces. Which types of cells appear to be most sensitive to such shear forces?

Chapter 6
Not always so simple – the batch process reconsidered

Batch processes seem to be quite simple at first glance: inoculation of the reactor, doing what we like to do namely nothing, and finally harvesting. However, there are some hidden pitfalls, which must be considered beforehand. This chapter is dedicated to description of some possible complications. These include switching between different limitation and inhibition conditions. To understand the interplay between the cells and the different medium compounds the topic of "kinetics" also has to be reconsidered, as well as its influence on the course of batch fermentation. Finally, this should lead to rational process development.

6.1 Formal kinetics – extrapolation from enzymatic reactions to cell growth

Complex reaction systems like the growth of cells in bioreactors must be described using strong simplifications. These can be educated guesses about the major steps and substances that reduce the system description accordingly. Other assumptions can be gained from global constraints like balances or thermodynamics of the system. We did already do this when assuming that only substrate uptake is decisive, and growth follows stoichiometric and thermodynamic constraints. This approach has been tried for decades in order to find easy manageable formulas for growth in batch and other process policies. The most important ones are given in Tab. 6.1. Nevertheless, for rational process development it is important to understand the assumed simplifications or observations. These can have their origin on the cell level or on the reactor level. In practice, not all compounds produced by the cells and accumulating in the reactor in small amounts are quantitatively detected nor is the medium ideally mixed. Neither are all metabolic pathways or intracellular control loops. That is why the notations are called "formal kinetics." They are a practical start to process analysis in the sense of a modular construction kit but include so many assumptions and simplifications that conclusions on underlying mechanisms are not allowed. Often, they are valid only for the process type for which they have been formulated, e.g., batch processes. In every case is simultaneous inspection of growth, substrate uptake, and product formation on demanding is indicated. To go a step further into a more mechanistic view, we have first to discuss what could eventually happen in a batch process demanding our attention.

Most of the kinetics is deduced from enzyme kinetics for substrate uptake as in the case of Monod kinetics. This is possible only with great care. The involved enzymes are often not characterized in their microenvironment with respect to limita-

https://doi.org/10.1515/9783110773354-006

Tab. 6.1: A collection of commonly applied formal kinetics to describe growth.

Author	Equation	Rationale behind it
Monod	$\mu = \mu_{max} \cdot \dfrac{c_S}{k_S + c_S}$	Enzyme kinetics for substrate uptake, Linear growth from substrate
Andrews, Haldane	$\mu = \mu_{max} \cdot \dfrac{c_S}{k_S + c_S + \dfrac{c_S^2}{k_{I,S}}}$	Enzyme, substrate inhibition
Hill, Moser	$\mu = \mu_{max} \cdot \dfrac{c_S^n}{k_S^n + c_S^n}$	Multiple binding sites, formal reaction order from chemical engineering
Contois	$\mu = \mu_{max} \cdot \dfrac{c_S}{k_S \cdot c_X + c_S}$	Biomass influences itself, diffusion limitation
Tessier	$\mu = \mu_{max} \cdot \left(1 - e^{-k_S c_S}\right)$	$\dfrac{d\mu}{dc_S} = k_S \cdot (\mu_{max} - \mu)$
Tessier-type with substrate inhibition	$\mu = \mu \left(e^{\frac{-c_S}{k_{I,S}}} - e^{\frac{-c_S}{k_S}} \right)_{max}$	
Blackman	$\mu(c_S) = \begin{cases} \mu_{max} \cdot \frac{c_S}{k_S} \\ \mu_{max} \text{ for } c_S > k_S \end{cases}$	Special case of enzyme cascade, hidden second step, more in microalgae paragraph
Aiba	$\mu = \mu_{max} \cdot \left(\dfrac{c_S}{k_S + c_S}\right) \cdot e^{\frac{-c_S}{k_S}}$	Monod plus Tessier type inhibition
Edwards, Webb, and others	Mixed terms from enzyme and Tessier type kinetics	Any kinetic construction kit

tion constants and turnover numbers. The amount present in the cell membrane is unknown. Most microorganisms exhibit more than one substrate uptake mechanism for different situations like lower or higher substrate concentrations in the medium. *E. coli*, for example, possesses five different glucose transport systems with three different transport mechanisms, and *S. cerevisiae* with 17 functional hexose transporters (seven for glucose) has the highest number. Nevertheless, a formal kinetics with two parameters has been found to work reasonably well. The background is that during a batch process substrate limitation occurs only during a short time interval, where only a few measured values are available. A second point is that the additive overlaying of several Michaelis-Menten kinetics looks quite like one with averaged kinetic parameters. The system is not observable with standard offline measurements.

Substrate inhibition is a clearly defined mechanism in enzyme kinetics. In a batch process high substrate concentration has a multivariate impact on the cell and reactor levels. This includes a high viscosity with side effects on mass transfer. High osmotic value is another effect to which the cells have to react by energy expenditure for water pumping or buildup of intracellular osmotic compounds. With respect to batch processes a long lag phase and slowly reduced inhibition effects during cultivation are the consequences, including changes in different growth characteristics. Thinking of substrates other than glucose, toxic effects may occur with multiple effects on the cells. The same holds for product inhibition, where approaches usually known from enzyme kinetics are applied. Extracellular compounds may not be the only ones to react with substrate uptake; intracellular steps may also attack the cell membrane as in the case of ethanol.

Contois kinetics looks a bit strange at first glance as biomass concentration appears on the right side of the equations. This means that the cells influence each other. That could actually be so in cases of quorum sensing, but often a hidden material transport limitation is the reason that the cells experience a lower local substrate concentration than is macroscopically measured.

The kinetics according to Tessier can be interpreted as an approach, where the difference of the actual growth rate and the maximum growth rate acts a driving force. The further the organisms are from the optimum the more effectively they can use additional substrate. Although the biological background is quite unclear the kinetics is frequently used because of its variability. For Tessier the kinetics is formally identical to the step response of a dynamic system with one time constant.

Substrate uptake is not always the limiting step over the whole range of substrate concentrations. Other steps further down the catabolic pathways can be limiting before the uptake is at its maximum. This case has been approximated by Blackman as a piecewise constant kinetic function. The linearly increasing behavior approximates the Monod kinetics for low substrate concentration, where the maximum represents a subsequent limiting step, which is masked for low substrate turnover rates. Indeed, Blackman kinetics fits better to the original data of Monod than the Monod kinetics itself. Apparent low k_S values, especially those lower than for the isolated enzymes, can be the effect of high overexpression of the enzymes involved in the transport system to allow the cell high turnover rates at low substrate concentrations. Nevertheless, the apparent μ_{max} is not given by the maximum of the transport system but by another intra- or extracellular limitation.

Other kinetics are valid for special cases, e.g., when an unknown limitation or inhibition is active during a special process pattern. Here our job is not simply to try out different formal approaches but to find out the mechanics behind it and set up equations for their precise representation. Before trying it out for our standard batch good guesses for reasonable values of the yield parameter $y_{X,S}$ are required.

6.2 Looking a step deeper – estimating aerobic growth yields from metabolic fluxes

Microbial metabolism is subject to the energy balance represented by ATP and the redox balance represented by $NADH_2$. Remembering Liebig's barrel model, ATP and $NADH_2$ can be visualized as the hoops keeping all the staves together. A more quantitative approach to find good values for yields as a link between substrate uptake and growth starts with evaluation of the main metabolic pathways as shown in Fig. 6.1 for aerobic growth on glucose.

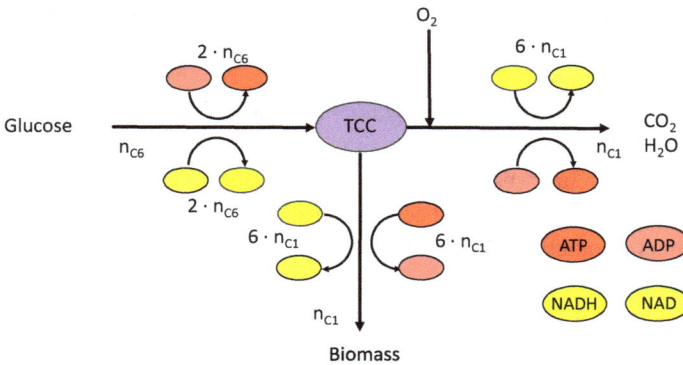

Fig. 6.1: Simple metabolic pathways diagram of aerobic growth for estimation of ATP and $NADH_2$ balance. The metabolic pathways are given in molar fluxes q for the C skeleton with carbon atoms indicated as Cn in the index. Assumptions: 2 $NADH_2$, 2ATP per C_6 in glycolysis, 1 $NADH_2$ per C3 pyruvate oxidation, 3$NADH_2$ + 1 QH_2 + 1 ATP per C3 acetyl oxidation in TCC, 1 $NADH_2$ > 10 H^+ membrane, 1 QH_2 > 6 H^+ membrane, 4 H^+ membrane > 1 ATP.

With the given simplifications we can set up the balance equations:

$$\text{ATP-balance:} \quad 2 \cdot q_{glyc} + 15 \cdot q_{resp} - 5 \cdot q_x = 0 \tag{6.1}$$

$$\text{NADH}^2\text{-balance:} \quad 2 \cdot q_{glyc} - (6-5) \cdot q_{resp} - \frac{1}{3} \cdot q_x = 0 \tag{6.2}$$

Normalizing q_{glyc} to 1 the solution is

$$q_x = \frac{2 \times 16}{5 \times 14 - 5 \times 16} \approx 3 \; ; \; q_{resp} \approx 1 \tag{6.3}$$

With the given simplifications it can be fixed that glucose can be completely oxidized yielding 32 ATP per mole. Glucose going into biomass utilizes 30 (6×5) mole ATP per mole glucose. From these relations follows a carbon flux allocation of about $q_{resp,c}: q_{x,c} = 1 \times 3:3 \times 1 = 1:1$ into respiration and into anabolisms. This value is indeed sensible, as in many cultivations a yield $y_{x,s} \approx 0.5$ g/g for aerobic growth on glucose is measured.

For the specific values of ATP and $NADH_2$ production in the diagram, growth and respiration are only coupled by ATP as the degree of reduction of biomass is the same as that of glucose. In principle, all glucose can be respired under ATP production without violating the redox balance. The same holds for allocation of glucose into growth provided there is another ATP source. Changes in the degree of reductions of biomass would change the situation in so far as the redox balance couples growth and respiration, making the oxygen consumption no longer follow a molar ratio of 1:1 to carbon dioxide production. Why we employed the ATP balance strictly, and did not consider something like decoupled AT-Pases, is an implicit suggestion of an internal mechanism of the cell that comes as close to optimum conditions as possible. In fact, that is not always the case.

The deduction of real yields from the values given in the metabolic pathway must be done with some care. The P/O ratio (mole ATP per mole O) is implicitly given here as 2.5 (15/6). Membrane leakage can lead to loss of protons needed to maintain the mitochondrial proton gradient driving ATP formation. In references a practical value of 1.75 mol/mol is often assumed. Similarly, uncertainty has to be faced for ATP demand fueling anabolic pathways for cell growth. This is, e.g., given as 100 mmol ATP/g biomass (calculation from Fig. 6.1 yields in 200 mmol/g). The lower estimations assume of a lower P/O ratio. Thus, the ratio, which we are basically are interested in, stays constant. Some attempts have been undertaken to account for the number of ATP molecules considering all known metabolic pathways from the synthesis of the metabolites up to the macromolecules. This gives only a part of the real demand. More growth-related energy is necessary for transport processes, repair, and other diffuse losses. Difficulties in interpretation of literature data comes from lumping vastly different reference points or allocation of losses to different pathways. Another point is biomass composition. Here it is assumed that the degree of reduction of biomass is the same as that of glucose. In fact, real values are lower, e.g., due to reduced compounds like fatty acids. Further, we neglected de-/carboxylation and de-/hydration processes distributed over many sites in the metabolism. More reductants given as $NADH_2$ equivalent are necessary to build up reduced compounds on the one hand but are in that case not available in the respiratory chain on the other hand. This leads to a reduced growth yield on a biomass basis. Nevertheless, lipids have a higher energy content, thus reducing the losses in yield on an energy basis.

6.3 Aerobic growth – case study heterotrophic plant cells

Plant cells can be cultivated under aerobic conditions on sucrose as substrate. This actually happens in growing plants where the cells not involved in photosynthesis are supplied with sucrose from the leaves. Cultivation data are shown in Fig. 6.2. After a lag phase of about one day the cells start to grow with a specific growth rate of about 0.21/day. Interestingly, sucrose is not taken up directly but cleaves to glucose and fructose due to strong invertase activity provided to the cells. This reaction is over after two days. Both monosaccharides can be taken up, while glucose is obviously the preferred substrate and is consumed first. Growth on two different substrates with different preferences is a common phenomenon during batch cultivations and is called "diauxic" growth ("aux-", Latin auxilium = help).

The reasons for preferences can be more active uptake systems, different yield coefficients meaning more ATP or carbon for the cells, or different catabolic pathways

(a)

(b)

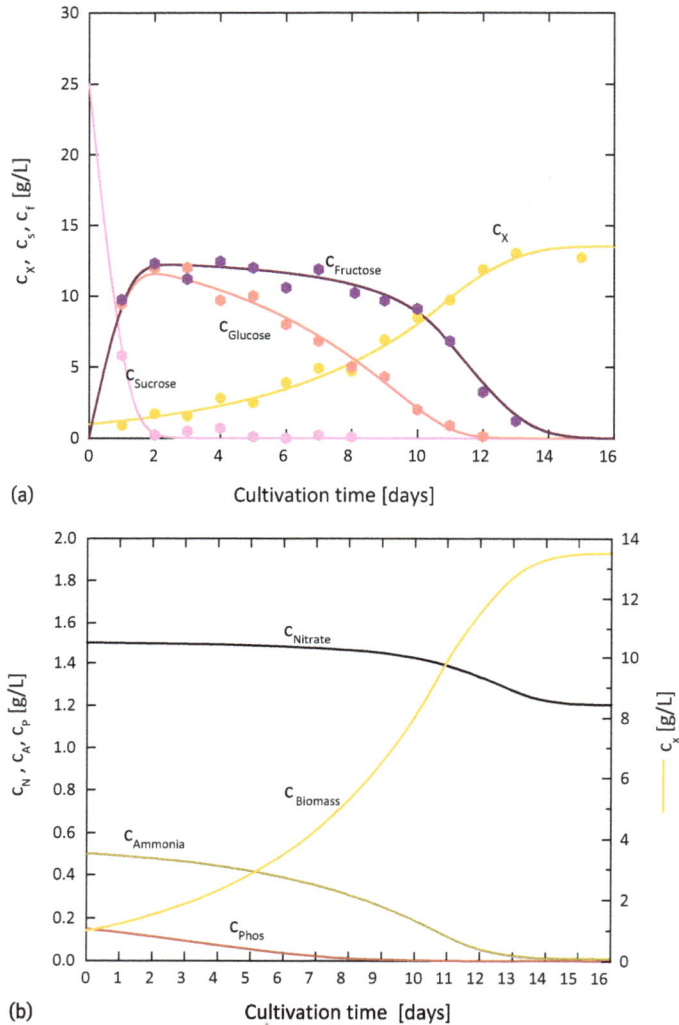

Fig. 6.2: Measurements of a batch culture with plant cells (*Echinacea*); continuous lines are simulations: (a) biomass and main substrates and (b) mineral nutrients.

meaning different expenditure with respect to necessary enzyme makeup. In many cases the physiological reasons are not clear. Fructose is taken up by fructokinase, while F6P enters glycolysis directly after G6P isomerase from glucose uptake. Thus, the energetic yield is the same for both sugars and there is even one enzymatic step less for fructose in comparison to glucose uptake. In wine making sucrose is hydrolyzed to glucose and fructose, which are fermented to ethanol with a preference for glucose. This existing biological preference is under investigation since this mechanism often leads to incomplete fructose fermentation during wine production, thus affecting wine flavor quality.

To find a way to describe the concept of a preferred substrate we assume that glucose is taken up following Michaelis-Menten kinetics:

$$r_{Glu} = r_{Glu,max} \cdot \frac{c_{Glu}}{k_{Glu} + c_{Glu}} \tag{6.4}$$

Fructose is taken up only in amounts to fill up the gap of the hexose flux induced by glucose limitation. For fructose uptake a limitation from low concentrations also has to be considered:

$$r_{Fru} = \begin{cases} r_{Fru,max} \cdot \dfrac{c_{Fru}}{k_{Fru} + c_{Fru}} & \text{for} \quad r_{Fru} \leq r_{Glu,max} - r_{Glu} \\[2ex] r_{Glu,max} - r_{Glu} & \text{else} \end{cases} \tag{6.5}$$

For the specific growth rate we get:

$$\mu = y_{X,Glu} \cdot r_{Glu} + y_{X,Fru} \cdot r_{Fru} \tag{6.6}$$

As expected, both yield coefficients have the same numerical value in this case as both hexoses deliver the same amount of ATP and carbon.

In this plant cell example, there are two nitrogen sources as well, namely ammonia and nitrate (Fig. 6.2). Here the same phenomenon of a preferred substrate can be observed. Ammonia is taken up on demand (nitrogen fraction of the cell), while nitrate is only used if ammonia uptake is reduced by limitation. This can be motivated by the necessity of using NADH for reduction of NO_3^-. Phosphate is taken up much faster than expected to cover the P demand of the cell. Phosphate can be stored in the large vacuole of plant cells. To store minerals in different chemical forms is a common behavior of microbial cells. For process development, that makes calculation of ideally balanced media difficult. Giving the total amount of, e.g., phosphate in the beginning leads to polyphosphate accumulation in the cell, but it is not certain whether this pool can be remobilized fast enough in a later stage of the culture. Kinetics for several substrates in their interplay is given in Tab. 6.2.

Multiplicative kinetics (Tsao, Hanson) is an approximation for the case of stoichiometrically coupled substrates. Examples are oxygen versus substrate uptake or nitrogen and phosphate uptake versus growth. In these cases, only one substrate can be limiting while the other is taken up on demand according to the cell's elemental composition or other intracellular stoichiometries. We already made use of this microbial behavior for media design. The approximation holds as in many practical applications only one substrate is in limiting concentrations, while the others may accumulate to saturation concentrations shifting the value of the respective Michaelis-Menten term to one. Nevertheless, the more structured formulation should be preferred.

Tab. 6.2: Three formal kinetics for several substrates and further structured approaches.

Referenced as	Formal	Applicable for	Structured
Interactive, multiplicative	$\mu = \mu_{max} \cdot \prod_i r_i(c_{Si})$	Stoichiometrically coupled essential substrates	$\mu = \min(y_{x,s,i} \cdot r_{s,i,\text{pot}}); \; r_{s,i} = \dfrac{1}{y_{x,s,i}} \cdot \mu$
Additive	$\mu = \mu_{max} \cdot \dfrac{1}{i} \sum_i r_i(c_{Si})$	Alternative substrates without preference	To be modified by physiological control considerations, e.g., one limitation and replenishment to maximum.
Noninteractive	$\mu(c_{S,i}) = \min(\mu(c_{S,1}), \mu(c_{S,2}))$	Alternative substrates with strong preference	
Coupled		Stoichiometrically dependent	One limiting, one on demand

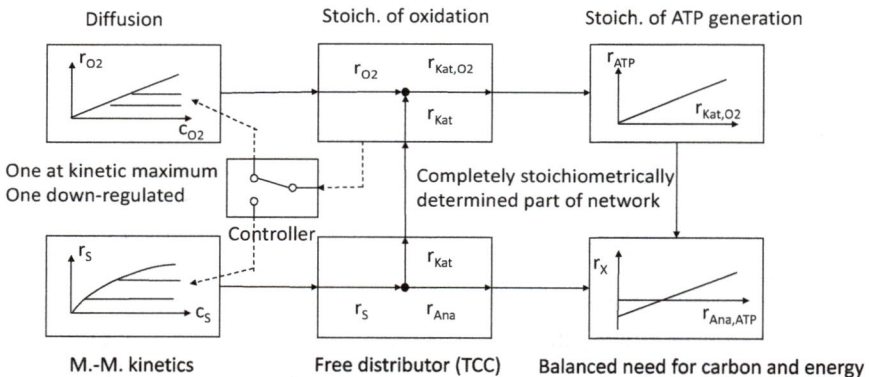

Fig. 6.3: Flowchart of the stoichiometry and uptake control for substrate and oxygen.

The additive approach is the most reasonable one, as the cells try to take up both substrates as long as they can further metabolize them. Prerequisite is that both substrates do not go through the same bottleneck.

The noninteractive approach (no happy phrase) stresses the concept of a preferred substrate. Only in rare cases it is a yes or no decision. It is a physiologically interesting task to find formulations for the intracellular control principle as given in the plant cell example. In cases where one substrate is absolutely preferred, a lag phase may occur, before the second one is used. In this "diauxic lag phase" the enzymatic set of the cells is adapted to the new substrate.

The case of substrate and oxygen uptake is represented as a flowchart in Fig. 6.3. Oxygen uptake and substrate turnover is stoichiometrically coupled by respiration, described by strongly coupled kinetics. ATP production in respiration is strongly coupled to growth. So basically, two growth patterns can be distinguished, namely substrate-limited growth and oxygen-limited growth, where the respective other substrate is taken up by demand. This is represented by the switch in the left box.

6.4 Looking a step deeper – estimating anaerobic product yields from metabolic fluxes

In anaerobic fermentations no oxygen is available as an electron acceptor. $NADH_2$ has to be used for reduction of another external electron acceptor or an intracellular metabolite. This is called endogenous respiration. The case of pyruvate reduction and ethanol formation is shown in Fig. 6.4.

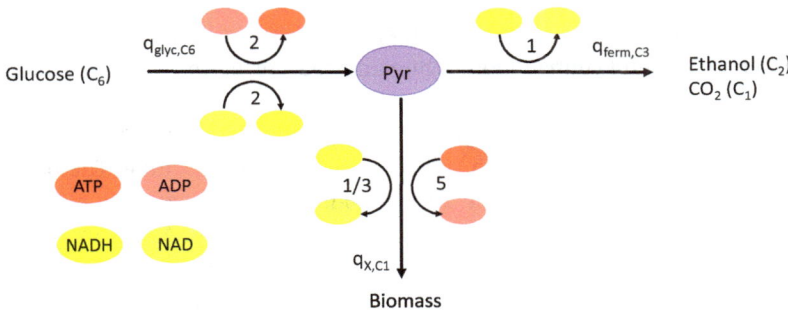

Fig. 6.4: Simple metabolic pathways diagram of anaerobic growth under ethanol formation for estimation of ATP and $NADH_2$ balance. The metabolic pathways are given in molar fluxes q for the C-skeleton with carbon atoms indicated as Cn in the index. Assumptions: 2 $NADH_2$, 2ATP per C_6 in glycolysis, 1 $NADH_2$ per C3 pyruvate reduction.

With the given simplifications we can set up the balance equations:

$$\text{ATP-balance:} \quad 2 \cdot q_{glyc} + 0 \cdot q_{form} - 5 \cdot q_X = 0 \tag{6.7}$$

$$\text{NADH}_2\text{-balance:} \quad 2 \cdot q_{glyc} - 1 \cdot q_{form} - \frac{1}{3} \cdot q_X = 0 \tag{6.8}$$

Normalizing q_{glyc} to 1 the solution is:

$$q_X = \frac{2}{5} \quad \text{and} \quad q_{form} = 2 - \frac{2}{15} = \frac{8}{15} \tag{6.9}$$

The allocation of carbon to growth is $q_{X,C} / q_{Glyc,C} = 2 \cdot /(6 \cdot 5) = 1/15 \approx 0.067$. Measured values for $Y_{X,S}$ are in the range of 0.03–0.05 g/g. This value is in good agreement with the "theoretical" value based on the metabolic scheme. Only a small part of the glucose is allocated to growth. In anaerobic cultivation a low biomass concentration can be expected. However, that is not a disadvantage as the intention is to get a high yield $Y_{P,S}$ of the end product, e.g., ethanol in the case described here. $q_{ferm,C}/q_{Glyc,C} = 28/15 * 3/6 = 14/15$ shows that nearly all carbon of the substrate is channeled into ethanol and CO_2. For ethanol as given in Fig. 6.4 $Y_{Eth,Glyc} = q_{ferm,C}/q_{glyc,C} \cdot M_{Eth}/(M_{CO2} + M_{Eth}) = 14/15 \cdot 44/90 \approx 0.46$. In terms of product formation, we have the very good result that about 95% of the theoretical maximum yield of glucose splitting to ethanol and carbon dioxide can be reached. The remaining part contributes to growth.

6.5 Anaerobic batch culture – case study ethanol production

Ethanol production for alcoholic beverages is probably the oldest deliberately performed biotechnological process in human history. Ethanol production for fuel, so-called bioethanol, works biochemically speaking in the same way, by anaerobic fermentation of sugars. Glucose comes in the case of wine from the grapes or other fruits. In case of starch an enzymatic step has to precede it: amylase is produced by the malts in case of beer brewing but must be provided in a separate step for production of bioethanol.

Before we go into details of "know-how" an engineer should think about "know-why" with respect to a useful contribution to human needs, or more strictly speaking to a viable market. Already Henry Ford proposed in the early 1990s: "The fuel of the future is going to come from fruit like sumac . . . – to almost anything." However, due to the emerging petrol industry this wish was not realizable at that time. The basic idea is to transform a diluted or otherwise difficult-to-manage substrate into a workable liquid fuel or drop-in fuel to be directly usable, e.g., for vehicles. As we are going to produce a fuel, the energetic efficiency should be the first step to check. This can be done by calculating the energetic efficiency $\eta_{C1,C2}$ of conversion on the basis of heat of combustion $H^0_{C,Comp}$ of the respective compounds:

$$\eta_{Eth, gluc} = \frac{H^0_{C,eth} \cdot 2 \cdot mol}{H^0_{C,gluc} \cdot 1 \cdot mol} = \frac{1,367 \text{ kJ/mol} \cdot 2 \cdot mol}{2,815 \text{ kJ/mol} \cdot 1 \cdot mol} = 0.97 \tag{6.10}$$

This is indeed good news, as the energetic efficiency is acceptably high. This is also the case if the 5% loss by biomass formation is accounted for, leading to $\eta_{Eth,Gluc} = 0.92$. Here we calculated with the higher heating value, which presumes a complete recovery of the heat of all side products being CO_2 and H_2O. This will not completely be the case as during cultivation cooling energy is required while for rectification heat is necessary.

The second question is directed to raw material availability. Plants as feedstocks are of course the ultimate renewable resources based on sun energy and carbon dioxide driving photosynthesis. The most abundant substrates are sugar and starch. Biofuels based on these substrates are called "first-generation biofuels." However, the amount of these carbon sources needed to substitute a remarkable part of fossil fuels is so high that a direct competition with food and feed is obvious as arable land is limited. Furthermore, the nexus between growing plants like sugarcane or corn and water and energy demand must be carefully evaluated. During the development of "second-generation biofuels" lignocellulosic materials, a waste product from agriculture comes into focus. As long as wastes from agriculture or food industry are considered, the conversion to a fuel seems to be uncritical. But hydrolyzation of lignocellulosics is much slower and more complicated compared to starch. Many life cycle studies are available to calculate the environmental footprint of bioethanol production. According to these results, bioethanol is not completely carbon neutral but about 70% of CO_2 released into the atmosphere can be reduced compared to fossil fuels.

The market itself is highly politically controlled. The Kyoto protocol in particular demands an increasing substitution of fossil fuels by biofuels. The United States and the European Union set road fuel targets at 5–10% in 2016 to meet these demands. The current energy production (2023) from fossil fuels is about 17.4 billion kWh worldwide. An annual increase of 5.8% is expected (CAGR 2023–2028). The UN Climate Change Conference 2023 in Dubai brought some cautiously optimistic signals. Nevertheless, cost issues are quite important for process design, as bioethanol is a very low-value product and still more expensive than fossil fuels, which is a big challenge for engineering. Even small changes of yield or energetic efficiency mean profit or loss for society. The production of bioethanol or biodiesel from crops remains questionable in view of the advancing drought.

Not all possible substrates are practically suitable. Molasses for instance needs water evaporation to come to more than 30% dry matter concentration necessary for high final product titer. Furthermore, several inhibitors including salts prevent highly intensive processes. A typical process for bioethanol production from starch-containing grains is shown in Fig. 6.5. The process starts with delivering the corn, which contributes a significant part to production costs. The corn is fed to the intake pit, from where it is conveyed to the precleaning section and finally to intermediate storage silos. After undesired materials such as sand or straw are separated, i.e., by ventilation, dry milling is the next process step, to make the starch accessible by the enzymes. Optimum and uniform particle size of the "flour" in the range of 2–3 mm is required to meet the optimum between milling costs and hydrolyzation time. This "mash" preparation is the third step namely a biochemical one. During liquefaction in a warm water mash, meant to reduce viscosity, α-amylase enzymes break up the starch molecules and gelatinize the broth. The final product is malto-dextrin, which is further converted in a saccharification step by glucoamylase to oligosaccharides and monomeric sugars, basically glucose. Free sugars make this process sensitive to bacterial contamination. Heat sterilization of the grains is practically forbidden to save energy and to prevent starch denaturation. Hygienic design and precise process control with respect to pH and temperature can contribute to a solution.

Now the real fermentation process of the "sweet mash" from the pre-saccharification can start. The preculture of the yeast is carried out aerobically on a side stream of the sweet mash and then transferred to the anaerobic fermentation as inoculum. Fermentation can lead to ethanol concentrations of up to 150 g/L. Reaching a high titer of ethanol

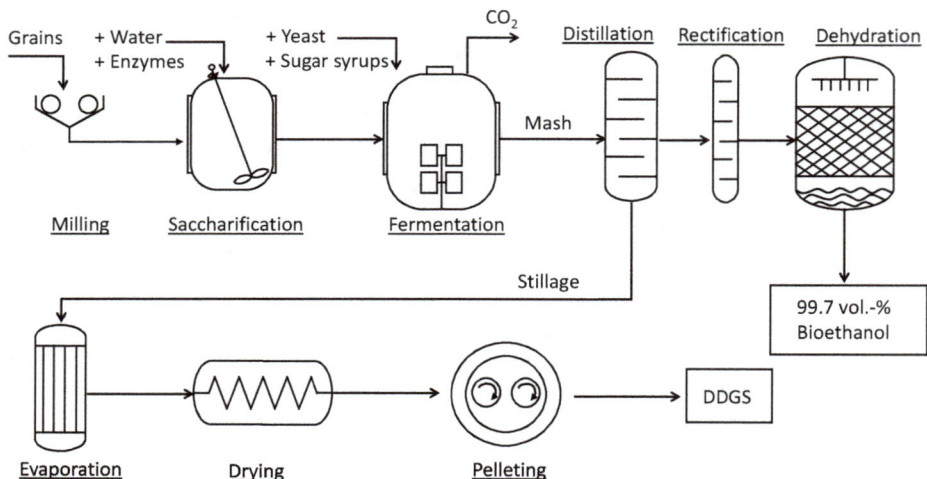

Fig. 6.5: Flow sheet of a bioethanol production process from grains to bioethanol and co-products.

during fermentation is crucial as it reduces the energetic costs of rectification. One limitation is the concentration of fermentable material in the flour and the mash. Another problem is the strong ethanol inhibition on substrate turnover, growth, and yield. One batch in fermenters (a typical word in industry for bioreactors) up to 2,000 m^3 lasts about three days. The fermentation system is designed in such a way that fermentation vessels can be operated in parallel. Such plants have a high degree of automation including cleaning by means of a CIP system.

The first task in this downstream operation is to separate a clear mixture of water and ethanol from the "fermented mash" (fermentation suspension), where the raw alcohol is withdrawn over the head of the column. This is done at moderate temperatures to prevent the proteins from denaturation and keep them valuable for feed purposes. Further rectification in a multicolumn system is done to reach ethanol concentrations close to the azeotrope equilibrium. Finally, the ethanol must be dried, e.g., by molecular sieve adsorption or pervaporation to meet fuel specifications, another constraint from the market. The remaining dealcoholized liquid, the "stillage" ("vinasse" liquid + grain residues + dead yeast) is pumped out for further processing including thermal and mechanical processing steps to allow for selling an additional product the dried distillers grains with "solubles" (DDGS), which is used as cattle feed. A modern bioethanol factory including DDGS production is shown in Fig. 6.6.

A high degree of heat recirculation between rectification and mash evaporation and stillage drying is crucial for minimization of energy losses. The attempt to make use of the whole value of a complex feedstock to obtain several products has found its expression in the concept of the "biorefinery." This type of process integration, which is typical for bioprocesses, will be further discussed in paragraphs about microalgae.

Fig. 6.6: Building a modern bioethanol production; in this stage the process equipment is already mounted including fermenters, rectification columns, and dryers (© GEA Wiegand GmbH).

To assess bioethanol production via anaerobic fermentation a look around at competing technologies is vital. The increasing amount of electricity retrieved by photovoltaics or wind power has brought up the idea of producing hydrogen by electrolysis from surplus energy. This hydrogen can be further processed on a biotechnological route by syngas fermentation (Chapter 10) or by chemical processes summarized as power-to-fuel or power-to-X technologies, where syngas (hydrogen and carbon dioxide) is converted to methane as fuel gas or used as educt for a "Fischer-Tropsch" process to yield liquid fuels.

6.6 Running example yeast production – growth in batch processes

Yeast is being produced for selling mainly to bakeries, but also for breweries or wine makers. The first trial of a production process is to do it as simply as possible – a reasonable batch process. Indeed, this has been the standard procedure for centuries. Looking at cultivation data in Fig. 6.7, we are surprised that the batch can be divided into several phases. In the first one, the yeast takes up glucose as expected but produces ethanol simultaneously even though enough oxygen has been supplied. A limitation in oxygen mass transfer would lead similarly to ethanol production. It would be visible by low pO_2 values, e.g., lower than 10% saturation. Furthermore, an increasing volumetric oxygen uptake rate (OUR) proportional to biomass propagation is observed, which is proof that mass transfer is not critical in this particular experiment. In the case of oxygen limitation, the simultaneous activation of the aerobic and the anaerobic glucose degradation pathway is called the "Pasteur effect." The yeast cells take up as much oxygen and glucose as they can. Respiration is then limited by oxygen availability. Many types of yeast do not downregulate glucose consumption to fit with oxygen uptake but use the excess glucose via the anaerobic pathway for ethanol production. After a while, glucose is exhausted, but the yeast keeps on growing after a short diauxic lag phase. Yeast is obviously able to take up the self-produced ethanol and use it for growth. This is a completely aerobic process as can be seen by the further increase of OUR.

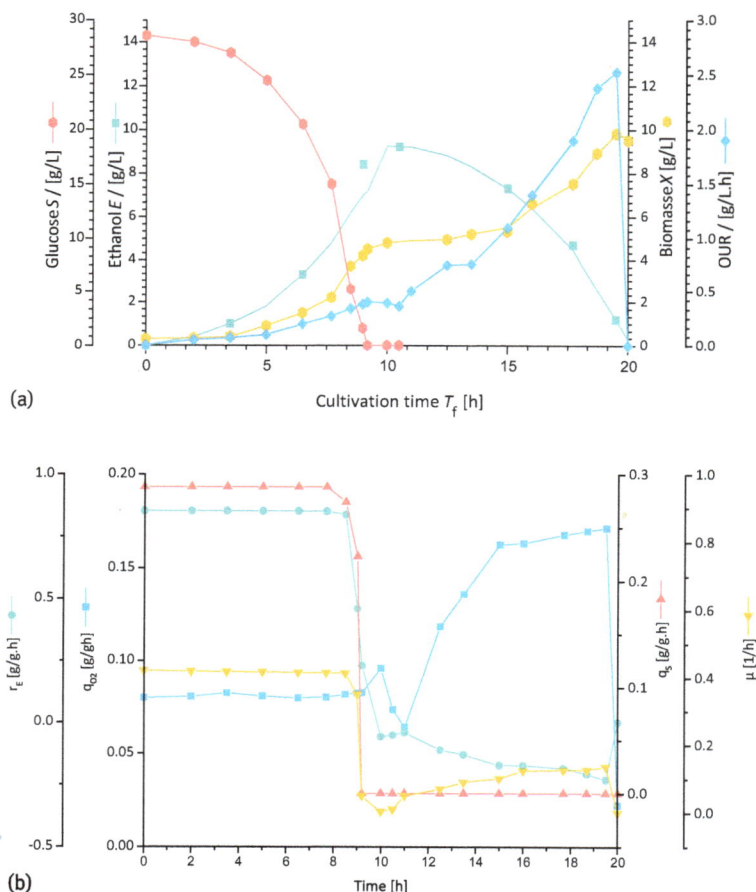

(a)

(b)

Fig. 6.7: Measurements during a typical yeast batch process: (a) substrate, biomass, ethanol, and the volumetric oxygen uptake rate; initial substrate concentration is 28 g/L; (b) calculated specific turnover rates of the main substrate and products shown in the yeast batch process.

To get a more detailed analysis of the data a look at the specific turnover rates of the batch experiment is informative (Fig. 6.7). The data show that the specific OUR remains constant during growth on glucose. Additionally the specific glucose uptake rate r_S is nearly constant, presumably close to $r_{S,max}$. As the substrate concentration reaches zero, substrate uptake decays to zero as well according to uptake kinetics. During growth on ethanol both the specific ethanol uptake rate and the specific OUR slowly rise proportionally. This is a hint at intracellular stoichiometry.

From molecular biological analysis it turns out that oxygen uptake as such is not the limiting step under high oxygen and sugar availability but a limitation in the respiratory chain. The capacity of the respiratory chain is thus limiting complete oxidation of the available sugar. This is called "overshoot metabolism" or more specifically the "Crabtree effect" (after Herbert Grace Crabtree, 1928). The maximum oxygen turnover rate is denoted as the critical specific oxygen turnover rate $r_{O2,crit}$ and the corresponding substrate turnover in the aerobic part of metabolism $r_{S,crit}$. In yeasts the onset of the overshoot metabolism may happen for glucose concentration higher than about $c_{S,crit} = 180$ mg/L, which corresponds to a substrate uptake $r_{S,crit}$ lower than $r_{S,max}$. At low substrate turnover rates $r_S < r_{S,crit}$ approximately one half $((1-y_{X,S}) \cdot r_S)$ of the substrate is oxidized while the other half is used for growth. Surplus of glucose is allocated to the anaerobic fermentation pathway of course with lower energetic yield. Assuming a given substrate uptake rate, growth, respiration, and ethanol formation are stoichiometrically coupled to substrate uptake under the constraint of an upper respiration limit as shown in Fig. 6.8.

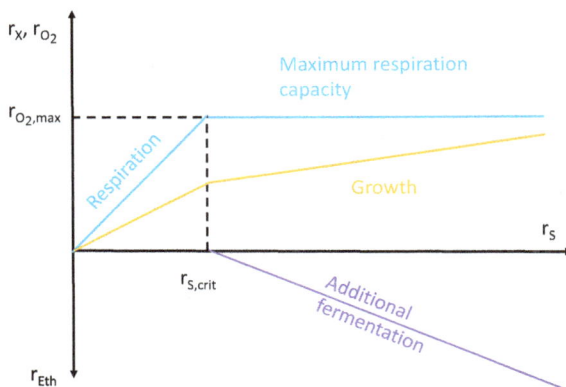

Fig. 6.8: Principal course of specific turnover rates as a function of substrate uptake; after reaching the limiting step for respiration, carbon is more and more allotted to ethanol resulting in higher growth rate but with lower yield.

Limiting steps in microbial metabolisms are often called "bottlenecks" resembling a bottle where pouring and drinking is limited by the neck of the bottle. A graphical representation of the "bottleneck model" is depicted in Fig. 6.9.

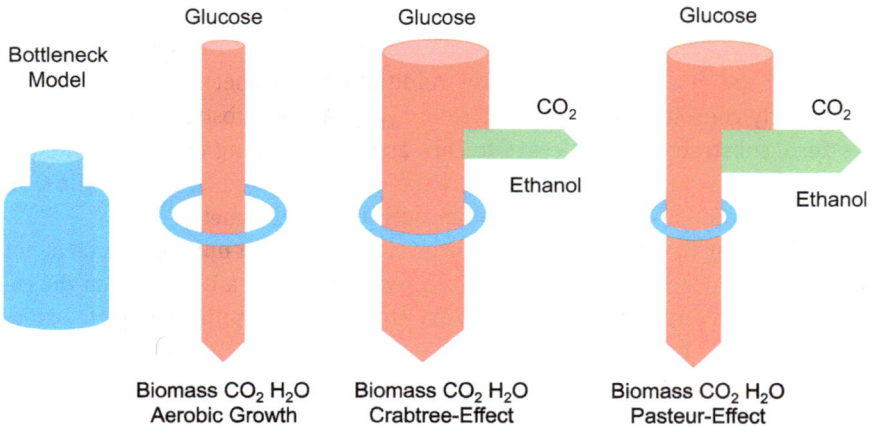

Fig. 6.9: Graphical bottleneck model of the Crabtree-effect; the blue ring represents the capacity of respiration, while ethanol formation occurs either by overloading with substrate (Crabtree) or oxygen deficiency (Pasteur). In the latter case the blue ring is smaller.

The rise of the specific OUR during growth on ethanol shows clearly that the yeast cells might well be able to increase respiratory activity. That the cells do not do it during growth on glucose is thought to be due to an intracellular optimization algorithm. On the one hand keeping the activity of the respiratory chain on a high level is a metabolic burden for the cell. On the other hand, the cells try to grow as fast as possible. This is finally possible at a specific respiration rate but at the cost of lost carbon in the form of ethanol. So, respiration is reduced in the presence of glucose, a phenomenon known as "catabolite repression." In the absence of glucose respiration is upregulated again because there is no alternative. Catabolite repression is also the mechanism for choosing preferred substrates. The best-known example is the preference of glucose over lactose in *E. coli* and may also play a role in the plant cell example. We will come back to this optimization aspect in the modeling chapter. Overshoot metabolism against the background of intracellular limiting steps is a common phenomenon in several bioprocesses, e.g., *E. coli* produces acetate at high glucose concentration (sometimes called the "bacterial Crabtree effect").

The Crabtree effect as a characteristic of yeast metabolism has obvious disadvantages in batch processes. Firstly, the overall yield is lower than expected as the cell cannot make up the low yield from the first growth phase during the second growth phase. Secondly, accumulation of ethanol – even to higher values compared to the example in the case of higher initial glucose concentration – leads to ethanol inhibition. Thus, an efficient yeast production in batch processes is prevented. In the next chapter we will see what to do in such cases.

6.7 Running example microalgae – how to deal with light as a nonmiscible substrate

After we have understood how light supports growth, we are now aiming towards a quantification of microalgal response to light. In analogy to Monod for the heterotrophic case, we start with having a look at growth during different light intensities. The specific growth rate shall be measured and plotted against light intensity $I_{hv,PAR}$, the photon flux density (PFD) in the PAR range. This leads to the so-called photosynthesis irradiance curve (PI curve), where a typical example is shown in Fig. 6.10. Such experiments are made understanding that all other nutrients including CO_2 are present in excess, and all cells "see" the same light intensity. Unlike an ideally mixed tank reactor a homogenous light distribution cannot be reached by ideal mixing but only by using low biomass concentrations or specific geometries. Timescale is also an issue. From a biology point of view, photosynthetic activity in terms of oxygen production is measured on a short-term basis in cuvettes. Measured PI curves from batch experiments is standard but gives only a snapshot valid during a given time interval. Microalgae react during long term light exposition by changing pigment furnishings, which is known as photo-acclimation. Continuous cultivation covers adaptation and acclimatization and gives the most reliable values for anticipated process design.

The PI curve looks remarkably different from Monod kinetics. The first linear phase reflects the physical mechanism of light absorption. Here the cells are light-limited depending on the passively impinging light energy without involvement of an enzymatic step. The slope represents the constant efficiency of growth in this intensity range. Microalgae also need maintenance energy, which is covered by a part of the energy gained by respiration. This becomes visible by the intercept of the kinetics with the μ- and $I_{hv,PAR}$ axis. At the compensation point $I_{hv,comp}$ the energy gained by photosynthesis exactly equals the maintenance energy leading to zero growth. The saturation point $I_{hv,sat}$ is usually in the range of 200–400 $\mu E/m^2 \cdot h$, so it can be found far below a typical value of sunlight even in middle latitudes.

For radiation values above the saturation point a constant photosynthetic activity, here measured as growth, is observed. Additional light can obviously not be used by the cells. So a bottleneck further down the metabolic energy flow is assumed. This can be directly after light absorption in the light reaction, e.g., during water splitting, in the dark reaction, e.g., CO_2 fixation, or even further down in the metabolism. So, we have the typical situation of Blackman kinetics (see Chapter 4) including two consecutive metabolic processing steps. Light energy being absorbed but not further processed is dissipated as fluorescent light and heat. This mechanism is known as nonphotochemical quenching (NPQ). The maximum specific growth rate μ_{max} depends strongly on the strain and can reach values of 1 g/g $\cdot$ day or even more than 2 g/g $\cdot$ day.

Above saturation, light intensity I_{sat} μ decreases. This situation is usually avoided during outdoor cultivation. However, some microalgae produce light-protecting pigments like carotenoids only in this intensity range, so it is applied to obtain high prod-

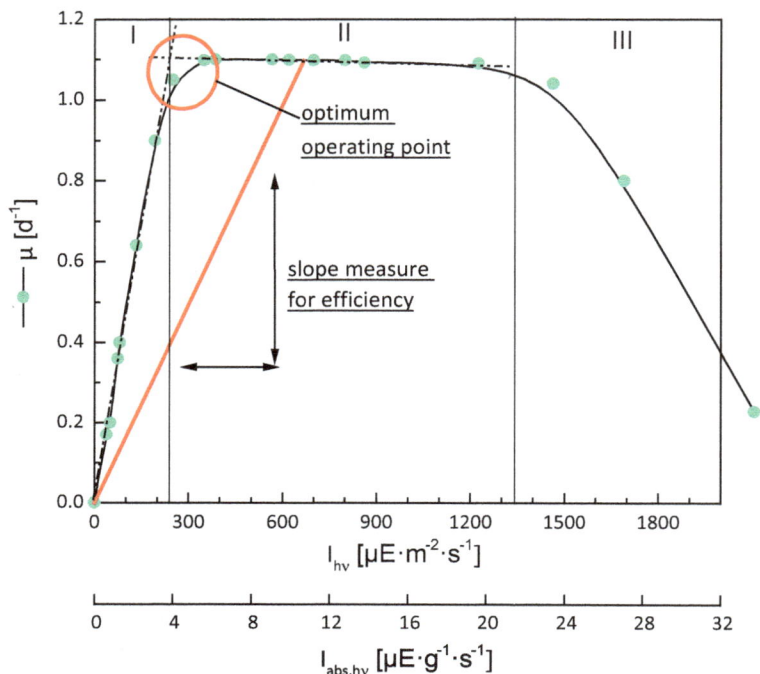

Fig. 6.10: Idealized photosynthesis irradiation response curve; the different intensity ranges are marked.

uct titers. The inhibition range is kinetically often not well defined and can be of different mechanisms and respective kinetics.

The quantitative description (eq. (8.8)) of light kinetics can now derived from what was said above and can be written down as

$$\mu(I_{hv,\,\text{PAR}}) = \begin{cases} k_{hv} \cdot I_{hv} - \mu_e & [I_{hv} < I_{hv,\text{sat}}] \\ \mu = \mu_{\max} & [I_{hv,\text{sat}} < I_{hv} < I_{hv,\text{inhi}}] \\ \mu_{\max} \cdot \frac{2 \cdot I_{hv,\text{inhi}}}{I_{hv,\text{inhi}} + I_{hv}} & [I_{hv} > I_{hv,\text{inhi}}] \end{cases} \qquad (6.11)$$

This kinetics is a black box approach similar to formal kinetics for heterotrophs and should be elaborated to get a transparent interface to physiological mechanisms. This includes firstly a transport step. This step is enzymatically enabled or facilitated in the case of heterotrophic microorganisms. In Chapter 4, we replaced at this point Monod kinetics $\mu(c_S)$ with Michaelis Menten kinetics $r_S(c_S)$. For phototrophs light absorption is the underlying mechanism for energy intake. Secondly, substrate is converted to biomass, which is a reaction step. Carbon is allotted to different metabolic pathways finally building up the cell mass. Due to the stoichiometric constraints this process is basically formulated by yield coefficients $y_{X,S}$ for heterotrophs. Speaking in system theory language this is a transport/reaction system. Now this structure will be applied to phototrophs.

In phototrophic organisms, light as the energy source impinges on the cells and is passively absorbed. Absorbance depends on biomass concentration, on pigment furnishing of the cells, and will be linearly dependent from light intensity. While the absorbed light flux $I_{hv,abs}$ represents the available energy flux for the cells, the value of interest is then the light flux absorbed by a given amount of cells. This value is given here as $r_{hv,abs}$ ($\mu E/g \cdot s$) with $r_{hv,abs} = I_{hv,abs}/c_X$ meaning the biomass-specific number of absorbed photons per time in analogy to the specific substrate uptake rate for the heterotrophic case. This is named "light availability." Referring to the dual character of the substrate as energy and carbon source, $r_{hv,abs}$ represents of course only the energy aspect of the substrate. The easiest formulation for this first step is to assume that $r_{hv,abs}$ is linear to light intensity $I_{hv,PAR}$ given as photon flux density:

$$r_{hv,\,abs} = \frac{I_{hv,abs}}{c_X} = \sigma_X \cdot I_{hv} \tag{6.12}$$

The proportional factor denoted as σ_X is the absorption cross-section known from Beer Lambert law here with the dimension (m^2/g). It represents the virtual area on which 1 g of spread algae would absorb the impinging light completely. Of course, this kinetic parameter depends on cell size or pigment content of the cells as well as on wavelength. A typical absorption cross-section spectrum is shown in Fig. 6.11. The parameter σ_X has to be measured for every algae strain and process condition; a typical value for *Chlorella* is 0.3 m^2/g for sunlight spectrum.

It is very important to note that the "substrate uptake," i.e., the light absorption, is a linear function and not hyperbolic like the M.-M. kinetics. Unlike heterotrophs, microalgae cannot reduce the amount of absorbed light in cases where growth is limited, e.g., by nitrogen or CO_2 availability, maybe only slowly in the acclimation process.

The typical shape of the curve reflects mainly the two absorbance peaks of chlorophyll (around 380 nm blue and around 680 nm red). The absorbance minimum between 500 and 600 nm is called the green gap. This makes microalgae look green but prevents them from using green light efficiently.

After considering the transport step of light absorption, the reaction step is to be formulated as rate of photosynthesis (here μ) as a function of photon absorption rate, sometimes called the "photosynthesis efficiency curve." This curve gives a good insight into the efficiency with which photons are utilized for biomass formation. Photosynthetic efficiency $\eta_{X,hv}$ can be read from the curve as $r_X/r_{hv,abs}$ for each given light intensity or in integrated form from data as $\Delta_{mX}/int(I_{hv,abs})$.

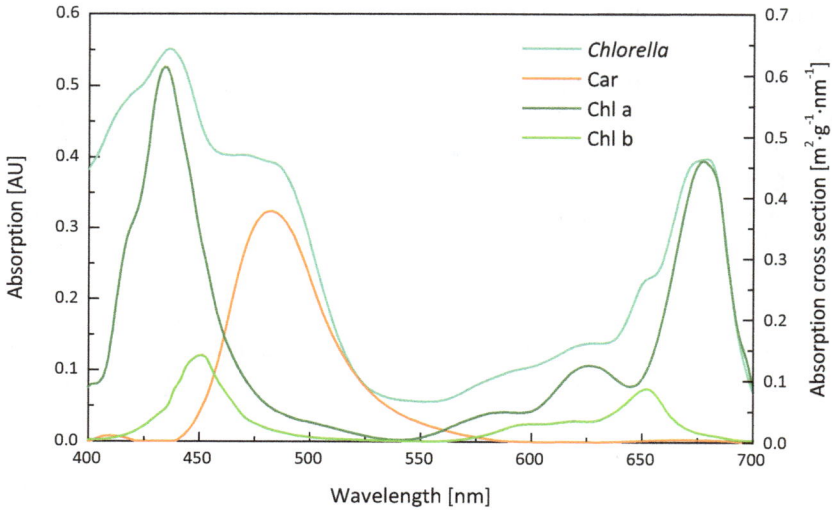

Fig. 6.11: Typical absorption spectrum of *Chlorella vulgaris* and the contribution of different pigments.

! The kinetic equations for phototrophic growth including eq. (6.11) now look like:

$$\mu(r_{hv,\,abs}) = \begin{cases} k_{\mu,hv,abs} \cdot r_{hv,abs} - \mu_e & [\mu \leq \mu_{max}] \\ \mu_{max} & [\text{environ factors}] \end{cases} \tag{6.13}$$

This is an example that kinetics is not always of M.-M type, but linear.

Besides the physiological model the reactor model equations for a batch process are to be set up. The basic difference here is the transport term for light, which is – as we already know – not miscible. In all particulate points in the suspension another light intensity is applied leading to different specific growth rates along the light path. That forces us to formulate the reactor equations as spatially distributed. Light transfer in molecular disperse systems obeys the Beer–Lambert law:

$$I_{hv}(l_{path}) = I_{hv,0} \cdot e^{-\sigma_X \cdot c_X \cdot l_{path}} \tag{6.14}$$

Here $I_{hv,0}$ is the incident light intensity and l_{path} (m) the light path length. However, this law has been developed for molecular dispersed systems. In the case of a suspension scattering also occurs. As forward scattering dominates, and side scattering is symmetrical for flat geometries a similar attenuation curve can be assumed. However, we cannot be sure that light attenuation is really linear with respect to biomass concentration. Basically, μ also depends on the location on the light path (Fig. 6.12).

For now, we assume that μ can be formulated as an average value μ_{av} valid for the whole reactor to set up the biomass balance:

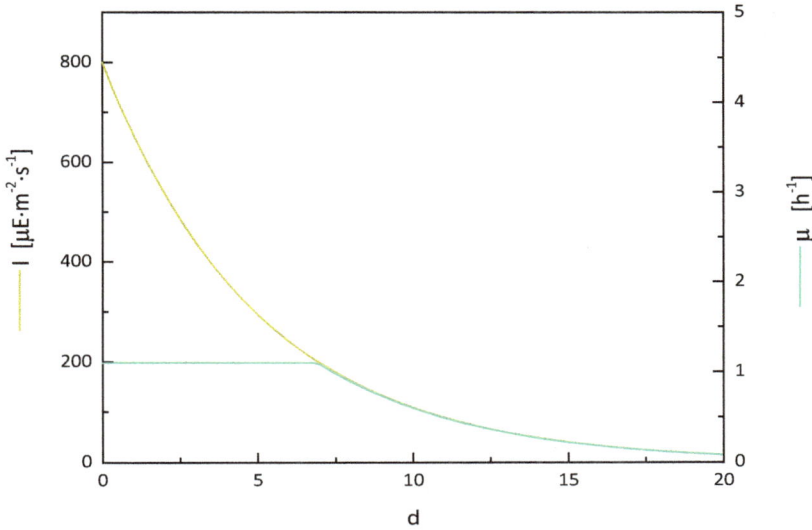

Fig. 6.12: Exponential decrease of light intensity along the light path; growth follows the light according to kinetics and is in the saturation range in the bright part and in the linear range in the darker part of the reactor.

$$\frac{dc_X}{dt} = \mu_{av}(I_{hv,0}) \cdot c_X \tag{6.15}$$

This second reactor equation is further elaborated for microalgae in Chapter 12.

Now we will test this approach for the case of the flat cuboid reactor and compare it with measurement data. The result is shown in Fig. 6.13. For simplicity light is applied only in the not light-saturated range.

Initially, for low biomass concentration, all cells experience more or less the same light intensity which does not change much for a few hours. This leads to exponential growth. In the longest middle phase of the cultivation all light is absorbed in the reactor. Light intensity at the dark side of the reactor with thickness D_R is nearly zero. With increasing biomass, the amount of photons absorbed per cell slowly decreases, so we expect a slowly decreasing μ_{av}.

The biomass balance can be further elaborated now as follows:

$$\frac{dc_X}{dt} = \mu_{av} \cdot c_X \approx \frac{1}{D_R \cdot \sigma_X \cdot c_X} \cdot I_{hv,0} \cdot \eta_{X,hv} \cdot \left(1 - e^{D_R \cdot \sigma_X \cdot c_X}\right) \approx \frac{I_0 \cdot \eta_{X,hv}}{D_R \cdot \sigma_X} \tag{6.16}$$

This is clearly not exponential growth as biomass increase does not depend on biomass itself but only on the constant flux of absorbed light. This phase is consequently called the linear growth phase.

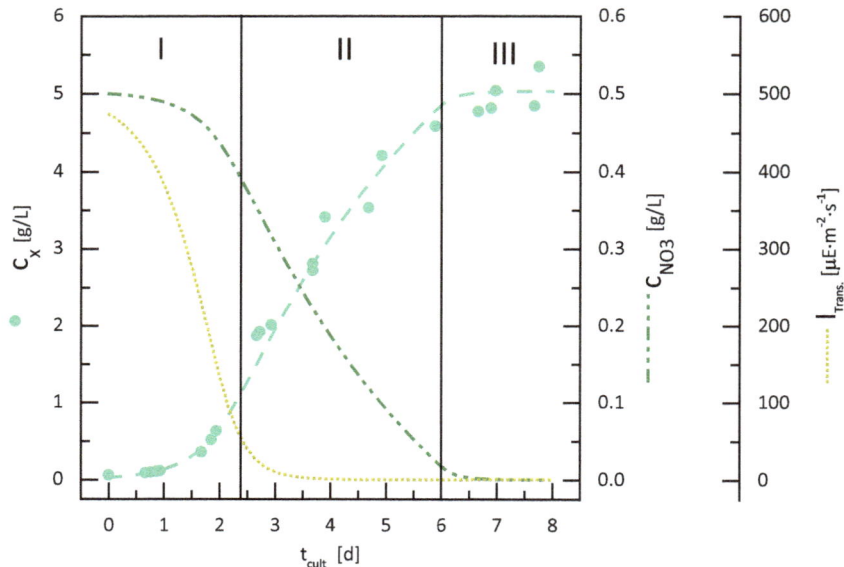

Fig. 6.13: Three different growth phases – exponential, linear, and limited – can be identified, marked by three sections.

Note the analogy with constant feeding in the fed-batch process for yeast production. It is also interesting to observe that dc_X/dt depends on reactor geometry. The thinner the reactor faster is the increase in biomass concentration. This can be understood from the fact that the total biomass produced depends directly on I_0, but is suspended in a volume dependent on D_R. In the late third phase either nutrients are limited, or the increasing dark volume increases the relative amount of maintenance losses.

6.8 Primary metabolites – products directly taken from central metabolism

We have already seen that product formation during anaerobic growth is directly coupled to energy metabolism. In aerobic processes product formation may be directly coupled to catabolic and/or anabolic pathways as well. Excretion of primary metabolites often has its kinetic reason in intracellularly limiting steps, either naturally or intentionally induced by genetic engineering. These are examples where substrate uptake limitation and a corresponding yield are not sufficient to describe growth kinetics, but in addition the assumption of another limiting intracellular step and a corresponding product formation are necessary. A general structure in terms of metabolic fluxes is given in Fig. 6.14.

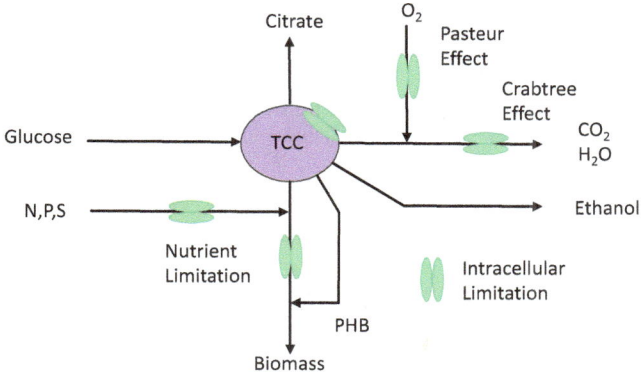

Fig. 6.14: The roughly simplified metabolic structure of aerobic microorganisms showing potential limiting steps and examples for related product formation.

Intentionally applied oxygen limitation in analogy to the Pasteur effect is applied in microaerobic processes like 1,3-propanediol production. This allows the cells to gain energy from respiration for maintenance purposes but forces them to produce an anaerobic product with high yields. The limitation can be somewhere in the metabolism. Animal cells produce lactate under glucose in excess. Excretion of citric acid in production processes is enforced by limiting the related step in the tricarboxylic acid cycle by deficiency of trace elements. This can potentially be reached by genetic engineering or by medium depletion of some trace elements. In other cases, substrate in excess is used by the cells to form intracellular storage compounds. These can be starch, PHB (polyhydroxy butyrate), or oils. To avoid changes in osmotic pressure and high cell volume such compounds are usually formed as water insoluble granules as depicted in Fig. 6.15. Such inclusion bodies can make up more than 50% of dry cell mass. This can be enforced, e.g., by introducing another limiting step again via the medium such as nitrogen depletion. Thus, growth is limited, leading to substrate excess.

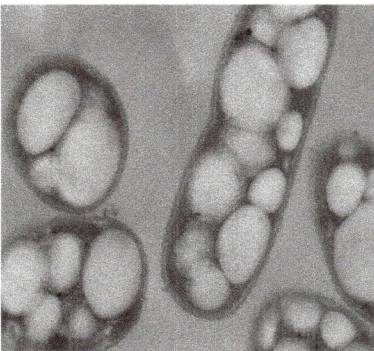

Fig. 6.15: Transmission electron microscope image of the bacterium *Ralstonia eutropha* producing polyhydroxybutyrate (PHB), which is accumulating within the cells as PHB granules (© J. Yu).

Now we look at simulation of batch processes under constantly high substrate concentrations and product excretions. For these types of products, the substrate uptake is constantly in the saturation range. Furthermore, the specific substrate uptake rate, the respiration rate, the specific growth rate, and the product formation rate are stoi-

(a)

(b)

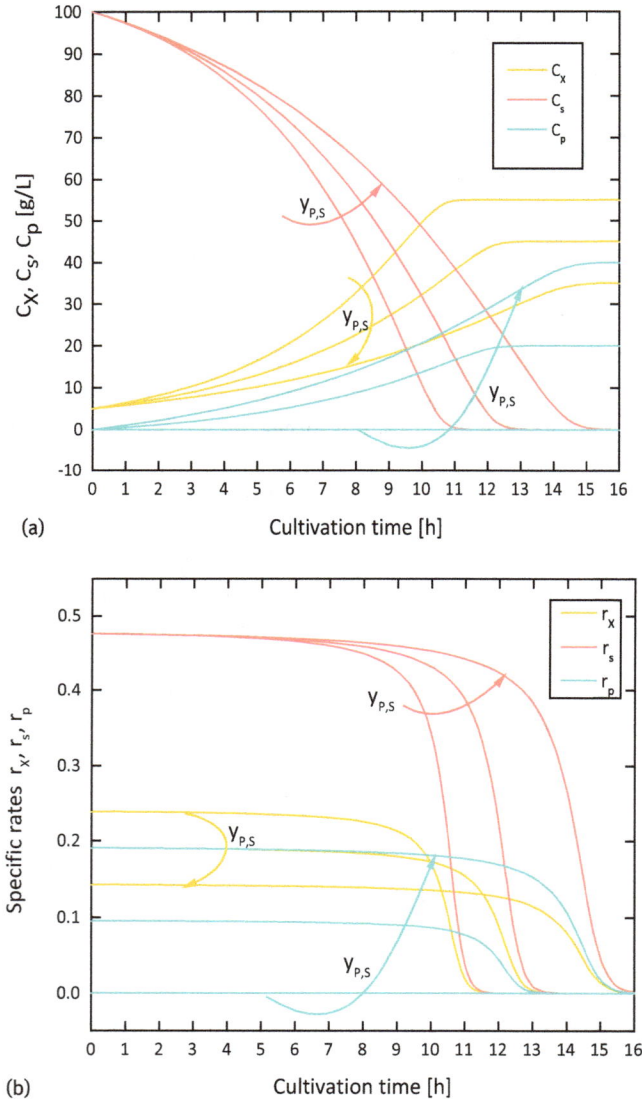

Fig. 6.16: Cultivation curves: (a) and the related specific rates and (b) for an aerobic batch process, where the product needs the same amount of energy as the biomass.

chiometrically coupled by energy, redox, and carbon balance, and are nearly constant during the batch process. Typical cultivation curves are shown in Fig. 6.16.

The simulations are carried out under the assumption of substrate uptake limitation ($r_{S,\text{max}} = 0.5$ g $\cdot$ g/h, $r_m = 0.0$). The product needs only a little extra energy for its synthesis, e.g., a polysaccharide. As a first trial the specific product formation rate r_P is related to the substrate uptake rate r_S with a constant yield coefficient $y_{P,S} = r_P/r_S$. The residual glucose is processed as usual via respiration and growth with a physiological yield coefficient $y_{X,S,\text{physiol}} = 0.5$ g/g. The specific growth rate r_X is lower for higher specific product formation rates due to constraint of the carbon balance. Consequently, the apparent growth yield $y_{X,S,\text{apparent}} = r_X/r_S$ is lower and depends on the product yield. Note that in such a batch process it is not possible to distinguish between different limiting steps unless an overshoot metabolism is active.

6.9 Serving mankind with basic needs – products needed in large amounts

Next to fuels the production of processed food or feed is the most important process in our society with respect to carbon flux. Since the invention of controlled usage of fire, people discovered many methods of food treatment for preservation and for increasing nutritional value. Among these operations, fermentation of vegetables, meats, or milk is an important factor in many cultures to broaden the food basis. Yoghurt is the most important example among dairy products. Lactose is degraded to lactate by homo- or heterofermentative bacteria; in the beginning fermentation was started by wild bacteria. Thus, digestibility is enabled or at least supported while decreasing pH increases stability against contaminations. Such processes can be regarded as batch processes, where no fresh substrate or microorganisms are added during maturation.

Remember that even today some cultivation processes like yeast production are performed at low pH values to prevent growth of rival microorganisms. Acidification as an important aspect of food stabilization has other side effects. As milk sours, proteins are subject to coagulation and the milk breaks down into curds and whey. The curds consist mainly of casein. The whey contains lactose, minerals, vitamins, and other small proteins. Separation of the solid and the fluid phase is the first step in cheese making. This involves pressing through a sieve or a cloth. Humans have domesticated milk-producing animals for over 10,000 years and made fermented dairy products from them. In many archeological museums in the Neolithic section, one discovers clay pots with holes in the bottom probably meant for cheese making. The first cheeses were not further fermented and are still today in trade as white cheese, curd cheese, or cottage cheese.

Presumably inspired by carrying fermented milk in vessels made of beef stomach, mankind developed the next step, which is interestingly an enzymatic one. Rennet from stomachs of young ruminants (cattle, sheep, goats, and others) can accelerate the separation process. This was further developed from ancient times to a standard procedure. Since the nineteenth century it has been known that the active agent is an enzyme. κ-Casein consists of a hydrophobic and a hydrophilic part, stabilizing casein micelles in the milk. The proteolytic enzyme chymosin (323 amino acids, 36 kDa) in the rennet breaks κ-casein at a specific point into the phosphate and calcium-rich para-κ-caseinate, which is the major component of cheese curd and glycol-peptide, finally found in the whey. During the cheese making process chymosin is added to the milk in cheese vats (Fig. 6.17) when enough lactic acid is formed during ripening. This clotting (coagulation) is the central step in curd making for hard cheese.

Fig. 6.17: Adding of chymosin to fermented milk (© Schönegger Käsealm).

Other bacteria or spores of molds are added for better taste, smell, or color. Then draining is again the next step (Fig. 6.18). After aging for months or even years and preserving the surface by a rind, grinding, and salting the cheese is ready to be marketed.

More than 30% of milk worldwide is used for cheese making in this traditional but nevertheless highly interesting bioprocess consisting of up to three bioreaction steps. Numerous different cheese products produced in a nearly equal number of process variations are on the market, a situation similar to the one in wine making or beer brewing. This reflects a complex interplay between local traditions, technical possibilities, consumer's expectations, and societal regulations. Small and domestic cheese makers or cheese makers in rural regions approach the same procedure as described above, starting with wild microbial strains for acidification and using rennet from calves. Larger commercial factories employ starter cultures and use pasteurized milk to avoid spreading of *Mycobacterium tuberculosis*. Pasteurization is obligatory in many countries and raw milk cheese allowed only under specific precautions. An ethical issue is slaugh-

Fig. 6.18: Cheese harp cutting the thickened (dehydrated) curd to ease water drainage to get a more solid cheese (© Schönegger Käsealm).

tering a huge number of calves besides their limited availability. In some religions and for vegetarians, animal rennet is not acceptable, so rennet substitutes are used. Recombinant chymosin is produced extracellularly based on bovine pro-chymosin DNA mainly by the filamentous fungi *Aspergillus niger* (GRAS) and the yeast *Klyveromyces lactis* (GRAS) but also *E. coli*. For better secretion the protein can be changed, or in the case of *Aspergillus*, be coupled to glucoamylase excretion. These products usually contain only one type of chymosin, while the natural product contains three main types. As chymosin is regarded as auxiliary production material and not as food supplement it does not have to be declared. Nevertheless, many people do not accept this as GM-free and ask for rennet substitutes based on plant or nongenetically modified microorganism. All the operations mentioned above have of course an impact on taste and smell and are therefore energetically debated among cheese lovers.

Besides lactate as the final product of anaerobic fermentation, other organic acids (carbonic acids) are employed as acidulant, chelating agents, or for flavoring in food products. Vinegar, with the main compound acetate, is next to ethanol one of the oldest biotechnological products in human history. The background is that (in earlier times airborne) microorganisms grow on alcohol-containing fluids (beer, wine) aerobically under conversion of ethanol to acetate. Nowadays industrial aerobic submerse cultivation (*Acetobacter* and *Gluconobacter*) is the standard process. The "substrate" (fruit juice, wine, spirit) is of interest insofar as it is decisive for the taste of the vinegar. The process is not fermentation in the narrow sense but a partial oxidation. Under low ethanol concentrations *Acetobacter* can gain its metabolic energy by complete oxidation of ethanol. So this process can be understood as overshoot metabolism, where under substrate saturation, it is more effective for the cells to bypass complete aerobic respiration.

Citric acid, a dicarboxylic acid, reaches a worldwide production volume of nearly two million tons and is therefore the most important biotechnological product with respect to market volume. It is not necessary to mention that such amounts can no longer be made available by citrus fruits. Since 1919 and industrially since the 1930s it has mostly been produced in submerse cultivation by *Aspergillus niger*. Citric acid is a central intermediate in the tricarboxylic acid cycle (TCC), meaning most organisms exhibit a large intracellular synthesis rate. The biologically interesting question is why the cells excrete it. The natural purpose is to act as chelating agent for iron ions to increase availability in the natural habitat. Remember that for exactly the same reason it is part of some growth media. This happens only in trace amounts. Consequently, an additional bottleneck has been introduced artificially. The overproduction of citric acid requires a unique combination of unusual nutritional conditions including excess of carbon source and suboptimal concentrations of certain trace metals and phosphate. The combination of high concentrations of glucose and ammonium represses the synthesis of α-ketoglutarate dehydrogenase, therefore inhibiting the catabolism of citric acid in the TCC and favoring its overproduction. Low manganese levels affect the cell wall enabling a large flux of citric acid into the medium. Other divalent cations influence the production as well. Furthermore, the inhibition exerted by citric acid on phosphofructokinase, a positive end effector, has to be suppressed. Citric acid production is therefore an example par excellence of how much control on microbial metabolism can be exercised by the medium here particularly by trace elements. As no recombinant genes are involved, producing strains have been developed by inducing mutations in parental strains using mutagens. Among physical mutagens, g-radiation is applied. The low pH value during the production phase (pH ≤ 2) reduces the risk of contamination by other microorganisms and inhibits the production of unwanted organic acids (gluconic and oxalic acids).

The theoretical yield is 112 g of anhydrous citric acid per 100 g of sucrose due to CO_2 fixation. Care has to be taken during gassing to avoid stripping of CO_2. However, in practice, the yield of citric acid often does not exceed 70% of the theoretical yield. The final yield after approximately one week is up to 140 g/L of citric acid and 10–15 g/L of dry biomass. Besides for food, other applications (20% of total production) are in the pharmaceutical industry as antioxidants to preserve vitamins, effervescent tablets, pH correctors, and blood preservatives or in the form of iron citrate as food supplements, ointments, and cosmetic preparations. In the chemical industry it is employed as a foaming agent for the softening and treatment of textiles. In metallurgy, certain metals are utilized in the form of citrate. Citric acid is also used in the household detergents as a co-builder with zeolites. This is supported by public regulations in order to act as a phosphate substitute avoiding eutrophication of the water body. Other biotechnologically produced carbonic acids are tartaric, itaconic, DL-malic, fumaric, succinic, and glucuronic acids.

In industry the "molecules" to be produced are divided into "performance molecules" and "platform molecules." The first group consists of typical biotechnological products like enzymes or vitamins (vitamin B or C). These are sold not only in the health or food sector for their biological activity, but also for technical applications. Enzymes for washing powder have already been mentioned. Another interesting case for technical use is hydrophobin, a small fungal protein able to organize itself in monolayers on surfaces, making them hydrophobic. During a walk in the forest, it can be observed that water easily drips off mushrooms. In particular, the spores are very hydrophobic. The second class, the platform molecules, are produced in larger amounts for further processing, e.g., polymerization. This group includes besides some of the carbonic acids mentioned above, 1,4-butanediol (BDO), "bio"-methyl methacrylate (MMA for making of PMMA), 1,3-propanediol, 3-hydroxypropionic acid, or 1-octanol (fatty alcohol for perfumes and flavorings). All these compounds can be either produced chemically or biochemically. So, a careful debate, "bio- versus petro-" has to be held. Decisive arguments are carbon and energy efficiency and arguments regarding the raw material (e.g., agricultural wastes) and production wastes. On the process level of course process intensity (productivity and final product titer) plays an important role. In downstream processing separation is an issue, as 1-octanol, e.g., is insoluble in water at concentrations above 0.5 g/L and can therefore comparatively easily be gained from a separate phase swimming on the medium. Of course, costs are in the focus, where the selling price is measured as \$/MMBtu (MM million, BTU British thermal unit, cost per heat of combustion). While ethanol is in the range of 28, 1-octanol is at 43. A high value is a good argument for "bio", justifying the greater efforts. Only a couple of molecules (about ten according to official organizations) have currently won the race towards "bio" or are candidates with good prospects.

Polysaccharides (PS) are integral parts of all cell walls (except animals) and fulfill functions such as storage compounds, mediators of spatial structure, and protective layers (mucus). PS are polymers of monosaccharides with the same or with different repeating units. They can either be linear or branched or can carry additional residuals. Some chemical structures are shown in Fig. 6.19. In aqueous solution they exhibit a linear or coagulated structure or can even form helices. Their mutual interaction via hydrogen bonds, water binding capacity via dissociated residuals, and their high molecular weight convey a high viscosity. For the same reasons strong physical binding to proteins can occur. Chemical synthesis is possible, but organisms can do it better or at least cheaper with respect to the monomer, the glycosidic linkage (α, β, etc.), and sequence.

As manifold as their role in nature is their application as products. Quite many PS are gained from plants. These include starch, paramylon (e.g., unicellular alga *Euglena*), pectin, gums, or cellulose. The macroalgae-derived polysaccharides agar, agarose, alginate, and carrageenan are used as thickeners, gelling agents (agarose gel), or as stabilizers in food. In particular, algae cell walls differ from those of terrestrial plants as they contain uncommon polyuronides and PS that may be methylated, acetylated, pyruvylated, or sulfated. As these are natural compounds, there are unconfirmed reports that, e.g., carrageenan exhibits adverse health effects. The latter one is an example of a sulfated PS. The unicellular red alga *Porphyridium cruentum* produ-

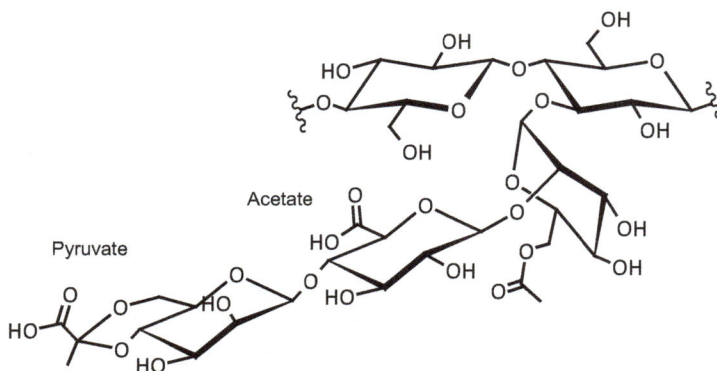

Fig. 6.19: Chemical structure of some polysaccharides with microbial and/or biotechnological origin.

ces a sulfated galactan exopolysaccharide with a sulfur ratio of up to 10%, which can replace carrageenans in many applications. Also sulfated is echinacin from the purple cone flower used as an anti-inflammatory agent and immunostimulant and was already applied by the indigenous peoples of North America. Some like chitosan (shrimp shells) and hyaluronic acid (cockscomb) used in cosmetics to rejuvenate the skin come from animals. Here an ethical issue applies, and some efforts are made to replace extraction from animal waste with microbial production. Already bacterially produced in batch (due to high viscosity) is xanthan (*Xanthomonas campestris*) as a food thickener. Harvesting is carried out by filtration subsequent to precipitation with organic solvents. Xanthan is not digested in the intestinal tract and is sometimes thought to cause allergic reactions. Dextran has medical use as anticoagulant and in affinity chromatography. Fungi-derived products are scleroglucan (moisturizing, sensory characteristics in personal care products) and schizophyllan, which is used in tertiary oil recovery, where it ousts the oil from deposits. Even cellulose, the most abundant PS in nature, is not only derived from plants but can be produced by bacteria to form thin layers for wound dressing. Here the unique selling point is spatial structure, which makes the difference between a low cost and a high value product.

Less emotional aspects compared to food safety apply in the discussion of bioplastics. At the center of endeavors are more aspects of practical applicability. The term "bio" means either derived from renewable resources, biodegradable, harmless in direct contact with living organisms, or in a narrower sense, produced via bioprocesses. The discussion on biodegradability has recently become more germane, as the tremendous problem of plastics in world oceans became more and more obvious. Between 5% and 10% of fossil fuels are deployed to produce plastic materials. This is a great motivation to seek alternative and renewable resources. Favored raw materials are wastes from agriculture containing lignocellulose or starch. The role of bioprocess engineering is here finding new possible routes of conversion from wastes to valuable products with the positive qualities mentioned above. Polyhydroxy butyrate (PHB) was one of the first commonly accepted products with some applications. Recently, polylactic acid has achieved wide application. It is produced from lactate via the cyclic diester lactide and subsequent polymerization and consequently called polylactide. It is questionable whether polylactide is biodegradable under environmental conditions.

6.10 Conclusions for batch processes – dealing with pros and cons

The batch process is still the most favored process due to its technical simplicity. In this chapter some aspects that make it more complicated have been mentioned. Examples are undetected limitations, diauxic growth or product inhibition. Other possible problems are collected in Tab. 6.3 and discussed as pros and cons. Some of the problems can be handled with careful observation and adjustment on the medium and the cell level. Other problems can be solved only by different process strategies as given in the next chapters.

Tab. 6.3: Advantages and disadvantages of batch processes.

Physiological aspects		Technical aspects	
Growth rate close to μ_{max}	+++	Simple and robust	+++
Permanent adaptation	−−	Average productivity	++
Substrate inhibition	−−	Average yield	+
Product inhibition	+/−	Only minor sterilization problems	++
Undesired coproducts	−	Low technical maintenance	+
Early uptake of co-substrates	−	Batch harvesting	+/−
		High viscosity	++

Medium conditions change permanently during the batch process. It starts from inoculation where the conditions in the pre-culture cannot be simply kept identical to the conditions in the fresh medium. Nevertheless, they should be as close as possible. With decreasing substrate concentration and increasing product concentration a permanent adaptation of the cells is required at the cost of yield and volumetric productivity, which is an intrinsic problem of the batch concept. As a batch culture is usually started with low biomass concentrations the volumetric productivity is low in the first half of a batch. This problem can be tackled in principle by a higher volume ratio of the inoculum. Instead of using the harvest from one scale as inoculum for the next larger scale, there is another idea. We harvest only a small portion, e.g., one-third of the volume and fill up with a corresponding volume of fresh medium in the same reactor. This approach is called "repeated batch" and partially overcomes the mentioned problems of substrate inhibition, low biomass at the beginning of the process, and adaptation.

Co-substrates and by-products can cause problems. Initial ammonia concentrations in stoichiometric amounts may cause inhibition. Uptake of ammonia leaves a proton back in the medium leading to a decrease in pH value. The same holds for production of acids as byproducts (see next paragraph). In cases where essential amino acids or other co-substrates like yeast extract are provided in the medium, the cells can take them up faster than necessary as carbon or nitrogen sources, leading to a depletion in a later fermentation stage. Such problems require repeated or continuous dosage of the respective compounds. Besides inhibition of the main products also other products, which are not all known, can contribute to inhibition. Experiments have shown that intentionally added ethanol to an ongoing ethanol fermentation leads to less inhibition than expected. The reason is that products like acetaldehyde or acetoin are produced in smaller amounts together with the main product ethanol. With respect to inhibition by the main product selection of robust strains showing fewer effects has been done for decades. Nevertheless, it is still an issue. For economic reasons high concentrations are mandatory. Here the batch is not that bad as it allows for production even at zero growth conditions or with dying cells as is the case in wine making.

The mentioned problems may be even stronger in the case of crude substrates like hydrolysates of wastes from the food industry or agricultural residues like straw. Modern approaches try to find a "robust" cell factory. This means maintaining cell factories in the presence of perturbations leading to less inhibition, better performance, and optimized costs: in general, better use of the cell's inherent potential. Action can be taken on the genetic and the metabolic level. Potentially inhibiting fractions in the medium can be metabolized by providing the cells with appropriate metabolic pathways. This is also in the interest of higher yields. For the suppression of potentially inhibiting by-products knockout of the related genes can be considered.

As in the example of yeast growth on glucose many other microorganisms show a so-called overshoot metabolism under high substrate concentrations. *E. coli* produces acetate, which also has a negative impact on growth. Mammalian cells convert a large part of the glucose to lactate, an effect happening also in the muscles of living animals during phases of vigorous exercise. To start with low concentrations to avoid substrate inhibition and byproduct formation would lead to low cell- and product concentration at the harvesting time and is therefore not feasible. Finding a remedy is reliant on keeping the substrate concentration low during the whole fermentation process. How this can be practically achieved is the topic of the next chapter.

6.11 Questions and suggestions

1. Microorganisms help make food more bioavailable and healthier through fermentation. To do this, however, they usually need organic raw materials. Can they also make a significant contribution to food itself?

2. Auf welcher Fläche müssten Mikroalgen angebaut werden, um 10% des von den Menschen benötigten Proteins zu produzieren und damit den Regenwald (Sojaanbau) merklich zu entlasten?

3. In the batch simulation (Fig. 6.16) it can be observed that a higher $y_{P,S}$ leads to a higher product yield and a higher final product titer but also to a longer duration of the cultivation. This reduces the volumetric productivity. Imagine you could change $y_{P,S}$ by means of genetic engineering but at cost to $y_{X,S}$. What would be an optimum value with respect to volumetric productivity $P_{V,P}$?

4. In Fig. 6.13 biomass increases constantly. What is the course of the specific growth rate?

Chapter 7
Little by little, one goes far – the fed-batch process

The drawbacks of batch cultivation could eventually be circumvented by measures on the process level. The basic idea should be to keep the cells in a physiological state – given as a defined working point in the kinetic space. This means constant specific rates and, consequently, constant medium concentrations. A sensible choice of such a working point could be to keep r_S smaller, but as close as possible to $r_{S,crit}$. In any case, it must be achieved via the respective substrate concentration being kept small and constant. We already know that starting with a low substrate concentration will help only for a few minutes. An appropriate feeding strategy of the related medium compound has to be calculated and implemented in the process.

The first idea is to measure c_S online and implement a controller to feed fresh substrate before c_S becomes too small, or stop feeding before it exceeds $c_{S,crit}$. This implies equipping the reactor with a substrate-feeding stream, $q_{S,f}$ (L/h), as the controlled variable. Controlling the substrate-feeding concentration $c_{S,f}$ could be another option in this direction. For this approach, a suitable sensor is necessary. In case this is achievable, there is a good chance of getting the required process without any overshoot metabolism. A similar idea is to measure the concentration of the unwanted byproduct, e.g., ethanol in the case of yeast production, and adjust the feeding rate to a value that keeps the byproduct concentration very low. Some yeast factories indeed have an ethanol sensor in the off-gas stream like the ones the police use in traffic control. This is a bit more indirect but it is at least an option to prevent an uncontrolled drift of the culture into the substrate, overshooting the metabolism. This danger is actually possible, as under complex media conditions, the kinetic parameters may not be known very precisely.

What about the idea of giving the cells exactly what they need to maintain the desired working point, so to speak, controlling r_S directly to keep it at an intended value, $r_{S,set}$, below $r_{S,crit}$? The total amount of fed sugar has to be balanced against the amount taken up. As this is the product, $r_{S,set} \cdot c_X$, the strategy requires knowing, in principle, the cell dry mass concentration. As measuring is not very reliable, the value can be estimated as shown below. The mentioned ideas need additional equipment for the reactor, compared to batch processes, i.e., the possibility to feed permanently fresh substrate. Therefore, such processes are called fed-batch processes. In this chapter, we will investigate fed-batch processes first, by calculation of the related reactor equations, based on kinetics and mass balances. Then, we will deal with process examples, e.g., the running example of yeast production, and finally discuss the necessary prerequisites as well as strengths and weaknesses of the fed-batch processes.

https://doi.org/10.1515/9783110773354-007

7.1 Setting up process equations – the formal deduction of a suitable feeding strategy

The structure of a fed-batch reactor is depicted in Fig. 7.1. Fresh medium, with the feed substrate concentration $c_{S,f}$, is fed with the volumetric flow rate, $q_{S,f}$ (L/h). As this changes, we distinguish here the nominal reactor working (reaction) volume, $V_{R,nom}$, and the real time-dependent liquid volume, $V_{R,liqu}(t)$. This has to be considered when setting up the balance equations.

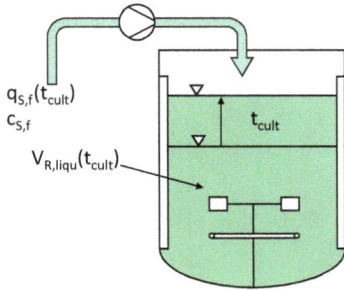

Fig. 7.1: Reactor configuration for a fed-batch process.

As no biomass is removed from the reactor, the biomass balance is the same as for the batch process, taking the biomass as a balanced state variable:

$$R_X = \frac{dm_X(t)}{dt} = r_{X,set} \cdot m_X(t) \tag{7.1}$$

Note that, here, the intended μ_{set} is used, assuming that we finally manage to keep it constant at the desired value.

For the substrate balance, an additional transport term counting the freshly fed substrate is necessary:

$$RS = \frac{dm_S}{dt} = -r_{S,set} \cdot m_X(t) + q_{S,f}(t) \cdot c_{S,f} \tag{7.2}$$

As the active volume in the reactor changes by the medium inflow, a third differential equation for the changing reactor volume $V_R(t)$ is necessary:

$$\frac{dV_R}{dt} = q_{S,f}(t) \tag{7.3}$$

Because the volume changes, we can no longer obtain differential equations for the concentrations simply by dividing the mass-related equations, through V_R. Nevertheless, m_X and m_S can be substituted by $m_X = c_X \cdot V_R$ and $m_S = c_S \cdot V_R$, leading to

$$\frac{d\big(c_X(t) \cdot V_{R,\text{liqu}}(t)\big)}{dt} = r_{X,\text{set}} \cdot c_X(t) \cdot V_R(t) \tag{7.4}$$

and

$$\frac{d\big(c_S(t) \cdot V_{R,\text{liqu}}(t)\big)}{dt} = -r_{S,\text{set}} \cdot c_X(t) \cdot V_{R,\text{liqu}}(t) + q_{S,f}(t) \cdot c_{S,f} \tag{7.5}$$

The next step is applying the product rule for differentiation, yielding

$$\frac{dc_X(t)}{dt} \cdot V_{R,\text{liqu}}(t) + \frac{dV_{R,\text{liqu}}(t)}{dt} \cdot c_X(t) = r_{X,\text{set}} \cdot c_X(t) \cdot V_{R,\text{liqu}}(t) \;\rightarrow \tag{7.6}$$

$$\frac{dc_S(t)}{dt} \cdot V_{R,\text{liqu}}(t) + \frac{dV_{R,\text{liqu}}(t)}{dt} \cdot c_S(t) = -r_{S,\text{set}} \cdot c_X(t) \cdot V_{R,\text{liqu}}(t) + q_{S,f}(t) \cdot c_{S,f} \tag{7.7}$$

Now, it is possible to normalize the equations to the volume, and reorder, yielding:

$$\frac{dc_X(t)}{dt} = r_{X,\text{set}} \cdot c_X(t) - \frac{q_{S,f}(t)}{V_{R,\text{liqu}}(t)} \cdot c_X(t) \tag{7.8}$$

Note that dV_R/dt has been substituted by $q_{S,f}(t)$, according to eq. (7.3). The second term on the right-hand side represents the dilution of the biomass already present by the freshly inflowing medium. Similarly, we can write for the substrate balance equation:

$$\frac{dc_S(t)}{dt} = -r_{S,\text{set}} \cdot c_X(t) + \frac{q_{S,f}(t)}{V_{R,\text{liqu}}(t)} \cdot (c_{S,f} - c_S(t)) \tag{7.9}$$

The balance equations themselves do not tell us something substantially new, but they can be employed to get a clearer view of what should happen in the reactor. We remember that the initial goal of the fed-batch process was to keep c_S small and constant. So, we can use these equations to calculate $q_{S,f}$ such that $dc_S/dt = 0$:

$$0 = -r_{S,\text{set}} \cdot c_X(t) + \frac{q_{S,f}(t)}{V_{R,\text{liqu}}(t)} \cdot (c_{S,f} - c_S(t)) \;\rightarrow \tag{7.10}$$

$$q_{S,f}(t) = \frac{r_{S,\text{set}} \cdot c_X(t) \cdot V_{R,\text{liqu}}(t)}{c_{S,f} - c_S(t)} \tag{7.11}$$

Because μ is a constant, $m_X(t) = c_X(t) \cdot V_{R,\text{liqu}}(t)$ is exponentially increasing. By substitution, $r_S = r_X/y_{X,S}$, and finally, the desired formula for the feeding rate is obtained:

$$q_{S,f}(t) = \frac{\mu_{\text{set}} \cdot m_{X,0}}{y_{X,S}\,(c_{S,f} - c_S(t))}\, e^{\mu_{\text{set}} \cdot t} \tag{7.12}$$

This looks reasonable, as an exponentially growing culture has to be fed exponentially to supply each cell with the same amount of sugar during the entire process duration. To set up the formula in a process computer, the yield and the initial biomass must be known. The initial biomass concentration and the initial volume are design parameters to tune the process with respect to the additional goals.

To test this exponential feeding strategy, it is advisable to first make a virtual experiment with the standard physiological model; the result is shown in Fig. 7.2.

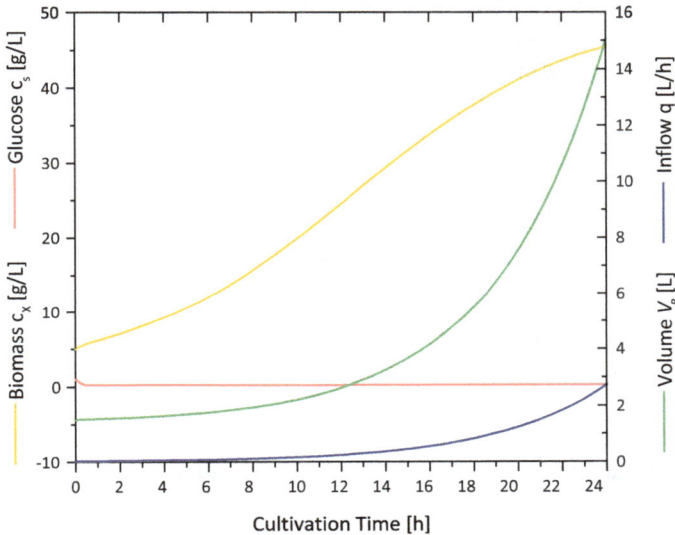

Fig. 7.2: Simulation of a simple fed-batch process with exponential feeding; the simulation parameters are $r_{S,max} = 2.0$ h^{-1}; $k_S = 1.0$ g/L; $y_{X,S} = 0.5$ g/g; $c_{X,0} = 5.0$ g/L; $c_{S,0} = 1.0$ g/L; $V_{R,0} = 1.5$ L; $c_{S,f} = 100.0$ g/L; $q_{S,f} = 0.0225$ L/h.

As expected, the feeding rate starts from a constant value and increases exponentially. Consequently, the filling volume of the reactor increases exponentially as well. After a short time, the initial substrate concentration stabilizes on a constant low value, which was the intention of this feeding strategy. The biomass concentration obviously does not follow in an exponential way. This is explained by a look at the biomass balance equation (7.8). The first term $r_X \cdot c_X$ is the growth term, which would lead to exponential growth, and the second term $-q_{S,f}/V_R \cdot c_X$ describes the dilution of the biomass present by freshly fed medium. This term becomes more and more dominant during the cultivation, leading to a weaker and weaker increase of the biomass curve.

To evaluate data from a fed-batch process and to calculate r_X and r_S, the total biomass and substrate in the reactor have to be entered into the appropriate formula: $m_X(t) = c_X(t) \cdot V_R(t)$ and $m_S(t) = c_S(t) \cdot V_R(t)$, respectively. The productivity is meant to give a measure of the cost efficiency of the process, e.g., with respect to the investment into the reactor. Therefore, for the fed-batch process, it is usually related to the nominal working volume.

$$P_{V,X} = \frac{m_{X,end} - m_{X,0}}{V_{R,nom}} = \frac{c_{X,end} \cdot V_{R,end} - c_{X,0} \cdot V_{R,0}}{V_{R,nom}} \tag{7.13}$$

The nominal working volume, $V_{R,nom}$ (L), of the reactor can be set equal to the final filling volume, $V_{R,end}$, as it is done in batch cultivations as well. This definition has to be evaluated for the whole process or for time intervals under consideration. It is spontaneously clear that in the beginning of the fermentation, this value is extremely low due to low biomass concentration and low filling volume. Nevertheless, it is a standard strategy for many products.

7.2 Running example, yeast production – an industrial feeding strategy

One example for a fed-batch process is the production of baker's yeast, where overshoot metabolism is to be avoided. This would occur typically at specific growth rates $\mu > 0.2$ to 0.23 h^{-1}. According to the specific demands and abilities of yeasts, some specific modifications can be found in industrial production processes. Process control policy needs careful integration of biological and technical knowledge to process control design.

7.2.1 Deviations from exponential – controlling physiology and obeying technical constraints

Looking at an industrial feeding strategy, as plotted in Fig. 7.3, several characteristic features can be observed.

The first thing to mention is that an exponential feeding strategy has not been applied but a piecewise linear dosing curve. This can be understood against the background of technical constraints, biological needs, and finally the desired product qualities. The first linearly increasing feeding phase is only an approximation of an exponential function. This is regarded as easier to realize under harsh process conditions. The specific growth rate is indeed not constant but increases in the beginning and declines at the end of the phase, reaching its maximum ($<\mu_{crit}$). A technical point is that the range between the lowest and the highest value of this phase (9–65) can be implemented with only one pump.

The second phase shows a constant feeding rate. This leads to a decrease of the specific growth rate. The reasons why constant feeding is nevertheless applied lies in a limitation of the reactor. In Chapter 6, it was already noted that sometimes, oxygen transport capacity is the bottleneck of productivity. This is indeed the case here. Why

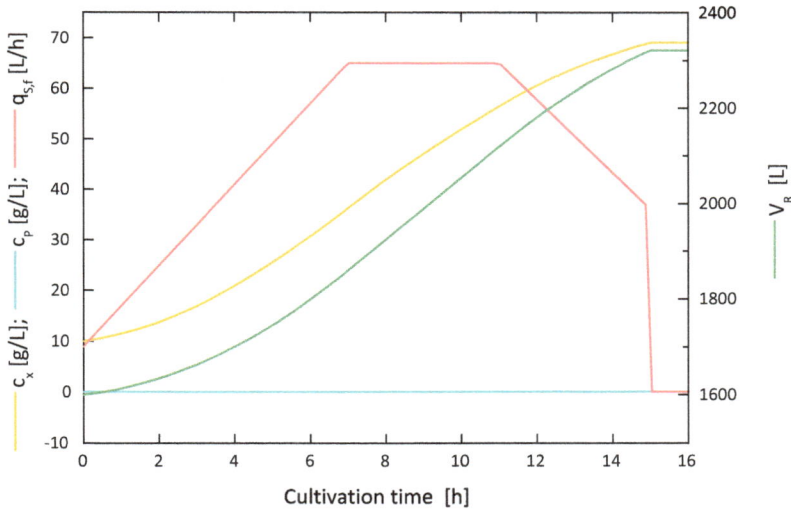

Fig. 7.3: Industrial feeding strategy for production of baker's yeast; the data refers to a reactor with 2,500 L working volume and a glucose concentration of 400 g/L in the feed.

substrate feeding has something to do with oxygen supply becomes clear when recalling intracellular stoichiometry:

$$R_S = q_{S,f} \cdot c_{S,f} = r_S \cdot c_X \sim r_X \cdot c_X \sim r_{O_2} \cdot c_X = R_{O_2} = OUR = OTR \qquad (7.14)$$

Consequently, the maximum oxygen transfer rate of the reactor means a maximum amount of substrate supply, which the cells can process via growth and respiration, and thus a constant feeding rate. Note that constant feeding leads to a linear increase of biomass concentration as the biomass integrates the feeding, so to speak. This means that in effect, the more biomass is present in the reactor, lower is the possible specific growth rate. This makes a reduction of μ unavoidable as soon the biomass reaches a given threshold, otherwise the Pasteur effect would occur.

The third phase is characterized by a linearly decreasing feeding rate, leading to an even lower specific growth rates. This stronger substrate limitation is brought about intentionally to force the cells to take up less preferred sugars and eventually produce traces of ethanol. Even small amounts of carbon sources in the molasses have to be used for higher yields to make the production process more cost effective. But the diminishing of the growth rate has another deeper meaning.

Not all yeast are equal. The presence of buds and mother cells is a characteristic feature of unequally budding yeast cells. The fraction of budding cells (FBC), also called budding index (BI), is actually a point of product quality for the bakery. Matured cells are more stable in the dough and show a higher fermentation activity than the buds. One target of the production process is to deliver a yeast product with a high ratio of matured cells. While high growth rates contribute to high productivities and lead to many buds, low growth rates favor matured cells. This is realized by a low growth rate toward the end of the cultivation. FBC shows values in the relevant range for production – 0.3 for $\mu < 0.1 \text{ h}^{-1}$ and 0.7 for $\mu > 0.2 \text{ h}^{-1}$.

The background is the cell cycle of budding yeasts, as schematically drawn in Fig. 7.4.

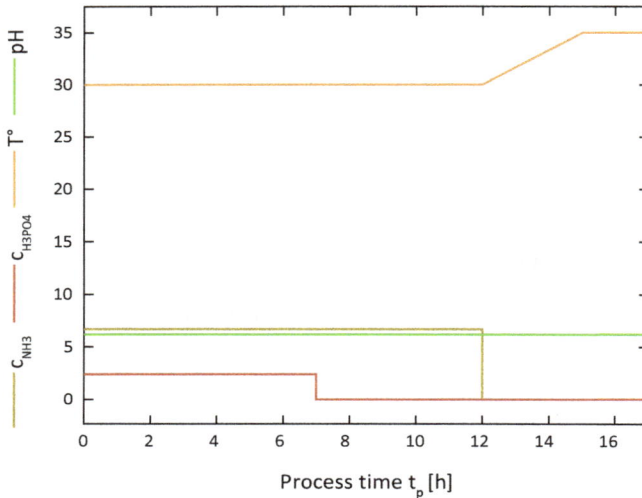

Fig. 7.4: Additional parameters of the industrial feeding strategy to be used to control yeast growth.

The haploid cell starts its life in the unbudded daughter phase by increasing the cell size. This takes the time, T_d, depending on the growth conditions. Then, the first checkpoint "start" is reached. Further action is not taken until the cell has measured its own size and the medium conditions. Only if both are satisfactory does the cell start the transition into the S phase. Here, nucleic acids are produced, the chromosomes divide, and the cell becomes diploid. The time Tp depends on the medium but also takes some time for the chromosomes to double. At the next checkpoint, correct doubling of the chromosomes is checked; the cell then decides to start budding. In the budded phase, the bud emerges and increases its size, firstly without having its own cell nucleus. The next checkpoint gives the green light for mitosis and the start of the meta-ana transition. This takes a defined time, independent of the medium conditions, and needs intracellular storage compounds like glycogen, the major intracellular storage carbohydrate in yeasts. After the bud has gained its own nucleus, it is separated from the mother cell, leaving it in the unbudded parent phase.

Consequently, low sugar concentrations can keep the cells resting in the G1 phase, and the fraction of budding cells decreases. High temperatures also stop further cell development at the "start" checkpoint and lead to accumulation of the disaccharide trehalose as a stress response, acting as heat protection to maintain the structural integrity of the cytoplasm. In process engineering terms, this is important for some steps in downstream processing like drying and freezing. Feeding of N and P (Fig. 7.5) is a further measure to support manipulation of the fraction of budding cells. Both nutrient components are supplied in fed-batch; here, at a constant rate. This prevents initial inhibition or precipitation. Stopping feeding at a determined point controls the protein concentration of the final product. This can vary between 40% protein at a C:N ratio (g/g) of 20:1 and 50% for C:N 10:1. The segregation of the cells into the different phases is also modified. Stopping feeding of P prevents nucleic acid synthesis and keeps the cells in the S phase. Concluding this process strategy determines not only the overall specific growth rate via substrate uptake kinetics, but also a measure used to influence cell morphology and product quality via medium concentration, measured by cell sensors.

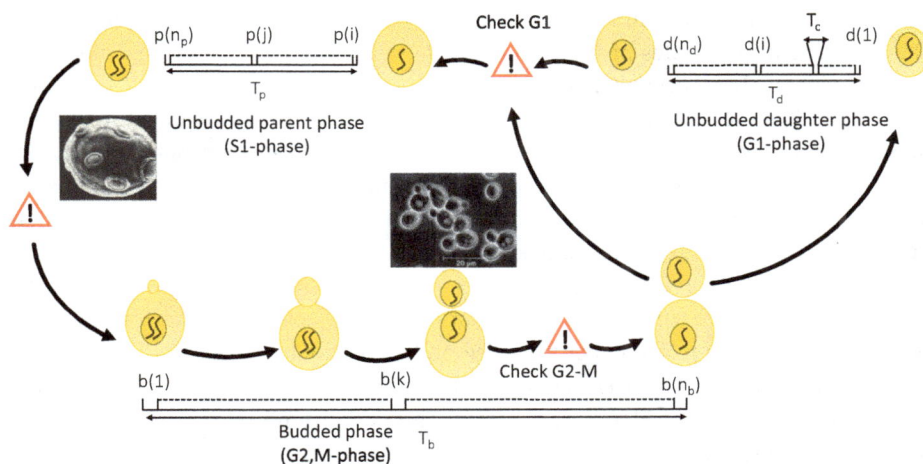

Fig. 7.5: Cell cycle model of yeast; checkpoints are used in production to keep the cells in their defined morphological states. The intervals give time points for simulation, as shown in the supplementary material.

This cell cycle is strongly conserved in eukaryotes and is therefore regarded as a model for animal cells and even humans. Transition into a new budding phase also depends on the number of buds a cell has already produced, visible by the budding scars. This can be taken as a first sign of individual aging in organisms. In this example, the integration of biological knowledge into process design is obvious. This has direct impact on product qualities, important for bringing the product into the market, depending on societal circumstances. A yeast for baguette baking is different from one used for bread in Caucasus, e.g., with respect to protein and glycogen content. The necessary time of dough rise can differ between 20 min and a few days, with different demands for yeast stability. Many ideas for genetic engineering can be thought of to make the process less complicated. One of them could be to induce etha-nol formation and the respective CO_2 production only in the dough, triggered by a compound therein. Another way might be manipulating the different checkpoint. However, regulations in many countries prohibit the employment of genetically ma-nipulated yeasts. On the other hand, production of "bio" yeast is an option. To get the "bio-label," strict demands are placed on the molasses: even ammonia from a chemi-cal plant is forbidden and natural protein hydrolysates are used as N source in some factories.

7.3 Production of secondary metabolites – case study on penicillin production

Penicillin was one of the first biotechnological products for medical use and one of the outstanding examples of biotechnology. Many books have been published and even films have been produced about the life and work of Alexander Fleming (1881–1955). In 1928, he saw "by pure chance" that a *Staphylococcus* culture on an Agar plate was contaminated with molds. Instead of discarding the plate like probably many sci-entists before him, he observed around the contaminated spot a zone free from bacte-ria. Further experiments revealed that the fungi indeed secreted a substance that could kill gram-positive bacteria, even in 800-fold dilution. The first thing to learn from him is to look carefully on the findings, even if it is not in the mainstream of the recent research. "One sometimes finds what one is not looking for," as Fleming said. Nevertheless, the published paper attracted no further attention. In 1938, a systematic study (Ernst Chain, Howard Florey) of antibacterial activity of microbial substances led to the rediscovery of Fleming's findings. The background was the Second World War where many soldiers did not die directly of their injuries but of secondary infec-tions. Consequently, further development was pushed by political authorities, mainly in the USA, and led to technical production of surface and submerse cultivation. This saved millions of lives during the war. This is another lesson to learn: scientific inter-est in terms of technology push is sometimes not enough for a useful development. In addition, a market pull is necessary; in this case, a high sense of affliction. The task of the bioengineer is to build a bridge over this gap and develop new processes as effec-tively as possible.

In 2015, sales of β-lactam antibiotics have reached over US $20 billion. They get their name from the β-lactam ring structure (Fig. 7.6). Product titers are with 60 g/L (=100,000 IU/mL), more than 50,000 times higher than the original strain of Fleming; reason enough to have a look at the mechanisms applied for strain and process im-provement. As a result of the first screening steps, production strains were changed from *P. notatum* to *P. chrysogenum*. Further remarkable development was achieved by undirected mutagenesis (e.g., X-ray, UV, yperite). These techniques cannot lead to overexpression of certain genes (unless intracellular control loops are deleted) and

Variable   ß-Lactam-   Thiazolidine-
side chain   ring   ring

NH   H

S   CH₃

N   CH₃

O

O   OH

O

Phenyl-Acetate   L-Cysteine   D-Valine

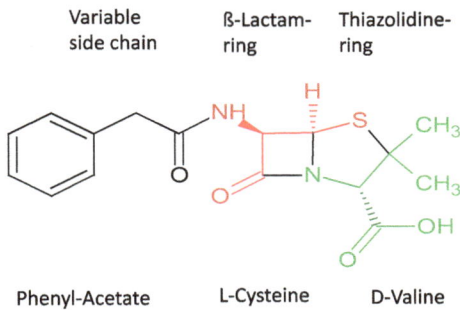

Fig. 7.6: Molecular structure of penicillin showing the eponymous cyclic amid structure and the precursors from which the cells synthesize the completed molecule.

certainly do not introduce new functional genes. Unlike primary products, which are limited by intracellular catabolic fluxes, determined by substrate uptake, stoichiometry, or intracellular bottlenecks, secondary products underlie a strong intracellular regulation. The fungus produces the desired products according to temporary physiological needs, stimulated by intra- and extracellular signals. The cells save the energy, which they would otherwise use. This could be, in the case of penicillin, a little bit speculatively, the presence of bacteria as competitors of glucose in the environment. The success of undirected mutagenesis is based on the destruction of the relevant intracellular control loops, leading to deregulation of negative feedback inhibition. For about 20 years, further process improvement has only been reached by employment of modern means of genetic engineering like overexpression of specific proteins, introduction of other promotors, and application of metabolic engineering, in general. Growing capacities go more and more to China and India.

Modern strains with high production rates have their limitations in ATP or NADPH demand and the presence of specific precursors, a situation of metabolic burden, not very different from the situation in recombinant protein production. So, only 10% of the consumed carbon is targeted to penicillin, while 25% is available for growth. The remarkably high remaining 65% is assigned to intracellular transport, product excretion, and hydrolytic loss of activated intermediates. Here, further progress is potentially possible despite the complex regulation network involved in penicillin formation.

Parallel to strain development, success has been accomplished by rational process design. This was also guided by insights into the biological circumstances, leading to high production rates. To understand some aspects of processes where filamentous fungi are involved, we have a look at their morphology. *Penicillium* and *Aspergillus* are not unicellular organisms like bacteria. Nonetheless, they rank among the microorganisms. They propagate by spores and form filaments of linearly connected cells, the hyphae (see Fig. 7.7) of a few μm diameter and up to the cm range in length. Growth occurs only at the tip of the hyphae, and cell material from the older cells is transported along the hyphae to the tip. Depending on the length of the filaments and the growth conditions, branching occurs by formation of new tips. The entirety of all hyphae is called the mycelium. The older cells form vacuoles of increasing volume and finally die. As in yeast, the individual cells have different age and physiology, and is to be described in terms of a segregated model.

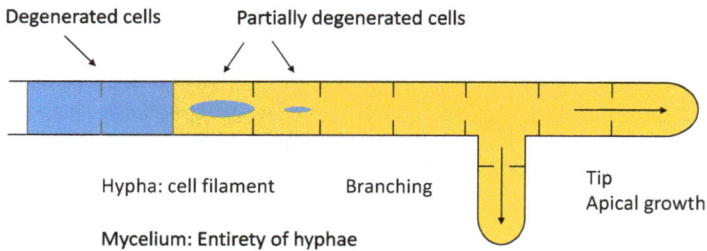

Fig. 7.7: Formation of a hypha and branches in filamentous fungi; depicted in golden color are the active cells in blue water-containing vacuoles.

In nature, the mycelium can cover an area of several m^2. As nutrients in the middle may be depleted, growth and fruiting body formation occurs mainly at the edge of the circular area, a phenomenon called a fairy ring. In the bioreactor, the hyphae are exposed to shear forces and vortices, bending them back to the center, finally forming a pellet of a few mm in diameter. A graphical simulation is shown in Fig. 7.8. This has repercussions on the process. Nutrients and oxygen are used up in the outer layer of the pellets, which form a diffusion resistance on top. The inner space depletes in nutrients, causing cells to die, leaving a hollow in the center. This is not totally a disadvantage. Pellets are easier to filtrate than single cells due to their size. Some degree of substrate limitation can also support product formation as outlined below.

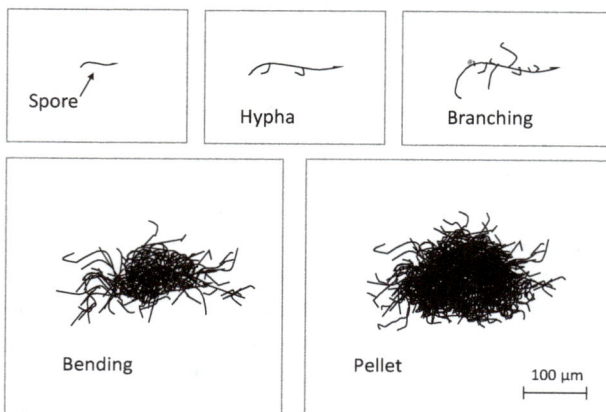

Fig. 7.8: Graphical simulation of pellet formation, starting from the spore and continuing over a growing mycelium.

The next observation is that *Penicillium* shows high production rates only under low growth rates. This is inverse, compared to products from primary metabolism. Of course, enough carbon has to be present, meaning that the peak of the specific production rate is observed at low specific growth rates (Fig. 7.9).

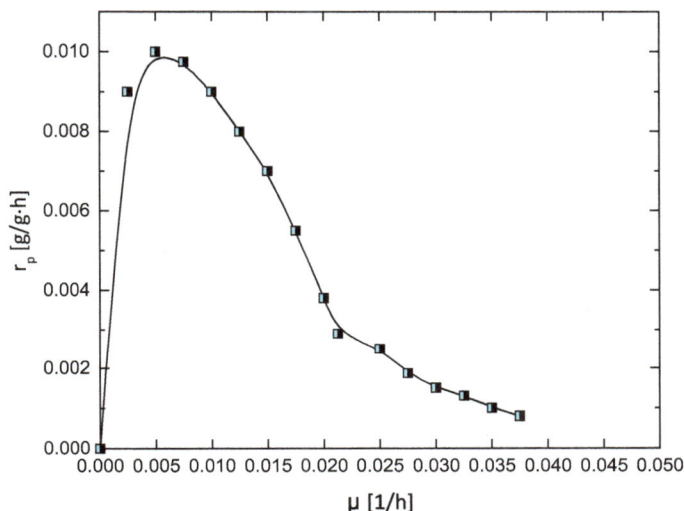

Fig. 7.9: The product formation kinetics has a pronounced maximum, which sensitively determines the optimum feeding strategy. It is of course strain-specific.

Glucose, in particular, suppresses product formation. This gives us the opportunity to install a two-phase fed-batch process, as in the case of enzyme production. During the first phase, mycelial biomass can grow up to a sufficient level. Here, glucose is the best substrate. In the second phase, production is induced by reducing the concentration of the carbon source to a level that allows only for low growth rates (Fig. 7.10). Here, lactose is preferred, as it does not suppress product formation. Both sugars can be in the initial medium, from which glucose is used up first. As penicillin is a comparatively cheap product, the process has to be based on cheap natural media. In the first phase, mainly molasses or other crude carbon sources can be chosen and nitrate as the N source. Feeding of corn steep liquor as additional C and N source is applied in the first phase and continued in the second phase. We remember that corn steep liquor contains mainly lactose as C source, while molasses contains glucose. Due to high product titers, nitrogen from corn steep goes to a large amount into the product. Ammonia would suppress product formation. The second phase can be prolonged by the so-called "mini-harvest" protocols, where a part of the reactor content, e.g., 30%, is harvested and the fed-batch can continue.

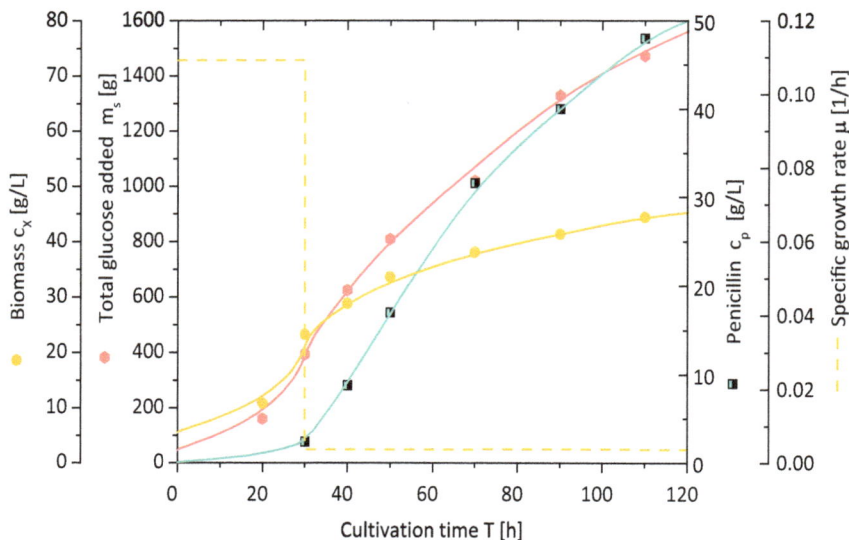

Fig. 7.10: Data from a penicillin production process showing the specific growth rate in the two phases, beside biomass and product concentration in the integrated substrate feed.

The next important aspect with respect to medium design is the additional feed of precursors to support the fungi in making the product. These are the amino acids, namely, L-cysteine and L-valine, and for penicillin-G, phenyl-acetic acid, so basically, the compounds from which the penicillin molecule is built up. This precursor leads to "direct synthesis," while other precursors lead to nonnatural "biosynthetic" antibiotics. For all side chain precursors, uptake must be ensured as well as induction of enzymatic activation.

Production fermenters reach volumes up to 100 m^3 scale, with scaling steps in the production train of 1:10. A flow sheet of the whole process is depicted in Fig. 7.11.

The first downstream process is cell separation by vacuum drum filtration. This is enabled by the pellets forming a quite porous filter cake. In modern processes, product titer is high enough to enable wet extraction without a prior solid/liquid separation step. Note that the mycelium, due to its content of residual antibiotic activity, cannot be used as cattle feed, but needs special disposal steps. Extraction occurs at low pH values to give a high partition coefficient. Unfortunately, the product is not very stable at this point, making a fast separation of medium and solvent necessary by the application of centrifugal extractors. Finally, the product runs through several purification steps. Crystallization has high ability to exclude impurities and to obtain a clean and dry penicillin salt.

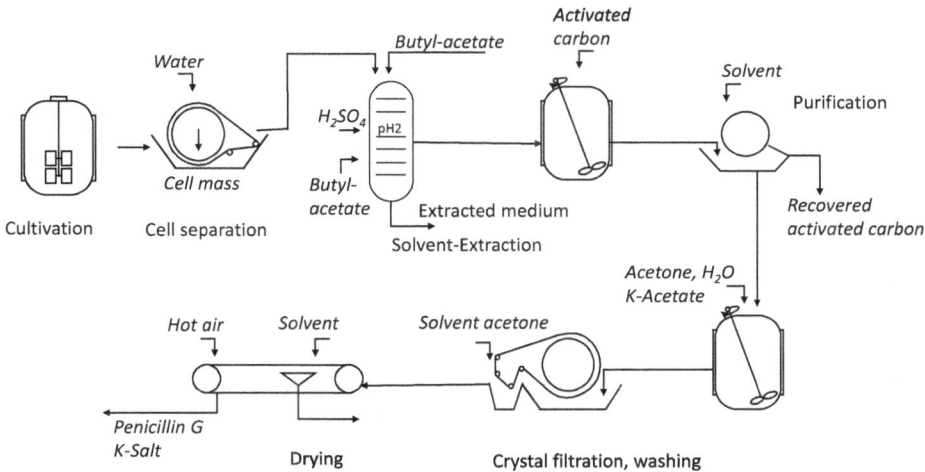

Fig. 7.11: Flow sheet of a penicillin production process, from cultivation through cell separation, and extraction to formulation.

7.4 Production of recombinant proteins – the cell factory

Until now, we considered whole cells and small molecules as products. While metabolites have a high turnover, they occur only in small intracellular concentrations. The cell itself is built up of macromolecules showing much lower turnover rates. Macromolecules carry the main functionality of a living cell. From a technical point of view, we find possible products among all groups of biological macromolecules. These groups are basically nucleic acids, proteins, polysaccharides, and lipids. Usually, they are built up from a long but specific sequence out of a couple of different monomers. This sequential order defines structure and function. In this paragraph, proteins are the focus and especially, process engineering aspects of their production. The complex biological machinery for the replication of nucleic acids, transcription, and finally translation into proteins is among the most amazing inventions of nature. In the meantime, this machinery is scientifically understood precisely enough for a success story on its own, to allow for a rational process design. Consequently, exploiting the bioactive capabilities of proteins in food production, for pharmaceutical use, or as catalysts in technical applications is a major field of bioprocess engineering.

7.4.1 Different ways to get hold of desired proteins – an overview of protein expression systems

The first option is to look for ready solutions made by nature. Many organic substrates are found in nature as water insoluble or even as denatured biopolymers. Nevertheless, they are degraded by microorganisms via hydrolytic enzymes, which are secreted into the environment. One potent exemplar of a microorganism with this capability is *Bacillus subtilis*. So quite early in the history of biotechnology, it was employed for the production of proteases (washing powder). To design a production process based on *Bacillus*, the same considerations with respect to the medium apply, meaning a suitable N source, not only for growth but also for the target protein. As in penicillin production, we need a biological signal forcing the organisms to start production and secretion. It is not really astonishing that protease is produced mainly during nitrogen limitation, while phytase is produced during phosphate limitation. As in the penicillin case, undirected mutagenesis or directed evolution are means to destroy intracellular feedback mechanisms between substrate availability and enzyme production. Consequently, the idea for the process is again a two-stage fed-batch process for getting high cell counts in the first phase and high product titer in a substrate limited second phase. An example is more closely described in Chapter 11 on fermentation.

It is desirable, not only the potential for production of large amounts of technical enzymes but also for its high ability to secret them efficiently into the environment or respectively the medium. This is an important example of integration between biology and downstream processing, as it enables easy separation of the enzymes from the medium without cell disruption and separation of the target protein from a few thousand other intracellular proteins. In fact, protein separation by chromatography after preceding mechanical cell separation and disruption steps is a sophisticated and, in the end, expensive enterprise. The idea of integration is to leave such things to the cells. Cell membranes have the most potential as selective separation devices. However, natural extracellular expression systems, which may be bacteria, yeasts or fungi, are employed mainly for technical enzymes as they are not very large. Another point of consideration is that recombinant proteins may be the target of natural proteases secreted by *Bacillus*, even if we do not want them to do it. Bacteria cannot produce proteins with human identical glycosylation or other forms of posttranslational modification. Such mainly pharmaceutical proteins can be provided only by eukaryotic organisms.

To make use of this property of eukaryotes yeasts, fungi and, finally, animal cells are applied. Phototroph production in moss or microalgae is also under development. Now, we collect some points how to support the "cell factory" along the way, from transcription to excretion. Firstly, we have to "convince the management" to produce the target protein and not the proteins needed by the cell itself. Geneticists will provide strains carrying the required gene in many copies, if necessary. Secondly, the promotor is to be defined to start production only at the beginning of the production phase. From a process engineering view, this is transmitting a chemical or physical signal to the cell to activate the promotor; so something like "placing an order" to the cell factory. Some commonly applied signals are collected in Tab. 7.1. Some of them are explained in more detail in the following paragraphs.

Tab. 7.1: Fed-batch processes, signals and induction (fill in further examples according to your own experiences).

Organism	Product	Signal for observation	Signal for induction	Mechanism
Bacillus			Substrate limitation	Intracellular control loop
E. coli	Rec. proteins	pO_2-peak	IPTG	Induction of lactose operon
			Heat	Heat shock proteins (chaperons)
			Light	Rhodopsin
Aspergillus	Penicillin	Target concentration	Low glucose	Intracellular control loop
Animal cells			pH	

The cells are optimized for self-reproduction, so forcing them to produce recombinant proteins will require resources of the cell factory, which are then no longer available for the cells to proliferate. This "metabolic burden" must be compensated as effectively as possible or tolerated only during the production phase. Gene expression should be balanced against the amount of available amino acids and cellular ATP generation. In fact, as soon as a high cell concentration is obtained in the growth phase, further growth is not obligatory but as much carbon as possible is allotted to the product. The necessary amino acids are the "parts" for further "assembling" the proteins. Feeding amino acids is a means to help the cell carry the metabolic burden. The translation process is located at the ribosomes, which can therefore be called the "work systems" or more precisely, the "assembly machine", stressing the cell factory picture. Their number depends on the type of the cell. Like in a real factory, short cycle times set under pressure of time can increase the scrap rate in the case of proteins – visible as incorrect sequence, like missing terminal amino acids or misfolding. Eukaryotic proteins are subjected to further posttranslational modifications, especially glycosylation, where oligosaccharides are attached to the proteins. These glycosylation patterns are very specific for different groups of organisms controlling enzyme activity, interaction with cell surfaces, or helping the organisms to distinguish between their own and foreign proteins. Incorrect glycosylation patterns can cause malfunction of pharmaceutical proteins in the human body. "Humanization" of the proteins makes production in mammalian cells necessary in many cases, although it is quite expensive. In other cases, glycosylation can be modified by subsequent enzymatic steps or even by genetically engineering the intracellular enzyme system. Finally, "parcel labels" have to be put on the products to allow the intracellular transport system to deliver the protein to the intended location, be it cytoplasm, periplasm, or extracellular space. These targeting tags are coded in the genome as well, detected by the cell and cleaved after delivery; see also Chapter 2.

7.4.2 High-density cultivation – a specific fed-batch for recombinant protein production

Recombinant proteins, mostly for medical use, are a big success story in biotechnology. Some of them like insulin can be produced by bacteria. Human insulin, produced by *E. coli*, was the first "golden molecule" of the biotech industry and has been on the market since 1982. This was only four years after the first insulin-encoding gene was transferred from mammalian to bacterial cells. Since 2000, insulin analogues with better pharmacokinetics have been available. In addition, to help millions of diabetics worldwide, effective production processes have been developed. In contrast to extracellular technical enzymes produced by *Bacillus*, intracellular proteins agglomerate as "inclusion bodies" (Fig. 7.12).

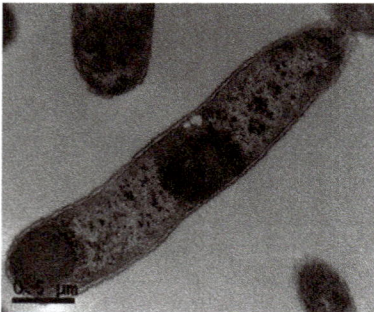

Fig. 7.12: Inclusion bodies of GFP (green fluorescence protein) by *Escherichia coli* BL21 as example of intracellular accumulation of recombinant proteins In this study, the incorporation of bacteriotoxic peptide (protegrin-1) into the GFP allows the expression of insoluble proteins in high concentration levels (© J. Yu).

The process runs through three phases as shown in Fig. 7.13. The first phase is a batch phase, where the *E. coli* cells grow with maximum specific growth rate to moderate cell densities. Concomitant acetate is produced due to the overshoot metabolism of *E. coli*, preventing further growth to higher densities. Increasing biomass concentration leads to exponentially decreasing dissolved oxygen concentrations. A controller prevents decrease to limiting values by increasing stirrer speed. At the end of this phase, all glucose is consumed. The idea is now to wait a short time to force the cells to take up the acetate. These events of glucose and acetate depletion cannot be measured directly. Now, an indirect signal is used, based on the substrate/respiration coupling. The complete lack of a carbon source, therefore, disables oxygen uptake. This is measured as a peak in pO_2. The pO_2 controller is parameterized in such a way that increasing pO_2 does not lead to falling stirrer speed. Sometimes, different peaks are observed, indicating glucose, acetate, and glycerin depletion. Glycerin is sometimes part of the medium coming from the deep-freeze cans in which the inoculum is transported. After the last peak, a classic fed-batch strategy is started. This first step shortens, in total, the long first phase of a fed-batch process, with low volumetric productivity.

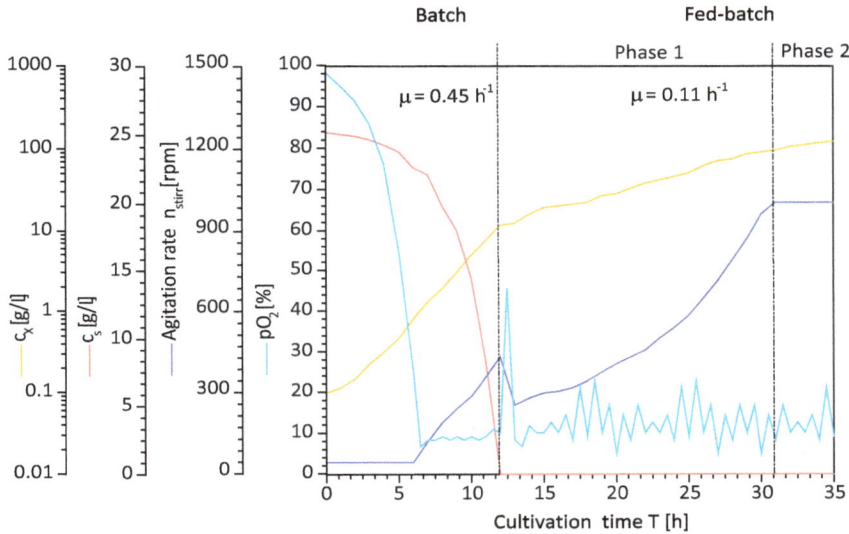

Fig. 7.13: Data from high-density cultivation showing the two phases, in particular, the gas phase related values of p_{O2} and agitator speed.

During the second phase, carried out in fed-batch mode, acetate production is suppressed by low growth rates and the culture propagates to high cell densities, up to 100 g/L. Dissolved oxygen is controlled at sufficient values, again by the agitator or higher gas flow rates. As we have seen in the yeast examples, oxygen transport limitation is possible, especially as shear stress induced by the stirrer should be avoided. Gassing with oxygen-enriched air is possible in so far as the high added value in contrast to yeast production is economically acceptable. During this phase, genes for product formation are not active.

Only when high cell densities are achieved, is production started by the induction of the respective genes in the third phase. Induction can be achieved by chemical signals (IPTG) or heat induction. In this case, chaperon in the form of heat-shock proteins are produced by the cells to prevent misfolding of proteins. The gene of interest has the same promotor as the chaperones. Overexpression of the recombinant proteins is a high metabolic burden for the cells. All carbon and energy are allotted to the product. Under these stress conditions, acetate is produced again as shown in the measurements in Fig. 7.14. This makes it clear that simultaneous growth and production is not a matter of choice. Low growth rates, before and during the induction phase, further reduces incomplete translation and incorrect folding.

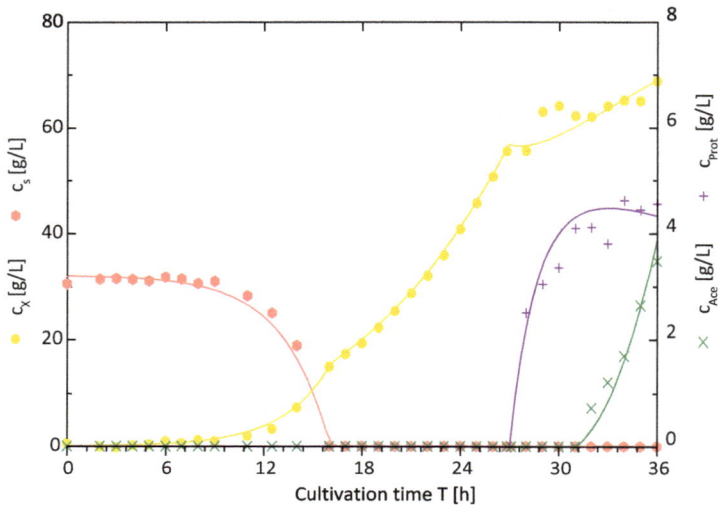

Fig. 7.14: Measurements and simulations of a high-density cultivation. Addition of inductor triggers the third phase, which is the production phase.

From HDC, we can learn two principles. The first one is to decouple physiological states into different phases of the process. Here, these states are high growth rate but acetate formation, lower growth rate but high cell density, and product formation without growth but good product quality. The second principle is to measure signals of the cells and to apply physiological signals to the cells for induction.

7.4.3 Production of pharma proteins with *Kamagataella pfaffii* – employment of a high-performance factory

Kamagataella pfaffii, formerly known and still sometimes termed as *Pichia pastoris*, is a methylotrophic yeast. It exhibits high growth rates and reaches high cell densities. It is able to grow on a simple, inexpensive medium, which is free from all compounds like proteins that could potentially interfere with production or product separation. *K. pfaffii* has been classified as GRAS by the American Food and Drug Administration (FDA), where one point is that no pyrogens or lysogen viruses are known. Furthermore, it shows high levels of secretion. So, it is an attractive host for the expression of recombinant pharmaceutical proteins. The yeast possesses two different alcohol oxidases (AOX), from which AOX1 has one of the strongest natural promotors. AOX1 is deleted and replaced by the recombinant gene sequence. Carbon uptake during growth and production relies on

the much weaker AOX2. Excretion is evoked by a fusion protein, e.g., the α-factor prepro-signal sequence.

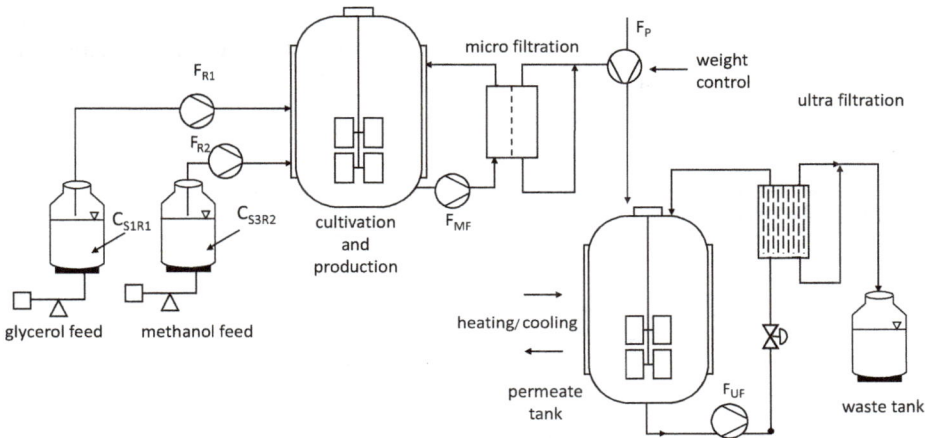

Fig. 7.15: Flow sheet of an experimental pilot plant for recombinant protein production with *Kamagataella pfaffii*.

The process requires two substrates, as visible in the process scheme in Fig. 7.15. During an initial batch phase on glycerol, the cells can grow with their maximum specific growth rate (approximately 0.15 h^{-1}). In the second phase, a glycerol-limited fed-batch phase is started, as in high cell density. Here, the biomass reaches predefined values of >50 g/L, while protein production is repressed. Then, the production phase is initiated. Methanol feeding leads to induction of the strong promotor and expression of the desired recombinant protein, which is finally secreted into the medium. Another advantage of *P. pastoris* is the tolerance for a high pH range, which allows pH adjustment to values favoring protein stability. For the sake of stability, a continuous separation of the proteins by filtration and cell retention can be foreseen.

7.5 Conclusions for fed-batch processes – dealing with strengths and weaknesses

The first idea of applying fed-batch feeding policies was to avoid initial substrate inhibition, overshoot metabolism, and the other disadvantages of batch processes. Media with high initial substrate concentrations can be applied without dissolution. Even if the main substrate is not applied in a fed-batch process, it can be applied for co-substrates. Ammonia is toxic only in comparatively low concentrations and is applied in fed-batch mode to reach high biomass concentration. The same holds for phosphate to prevent precipitation or intracellular accumulation. Furthermore, the fed-batch process turns out to be a flexible means of controlling cell physiology (e.g., age and cell composition) and designing processes in different phases, especially the growth and production phase. Therefore, fed-batch processes are widely applied in different kinds of (pharmaceutical) processes due to their flexibility and strict physiological control. Some of the pros and cons are collected in Tab. 7.2.

Tab. 7.2: Pros and cons of fed-batch processes.

Physiological aspects		Technical aspects	
Growth rate controlled	+++	Control necessary	–
No substrate inhibition	++	Average productivity	+
No overshoot metabolism	++	Average yield	++
Application to co-substrates	+	Changing volume	–
High substrate concentration	++	High technical maintenance	–
Product induction	+++	Batch harvesting	+/–
		Flexible process design	++

Technical problems include changing volume during the process and low productivity in the initial phase. This can be avoided by a preceding batch phase or by choosing high substrate concentrations. The ideal case would be to employ pure substrates with only little water, which may be gaseous (methane), fluid (methanol, glycerol), or solid. This last option is of course hardly practicable. High technical maintenance demand and necessity of a good measurement and control mechanism could be a hurdle to be managed, based on existing technology and experiences. Batch-wise harvesting can be a tribute to batch-wise downstream technologies and especially allows for charge-wise quality control in the pharmaceutical industry.

7.6 Photobioreactors – the interface between algae and sun

Photobioreactors are the centerpiece of each photo-biotechnological process. They can be artificially illuminated or as outdoor reactors, illuminated by the sun. In any case, the physiological requirements of microalgae define the construction concepts of photobioreactors. So, the work of an engineer designing a photoreactor resembles the work of an architect planning a house for a family. The first step is to ask the family what they need, and a second step is considering the environmental conditions and make compromises. We already collected the necessary information, which are kinetics for the microalgal needs and light conditions at different sides. The reactor has to transform these conditions into suitable conditions inside the reactor.

In terms of hydrodynamics, a photobioreactor represents a four-phase system, with the liquid phase providing water and nutrients. The gaseous phase supplies carbon dioxide and removes excess oxygen from the system. Eventually, the solid phase is given by the cells. The fourth interacting phase is the superimposed light radiation field.

7.6.1 Starting from natural conditions – lakes and open ponds

Microalgae live naturally in lakes or in the sea but also in other wet environments, and are adapted to these conditions. So, a start is to mimic these natural environments; of course, improving some factors for the algae in view of better productivity. *Arthrospira* (brand name Spirulina) is produced, e.g., in volcano lakes in Myanmar or soda lakes in China. The high pH value protects the culture from predators or competing microalgae. Harvesting is performed simply by filter nets, which is possible, thanks to the filamentous growth of these cyanobacteria. Cost is low for this extensive approach and so is productivity. Reasons for these low values are bad mass transfer for CO_2 and suitable nutrient supply. Other problems are seasonal fluctuations and last but not least, lack of global availability.

To tackle these problems, open ponds are constructed, especially for better mixing and CO_2-supply and, in general, to maintain better and constant environmental conditions. These are usually constructed as recirculation ponds and are recirculated by paddle wheels. Such "Raceway Ponds" (Fig. 7.16) are available in different configurations. While a single pond can be 100 m long, production plants with many ponds operated in parallel can cover several ha. Actually, the by far greatest share of microalgae production comes from raceway ponds.

The mean flow velocity v_f (m/s) follows from the need to prevent sedimentation and to mix the lower with the higher water layers. In practice, often 0.3 m/s is adjusted. The necessary power input as a contribution to costs is calculated as

$$P_{W,pond} = 1.59 \cdot A_G \cdot \rho_L \cdot g \cdot u_f^3 \cdot f_M^2 \cdot d_h^{-\frac{1}{3}} \tag{7.15}$$

This includes the ground area A_G (m²), the density of the medium ρ_L, which is practically 1 (kg/L), the gravitational acceleration g (m/s²), and the Manning factor f_M (s/m$^{1/3}$) describing the friction at the walls, which depends on the roughness, which is between 0.01 and 0.016. The hydraulic diameter d_h for the given geometry is $4 \cdot w \cdot h/(w + 2 \cdot h)$. The mechanical energy input calculated in this way is with 0.5 W/m² or 1 W/m³, which is relatively low. The energetic efficiency of the paddle wheel (Fig. 7.17) itself can make up another 50%; the solution is to dip the blades dipping vertically into the water.

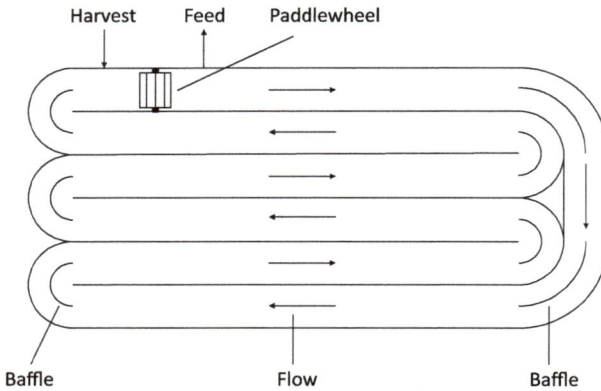

Fig. 7.16: Ground plan of a typical raceway pond, the most important gauges and peripheral devices are delineated.

Fig. 7.17: Picture of a raceway pond; the paddle wheel is a CAD drawing showing that the paddles reach nearly the ground to care for a uniform flow over height and to prevent sedimentation (© a4f).

Now, we compile some pros and cons of these "open reactors." Areal productivity P_A is with 10 g·m^{-2}/day ≈ 35 t/ha·year, which is relatively low but at least remarkably higher than what one gets from terrestrial plants. CO_2 is applied via bubbles behind the paddle wheel but will evaporate to a large portion from the free surface to the atmosphere. Also, water evaporates (up to 10 mm/day of water column) and has to be replaced, preferably by sea water, not to mention the necessary pumping energy. The salt content will rise but to an acceptable level, e.g., when the pond is operated near the Mediterranean Sea. In some desert areas, ground water is used, which is of course against the sustainability idea. At last, water evaporation helps in keeping the temperature quite constant, rising not more than typically 10 °C above the ambient air temperature. Evaporation can be diminished by spanning plastic foils over the then so-called "covered pond." Contamination by predators (daphnia, paramecium), fungi, or competing microalgae is an important issue.

While large ponds are around 30 cm deep (more than optimal) and are made simply by gasketed lacunae, modern concepts employ stainless steel trays that are only 10 cm deep mounted in green houses. A special reactor type, already known for some time but attracting new attention, is the "inclined surface reactor" (Fig. 7.18). It is based on raceway ponds but with an inclined tray, where recirculation is managed not by paddle wheels but by a volumetric pump. The gravity driven suspension film is in this way only a few mm thick. What that is good for we will learn in the next paragraph.

Fig. 7.18: Inclined surface reactor; the algae suspension flows down into a collection channel for gas exchange (© TU Munich).

7.6.2 Flat plate reactors – the bubble column in an "enlightened" guise

The next step is to replace all open surfaces by transparent enclosures to gain a "closed photobioreactor" and analyze which additional design features are possible, compared to the open reactors. The transparent walls could be made of glass, PVC, PE, or PMMA. The first design aspect is to let in as much light as possible into a given reactor volume V_R, meaning that the reactor-surface-to-reactor-volume ratio, SVR = A_R/V_R, should be as high as possible. For a rectangular geometry, counting both sides, yields:

$$\text{SVR} = \frac{2 \cdot L_R \cdot H_R}{L_R \cdot H_R \cdot D_R} = \frac{2}{D_R} \tag{7.16}$$

For e.g., $D_R = 2$ cm SVR = 100 m^{-1}. For a pond of 25 cm depth, the value is only 4 m^{-1}. This is equivalent to making one dimension, referred to here as thickness D_R, as small as possible. The idea has its equivalence also in eq. (7.17). The result is the "flat plate" reactor (Fig. 7.19). During the process, biomass concentration can be kept so high that the ambient light impinging on the reactor and being distributed inside the reactor serves all algae cells with an optimal amount of photons, e.g., 50 µE/g/s.

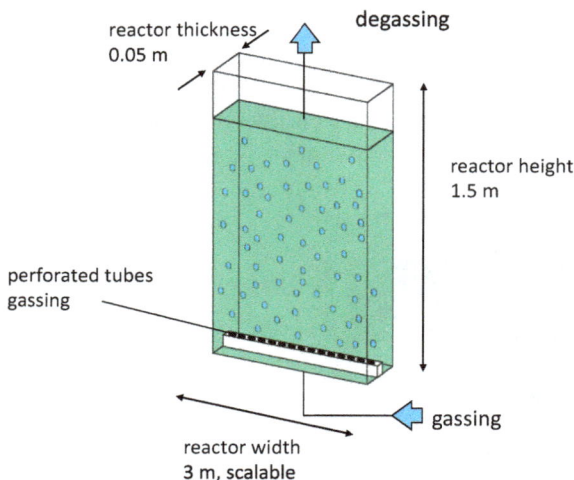

Fig. 7.19: The principle of a flat-plate photobioreactor with peripheral devices. Mutual shading contributes to light dilution (©KIT).

To mount such a flat plate horizontally on the ground to make a water body like an open pond is not a good idea. The direct sunlight is much stronger than what the algae need, at least in the upper layer. Energy loss by NPQ or even light inhibition will be the result, which is an additional problem in open ponds. The second aspect is to dilute the sunlight to values near the linear range of the PI-curve. This "light dilution" can be obtained by mounting the plates vertically, as shown in Fig. 7.20. Decisive for overall productivity is the vertical light component $I_{hv,horiz}$ (global horizontal irradiance), but physiologically active and penetrating the reactor is only the component $I_{hv,norm}$, perpendicular (normal) to the reactor surface. The light balance is therefore $I_{hv,horiz} \cdot A_G = I_{hv,norm} \cdot A_R$.

To summarize this second design aspect in a formula, it can be stated that the reactor surface to reactor footprint (ground) area $SFR = A_R/A_G$ should be as high as possible. For a flat-plate reactor installation (Fig. 8.18), both transparent sides count, making:

$$SFR = \frac{2 \cdot H_R \cdot L_R}{L_R \cdot d_R} = \frac{2 \cdot H_R}{d_R} \quad (7.17)$$

Here, d_R is the distance between two reactor panels. Mutual shading of reactor panels and reflection on the surface contribute to light dilution. In technical installations, a factor of up to 10 is realistic. A pond has a value of only 1. Also, vascular plants like

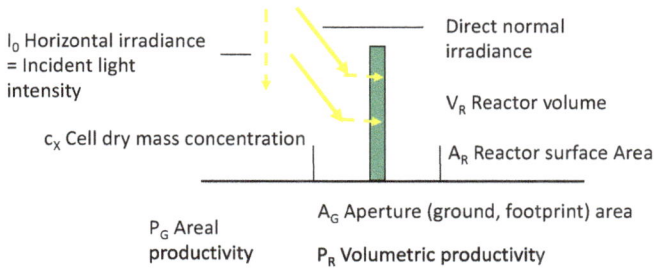

I_0 Horizontal irradiance = Incident light intensity

Direct normal irradiance

V_R Reactor volume

c_X Cell dry mass concentration

A_R Reactor surface Area

A_G Aperture (ground, footprint) area

P_G Areal productivity

P_R Volumetric productivity

Fig. 7.20: The principle of light dilution of ambient light on a large reactor surface and different characteristic values for design and evaluation of photobioreactors.

trees use this principle. Leaves are very thin but span a high total surface of leaves (e.g. 2,500 m²) over the ground area, shaded by the tree (e.g., 100 m²). A lime tree can outnumber a value of SFR = 20 (in botany, only one side of a leaf is counted). The two aspects shown above also mean that the fluid volume per ground area has to be low (10–50 L/m²). This is sometimes not understood intuitively as "the medium is the reaction volume" is a rule of thumb in reaction engineering. Here, the algae cells make up the real reaction volume.

Fig. 7.21: Arrangement of flat plates, the so-called "green wall panel," where plastic bags are mounted between wire fences for stabilization.

Mass and mechanical energy transfer in flat-plate reactors is ensured by bubble aeration from a sparger, similar to what happens in bubble columns. As gas transfer is essentially lower, e.g., by a factor of 100 than in heterotrophic processes (lower μ, lower c_X), flat plates are operated in the laminar regime without bubble dispersion and coalescence. Typical aeration rates are 0.05–0.15 vvm. This is with, e.g., 50 W/m^3 or 1 W/m^2 – a considerable part of the chemical energy is produced as algae biomass. A careful control of the aeration rate according to the needs, e.g., lowering at low light conditions or during the night, can reduce the costs for energy transfer. Fluctuating CO_2 consumption due to light fluctuations is usually compensated by controlling the CO_2 fraction in the gas.

To reduce pressure loss and therefore energy expenditure, the height of the reactor can be lowered and the number of plates per ground area increased accordingly. This also reduces material expenditure as the lower hydrostatic pressure at the bottom of the reactors allows for the application of thinner reactor materials. This reflects a current trend in photobioreactor design.

7.6.3 Tubular reactors

To further increase SVR and SFR, employment of transparent tubes is a geometrical means (Fig. 7.22). They are installed horizontally, with the length axis being the length axis of the reactor. The length dimension L_R can be several hundred meters. Several tubes, n_{tube}, are mounted above each other and are connected by bends in this way, forming the so-called "fences." Tubular reactors belong to the largest closed reactors.

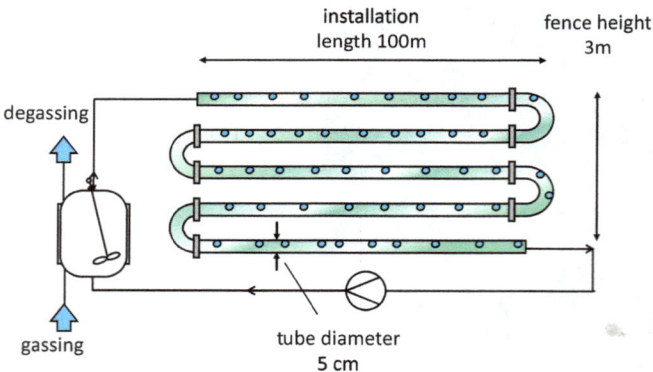

Fig. 7.22: Principle of tubular reactors with peripheral devices.

Light capturing, given as SVR, results in

$$SVR = \frac{n_{tube} \cdot L_R \cdot 2 \cdot \pi \cdot R_R}{n_{tube} \cdot L_R \cdot \pi \cdot R_R^2} = \frac{2}{R_R} \tag{7.18}$$

This is quite similar to the flat-plate reactor. Especially for light distribution, there is an advantage. Light can penetrate the tube in two perpendicular (horizontal and vertical) directions. This leads to a homogeneous light distribution as exponential light attenuation ($I(r) \sim e^{R-r}$) is partially compensated by the "focus effect." This term describes that light beams approaching the suspension in the radial direction, coming closer and closer to each other. This leads to a virtually higher local irradiance, as the photons penetrate a decreasing circumferential surface area closer to the axis ($A(r) \sim 1/r$).

Light dilution over the surface is another big advantage as interpretation of eq. (8.16) shows:

$$SFR = \frac{n_{tube} \cdot L_R \cdot 2 \cdot \pi \cdot R_R}{L_R \cdot d_R} = \frac{n_{tube} \cdot 2 \cdot \pi \cdot R_R}{d_R} \qquad (7.19)$$

The number of tubes corresponds to the height of a flat panel, but the circumference term ($\pi \cdot R_R$) increases the value remarkably.

Tubular reactors are operated as plug-flow reactors, with a flow velocity between 0.2 and 0.5 m/s. Cycle times are of course much smaller than specific growth rates; so with respect to biomass, they behave like ideally mixed tank reactors. The flow can be driven by a centrifugal pump or via the airlift principle, where the reactor itself is the downcomer.

To operate tubular reactors, an external pressure difference must be applied. This is done either by centrifugal pumps or via the airlift principle, where the reactor itself is the downcomer. The pressure difference can be calculated according to the Hagen-Poiseuille equation:

$$\Delta p = \frac{q_L \cdot L_R \cdot 8 \cdot \eta_{med}}{4 \cdot R_R^4} = \frac{v_L \cdot L_R \cdot 8 \cdot \eta_{med}}{R_R^2} \quad \text{with} \quad q_L = \pi \cdot v_L \cdot R_R^2 \qquad (7.20)$$

Here, η_{med} is the dynamic viscosity of the medium ($\eta_{H2O} = 0.9$ Pa $\cdot$ s at 25 °C) and v_L the average medium velocity. What looks a bit frightening is that the radius in the denominator scales with a power of 4, which is directly opposite to the intention of making it small. But to drive the volume flow through a smaller tube, it is necessary to keep the velocity constant. This reduces the problem to a power of 2. Nevertheless, the high pressure difference has an impact on the mechanical stability of the tubes and also on the microalgae experiencing a pressure jump during each cycle.

The same holds for the power input:

$$P_{tube} = \frac{8 \cdot \eta_{med} \cdot L_R \cdot q_L^2}{\pi \cdot R_R^4} = 8 \cdot \eta_{med} \cdot L_R \cdot \pi \cdot v_L^2 \quad \text{with} \quad q_L = \pi \cdot v_L \cdot R_R^2 \qquad (7.21)$$

At constant liquid velocity, power input is constant for different tube radius, but volumetric power input increases with decreasing radii. In practice, tube diameters of 4 cm are commonly applied, leading to a power input of more than 100 W/m^3 (Fig. 7.23).

Gas transfer is foreseen by agitated vessels being coupled to the liquid cycle. During their way through the reactor, the microalgae take up CO_2 and produce O_2. This leads to pH, pCO_2, and pO_2 gradients. When all dissolved CO_2 is used up, the medium

Fig. 7.23: Large-scale installation of tubular reactors in parallel double fences, making bending easier (© a4f).

must again reach the gas exchange vessel, at the latest. This is usually after about 2 min. Besides the pressure drop along the longitudinal axis, this limits the lengths of tubular reactors. A little help makes the gas bubbles drift along the tube. This is the price we have to pay, taking two spatial coordinates for light transfer.

7.6.4 New developments and ideas

Flat plates can be pronounced as tall bags or long bags submerged in a "waterbed" (may also be open seas) to increase the mechanical stability and reduce the impact of heat accumulation. Better flow patterns or light-conducting structures for better light distribution are other trends. Even the horizontal reactor (KIT) comes back but with corrugated surfaces to cope with the basic parameters mentioned above. The sun energy ends up nearly completely as heat inside the suspension. Besides contamination, this last item is one of the main problems in outdoor cultivation. Modern concepts go away from these basic types. Horizontally arranged reactors would offer advantages like low pressure and no need for racks. However, light has to be guided into the suspension, e.g., by corrugated surfaces or light conductors. Further, such concepts allow the collection of IR outside the photoactive volume, thus decreasing the heat uptake and the further application of "transparent" photovoltaics.

7.7 Questions and suggestions

1. Read more about the history of Fleming to understand the dynamics of science and process de-
    velopment. Find other examples where "lucky findings," technology push, and market pull are
    driving forces of our endeavors.
2. Does the strong maximum of the penicillin production kinetics possibly have a biological
    meaning?
3. Sometimes, constant feeding is chosen (with limitations in OTR, pumping rate, gaseous substrate,
    and light). How do the biomass and the specific growth rate develop?
4. What is the theoretically maximum final biomass concentration that can be reached in a fed
    batch? What happens during several cycles of mini-harvest and filling-up of the harvested volume
    with fresh medium?
5. Calculate r_X, r_S, $y_{X,S}$, $P_{V,X}$ from the measured data in Fig. 7.2: $t_0 = 0.0$ h, $c_{X,0} = 5.0$ g/L, $c_{S,0} = 2.5$ g/L,
    $V_0 = 1.5$ L, $t_{end} = 24$ h, $c_{X,end} = 45.4$ g/L, $c_{S,end} = 2.5$ g/L, $V_{end} = 15.0$ L. The feeding rate $q_{S,f}$ is given as
    $0.0225 \cdot \exp(0.2 \cdot t)$ L/h, $c_{S,f} = 100$ g/L.

Chapter 8
Bioseparation – the art of selectively directing particles and molecules

After the actual cultivation, the next task of bioprocess engineering is to separate the final product from the residual suspension. This typically requires a set of several consecutive "bioseparation" steps. The whole process chain is called "downstream processing," like following a river along several cascades. Consequently, preparation of a process in the lab is referred to as "upstream processing" (Section 1.4). Specific problems result from the biological nature of the complex material matrix to be treated. An idealized look of the suspension is presented in Fig. 8.1.

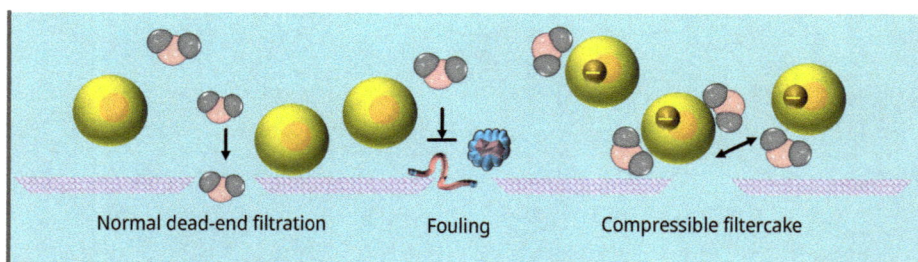

Fig. 8.1: The initial situation, after cultivation; beside the cells, macromolecular substances may be proteins or polysaccharides that are present, partially intended, partially not. Low molecular substances could be residuals from the initial medium or have been produced by the cells. Solution in water is mainly mediated by the dipole character of the water molecule, salt ions, and surface charges of the organic matter.

Whole cells are initially the most visible fraction of the suspension. They may be the product themselves or they may contain an intracellular product. In the third case, the cell product was excreted into the medium during cultivation. Macromolecules, especially proteins and polysaccharides, are typical representatives. The cells usually excrete not only the target product, but also other metabolic products that must be removed later. Smaller molecules can be monosaccharides or bioactive substances such as penicillin or cortisone. Even in this group of substances, not all molecules are products.

The problem now is to separate similar particles or molecules, cleanly. Furthermore, the individual components can interact with each other. The surface charges are often responsible for this behavior. Polysaccharides, for example, show a strong bond to water. This results in a strong increase in viscosity with negative consequences for some processes. Furthermore, polysaccharides attach themselves to proteins to such an extent that they are almost impossible to separate. In the area of extracel-

https://doi.org/10.1515/9783110773354-008

lular proteins, it is to be expected that the producing cells also excrete proteases that destroy the product. The list could be continued indefinitely. The first task of the development engineer is therefore to identify such problems. Speed, as well as temperature and pH control, are useful tools in avoiding them.

At the lowest level, bioseparation means that forces are applied to specifically act on the target product, causing it to spatially change its direction of motion. This finally enables segregation and separation. Such forces can be gravitational forces and steric hindrance, electrostatic or hydrodynamic forces, or directed diffusion or molecular interactions. In each case, properties must be defined that distinguish the target molecule or particle from all others present in the suspension. Specific problems result from the biological nature of the substance matrix to be treated. Consequently, downstream processing in bioprocess engineering requires specifically elaborated operation steps. However, it also shows that biology offers specific solutions. The "art" of process development now consists in creatively "playing" with these possibilities. See in the picture how many mechanical and thermal steps a downstream process can consist of. In the following chapter, some typical bioseparation steps are introduced in more detail. Also, problems are pointed out and possible solutions are mentioned.

8.1 Identification of separation tasks – things to do around the whole process

Firstly, different separation tasks are identified (see Fig. 8.2). In any case, the cells must be separated from the suspension. This step is commonly referred to as cell harvesting. In fact, the specific realization may depend on whether the cells contain the product or whether the medium contains the valuable substance. This dual view also applies to the other steps. In the case of a substance dissolved in the medium, a further step can separate small (e.g., sugar) and large molecules (e.g., proteins). Again, one of the two types of molecules can alternatively be the actual product. This pattern can be employed several times in the production lines, with several particle or molecular sizes to be separated and different unit operations employed. During the "downstream," individual "river" sections are defined. This definition is known as the "RIPP" model, according to the typical character of the consecutive separation tasks – removal, isolation, purification, and polishing.

The first step "R" means cell recovery or cell removal, depending on whether the cells are the product. This biomass harvesting step includes the main dewatering step. In this first step, large volumes are processed. The concentration of the valuable product, on the other hand, is often low. Therefore, attention must be paid to the lowest possible energy input and good scalability. However, harvesting should be fast, since prior cooling to suppress ongoing reactions is expensive. Typical methods employed are filtration and centrifugation. The isolation step "I" is intended to separate the target product as well as perhaps the interfering molecules or particles. Cell disruption also

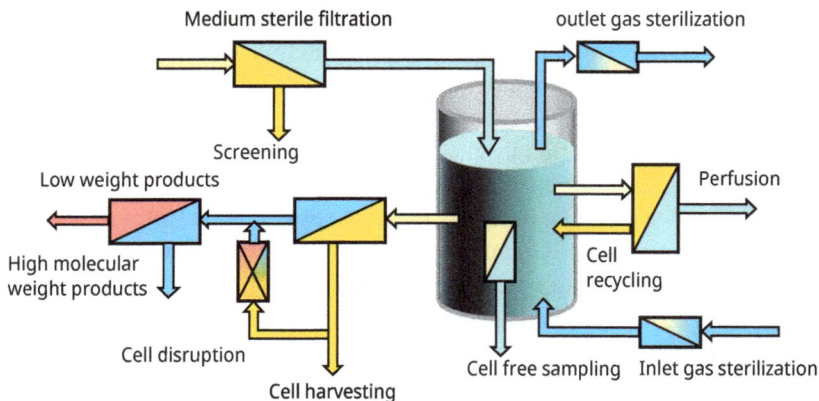

Fig. 8.2: A dual view of the different bioseparation steps during a bioprocess; the process icons mark the specific separation steps where the diagonal line can be regarded as filtration membrane. Water is shown in light green, air in blue, and other colors indicate the respective concentrated material.

belongs to this group. This is accompanied by a further concentration and thus a further reduction in the volume of the dispersion. Attention must also be paid to the fact that if concentration is carried out without prior separation of any contaminants, undesirable side reactions may increase, according to the mass action law. This can be a problem, especially when working with disrupted cell material. Besides special filtration methods, extraction plays an important role in isolation. Also, precipitation may be an option. Purification – the first "P" – is especially important in bioprocesses when the product is used in food or pharmaceutical applications. These are in fact the standard cases. Even an extreme low contamination with a misfolded protein could trigger an allergic reaction. On the one hand, the material flow to be processed is now lower than at the beginning of the chain, but the processes must be much more specific to the product molecule. Therefore, methods based on specific adsorption to a matrix, such as electrophoresis or chromatography, are used. However, purely physical, or simple chemical criteria are now often no longer considered as characteristic properties. The decisive factor is exclusively the efficacy, which then has to be demonstrated by binding to a specific receptor. For this purpose, affinity chromatography can be considered, which, for example, shows analogues of the receptors as binding sites on the matrix. Polishing as the second "P" removes the last impurities and brings the isolated substance in a suitable configuration to be sold and applied. This last action is usually referred to as formulation. Last residues of the solvents have to be removed, e.g., by drying, or mechanical properties have to be adjusted by crystallization. Incidentally, crystallization is one of the most specific operations because foreign molecules in the crystal are excluded.

While harvesting is the most prominent unit operation, there are many other separation tasks to be solved in a bioprocess. Starting with medium preparation, a possible bioseparation step is the removal of contaminants by sterile filtration. That holds, especially for vitamins or trace element solutions, which cannot be heated without problems. The filtrate is then the final product to be used in medium preparation. The other way around, there may be an ecological task to find a few bacteria in a lake. Then, these cells as residue are the valuable product of filtration, while the huge amount of filtrate is discharged. The next operation step is the cultivation. Here, it may be necessary to take cell-free samples. This is to avoid changing concentrations, which otherwise could be metabolized by the cells even during the short time, until the sample is frozen. Larger amounts of cell-free medium have to be removed during continuous cultivation (see Chapter 9). Cell recycling or cell retention are measures to keep the cell concentration high and to change the medium. This is called perfusion culture. It cares either for supplying fresh substrates or to remove inhibiting products.

8.2 Filtration – a smooth but not always simple approach

Process engineering processes were invented by mankind in parallel to mechanical things like hand axes or spearheads. However, since they are not so well representable, they are less present in public perception, e.g., for filtration. The availability of clean water has always been critical to human survival. Simple filters, dating back thousands of years, used plant materials. Porous ceramics and cloths can be traced back for centuries. Also, for the preparation of food, such methods were certainly used. In any case, a liquid loaded with (unwanted) solids is passed through a filter medium. Do not mix up this term with the chemical growth medium. The solid medium here has pores that allow the water to pass through but represent a steric hindrance for the particles. The particles – usually solids – accumulate on the filter medium, thus forming a solid layer. This slowly building layer forms a porous pile, called the filter cake; the process as such is dead-end filtration. Even today, filtration is a central step in many process engineering procedures, including biotechnology. For example, see Fig. 7.11.

Harvesting, and also of many other steps, belong to mechanical process engineering, where they are referred to as solid/liquid separation. Here, the cells represent the solid, but at the next stage, the solids can also be defined by the harvested enzymes. Of course, neither cells nor proteins are really solid. The most direct and simple operation for cell separation is filtration. The cells or particles, in general, are considered as spheres of a given diameter. A filter membrane contains pores with a much smaller diameter – keeping the particles from passing through the membrane. Slowly, more and more cells settle on the particle layer already there, forming by and by a so-called filter cake. This has also to be passed by the fluid, driven by a pressure difference, Δp (N/m^2), over the filter cake and the membrane. Particles experience a steric hindrance at the membrane, especially at the pores. The filtration process can be interpreted in terms of a force acting on the particles, but not on the fluid. This mechanical force is transmitted through the pile of the filter cake and the membrane. The basic linear approach is mathematically described is analogous to a current flow through an electric resistance. The reciprocal permeability of the filter cake itself is formally a resistance, $R_{f,cake}$ (m^{-2}), while the filtrate flow Q_{fluid} (m^3/s) can be analogous to the electrical current, driven by the pressure – resp. the voltage. This relation is known as Darcy's law (Henry Darcy, 1856):

$$Q_{liquid} = \frac{1}{R_{f,cake}} \cdot \frac{A_{cake} \cdot \Delta p}{d_{cake} \cdot \eta_{liquid}}$$

(8.1)

This equation includes three design parameters, namely the cross-sectional area, A_{cake} (m^2), and the current thickness, d_{cake} (m), of the cake as well as the viscosity, η_{liquid} (N/m$^2 \cdot$ s = Pa $\cdot$ s) of the liquid, being, e.g., the medium. The resistance can be referred to the cake thickness – indicated as the specific resistance, $r_{f,d}$ [] = R_{cake}/d_{cake}. As the thickness of the filter cake may be very small, especially in bioseparation, a mass specific filter cake resistance is defined as $r_{f,m}$ (m/kg) = $R_{f,cake} \cdot A_{cake}/m_{cake}$. The filter cake mass can be estimated from the filtrate mass as $m_{cake} = m_{filtrate} \cdot c_{solids}$. This approach assumes that the particles in the original suspension take only a small part of the volume and that they settle quantitatively on the filter membrane. This mass specific resistance is independent of the filtration area and the current height or mass of the filter cake. Consequently, it describes the intrinsic property of the cake, and it is convenient to be extrapolated to comparable situations. This mass specific resistance basically depends on the particle size and shape as listed in Tab. 8.1. However, even for a heap of monodispersed spheres, an a priori calculation of the filter cake resistance is only roughly possible. For large particles, the Hagen-Poiseuille law (see Chapter 7.6.3) gives a rough estimate, describing a laminar flow through small capillaries.

Tab. 8.1: Types of filtration, ordered by particle size.

Approach	Microfiltration	Ultrafiltration	Nanofiltration	Reverse osmosis
Pore size (μm)	0.1–10	0.001–0.02	0.0005–0.002	0.0001–0.001
Particle size (μm)	>0.1	>0.01	>0.001	>0.0001
Transmembrane pressure (bar)	0.1–2.0	0.1–5	2–20	10–100
Typical particle	Cells	Macromolecules, proteins	Small molecules, ions,	Small molecules

In the literature and in practice, several devices are described with which filtration can be performed. These differ mainly in the way in which the membrane is stretched and how the transmembrane pressure is generated. The simplest variant is the filter press. Two filter cloths are clamped to a frame. Both frames are mounted opposite to each other, and the suspension is fed between them. While the filtrate passes through the cloths to the outside, the filter cake is built up between them. The pressure required for this can be built up via the suspension with pumps or by squeezing the plates together, may be in a larger stack. After that, the filter cakes are removed, optionally with the help of pressed air. Typical pressures in this discontinuous process are 10–20 bar. A more elegant and very common device is the rotary vacuum-drum filter, as depicted in Fig. 8.3.

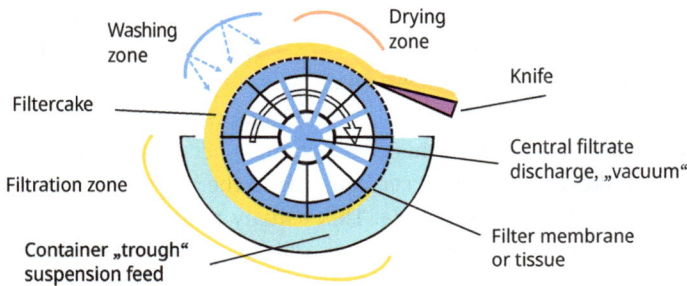

Fig. 8.3: Sketch of a rotary vacuum-drum filter; The suspension is given in light blue, the filtrate in vacuum in blue, and the filter cake in dark yellow.

This apparatus is basically a rotating drum on which a filter cloth or membrane is stretched. The transmembrane pressure is realized by applying a vacuum (under pressure) from below the cloth. To prevent the entire drum from being under vacuum, there are separate chambers under each membrane segment. Following the pressure difference, the filtrate is sucked through the membrane and discharged via channels and the central axis. The function can best be understood by following the course of the biomass. The filter cloth dips into the container with the suspension, shown on the right in the picture. The prevailing vacuum sucks the suspension through the submerged filter section. While the filtrate is being drawn off at the end of the filtration zone, the solids, i.e. the microorganisms, remain on the filter cloth. In this way, the filter cake slowly builds up. When a segment leaves the vessel again, shown on the left in the picture, the filter cake reaches its maximum thickness. In the washing zone, the filter cake can be washed using spray nozzles. In the drying zone, vacuum removes the residual liquid from the pores and from the gussets of the pile. A blade, also called a knife or scraper, then lifts the filter cake off the filter cloth, partially with the assistance of compressed air. This completes one cycle.

The filtration process sounds quite easy, but in the case of bioseparation, there are some pitfalls. Firstly, the biological "particles," meaning cells or macromolecules, are much smaller than particles in classic filtration. The capillaries between the particles through which the medium has to flow become smaller and smaller with smaller particle diameter. Further, the suspension may contain additional material, which can block the pores of the membrane, as shown in the middle of Fig. 8.3. This process is quite common and is described by the term "fouling" or "filter clogging".

According to Darcy's law, the filtrate flow should increase with an increase in pressure. In biological suspensions, however, only a sub-proportional increase in filtrate flow or even no increase at all is observed. The specific filter cake resistance therefore increases. One then speaks of "compressible" filter cakes. Actually, a physical compression cannot directly be observed, as the filter-cakes in bioseparation often exhibit a thickness of less than 1 mm. Apparently, the capillaries assume an increasingly smaller diameter. This can be caused by several reasons. First, small particles can slip into the spaces between the large particles due to the pressure build-up. This may be true, for example, for cell debris between whole cells. The cells or even macromolecules could deform due to the pressure and then become closer to each other. However, in the example of yeast, see below, this does not apply because the cells are practically non-deformable, like an inflated soccer ball, due to the high turgor. However, the surface charge plays an important role. This usually takes on a negative value and keeps the cells and the other components at a distance. Furthermore, the dipoles of the water molecules are aligned, which leads to a lower mobility and thus to a higher viscosity of the water, i.e., the continuous phase. However, increased pressure can partially overcome this electrostatic repulsion, thus reducing the radius of the capillaries.

8.3 Centrifugation – a powerful and effective solid-liquid separation step

Employment of gravity is an obvious idea to separate cells from the suspension. This is known as sedimentation. To calculate the sedimentation velocity, two different forces have to be respected. The mass of a cell m_{cell} induces a gravity force F_{grav}. In the opposite direction, buoyancy induces a force F_{buoy}, depending on the volume of the cell v_{cell} and the density of the medium ρ_{medium} (kg/L). The total force to move a particle in a liquid is then

$$F_{tot} = \left(\rho_{part} - \rho_{liquid}\right) \cdot g \cdot v_{part} = \left(\rho_{part} - \rho_{liquid}\right) \cdot g \cdot \frac{4}{3} \cdot \pi \cdot r_{part}^3 \tag{8.2}$$

Often, cells have only a slightly higher density ρ_{cell}, e.g., 1.1 kg/L than that of the surrounding medium, appr. 1.0 kg/L. Actually, this is the critical value for all separation steps based on gravity. In many cases, the density difference approaches zero. This may happen when the cells contain many lipids or when the medium has high substrate or salt concentrations. The radius r_{cell} of a cell, e.g., 2 µm, is much smaller than that of other particles in the technical solid/liquid separation steps. This value enters the formula with the third power and is therefore also critical.

After the cell – i.e., the particle – has started to move, it reaches a constant final velocity v_{part}. The reason is an increasing resistance of the fluid with increasing velocity. Stokes law gives the relation for this drag force F_{drag} (N) on a small sphere as

$$F_{drag} = 6 \cdot \pi \cdot r_{part} \cdot \eta_{fluid} \cdot v_{part} \tag{8.3}$$

Compare this equation with the drag force calculated in eq. (5.2). Setting the sum of all forces $F_{grav} - F_{drag} - F_{buoy} = 0$ for constant conditions delivers the sedimentation velocity, known by the Stokes equation:

$$v_{part} = \frac{2}{9} \cdot \frac{r_{part}^2 \cdot g \cdot \left(\rho_{part} - \rho_{liquid}\right)}{\eta_{fluid}} \tag{8.4}$$

This also depends also the viscosity η_{liquid} [N/m$^2 \cdot$ s = Pa $\cdot$ s] of the liquid, where the numerical value for water is 1.01 mPa $\cdot$ s. The gravity of Earth can be given as g = 9.81 m/s. Calculating this velocity with the practical values given above shows a disappointing result, $v^{part} \approx 0.5$ µm/s. That means only 1.5 cm sedimentation can be observed within 8 h. That is of course not enough for a technical process, and is hardly applied without additional measures.

The next step in developing a more intensive process is to apply centrifugal forces, which are also based on the density differences but are not limited by the gravity of the Earth. Basically, the suspension and thus the particles are forced into a circular path in an apparatus. The inertia then leads to an apparent outward force, the centrifugal force:

$$F_{Cf} = m_{part} \cdot v_{circ}^2 / r_{circ} = m_{part} \cdot \omega_{circ}^2 \cdot r_{circ} \tag{8.5}$$

And consequently, the centrifugal acceleration:

$$a_{Cf} = r_{circ} \cdot \omega_{circ}^2 \tag{8.6}$$

Here, r_{circ} is the radius of the circle path, v_{circ} (m/s) is the path velocity and ω_{circ} (rad/s) is the angular velocity. It is important to note that the highest values of the centrifugal acceleration appear at the outer radius. The centrifugal acceleration is often normalized to the multiple z of g. This centrifugal number z can reach values of 10 × 10^3 in technical centrifuges for cell separation and for protein separation ($r \approx 5$ nm!) up 200 × 10^3 but not in large scale. In practical applications, often the rotation frequency n, revolutions per time (e.g., "rounds per min"), is given. It is not enough to give only this value in publications, because it cannot be translated to other devices without knowing the radius. The bad news is that the problem of the small density differences can be mitigated but not really solved because of the opposite lift force.

Small values of the sedimentation velocity make the development of bio-specific centrifugal devices with sedimentation path lengths of only a few mm necessary. The most frequently used device for cell separation is the disc stack separator, as shown in Fig. 8.4.

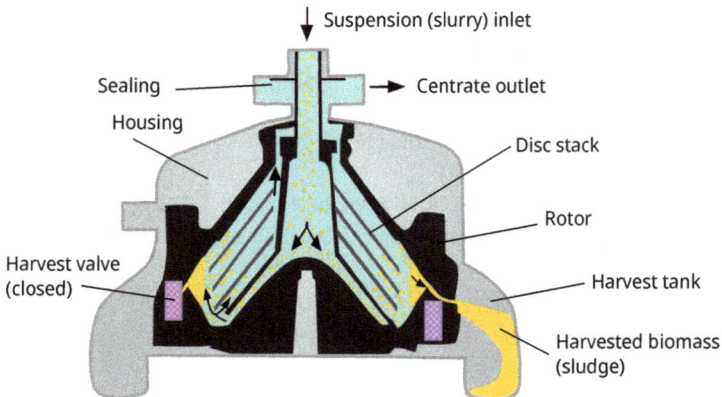

Fig. 8.4: Technical drawing of a disk stack separator; note that the left and the right side represent different radial cuts of the centrifuge; the discs, e.g., are fixed to the central axis only at some points to give room for the centrate to leave the machine.

Instead of a cylindrical beaker, the rotor contains a stack of several so-called conical discs in a few mm distance. The suspension is pumped into the operation volume between the discs, where centrifugal forces bring the cells in radial direction to the underside of the upper disc. Here, a film is formed, sliding along the disc to the collecting chamber. To avoid cell damage, the paste is slowly decelerated.

The suspension is pumped into the process chamber through the rotating inner axis. On the underside, it then flows outward, following the centrifugal force, and is further accelerated to the peripheral speed. On the outer circumference, the somewhat heavier cells are separated, and accumulate as slurry. Some particles follow the path of the suspension, back inward, but sediment to the nearest disc. There, on the downside, a film of cells collects, which can slide outward again and is deposited there. The now clarified centrate flows back inside and leaves the apparatus through the outlet. A seal (caulk) prevents the fresh suspension and the clarified liquid from flowing together. Incidentally, this seal is also the transition from a necessarily static inlet to the rotating central tube. For final harvesting, a valve can be opened while the machine is running, which discharges the slurry through a nozzle into the harvesting chamber. During this process, the biomass is slowed down again. This must be done carefully over a longer distance to avoid creating mechanical stress.

Another way to realize short sedimentation path lengths is the decanter centrifuge. This device is basically a horizontally oriented elongated centrifuge, as depicted in Fig. 8.5. The suspension ("slurry") is pumped into the centrifuge through a feeding tube along the central axis. There are nozzles in the feeding chamber through which the suspension is thrown outward to the inner wall of the rotating bowl. Here, the slurry is accelerated, and centrifugal forces begin to act. The cells settle and deposit at the inner wall of the bowl as a biomass film, while the clarified liquid forms a "lake".

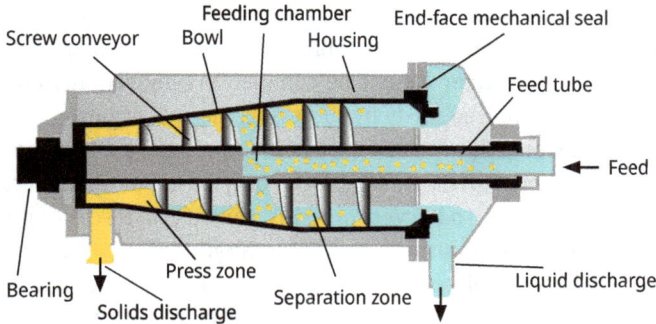

Fig. 8.5: Technical drawing of a decanter centrifuge – basically a horizontal centrifuge. The cake that collects on the inside of the outer wall is discharged by the screw. This moves at a low differential speed to the rotating outer shell.

The special feature of the decanter centrifuge is the way in which the product is further transported. Inside the rotor is a scroll. This screw conveyor rotates at a low differential speed to the bowl. This movement scrapes off the biomass and transports it toward the conical part. Here, it is further compacted, dewatered, and finally discharged. The clarified liquid flows in the opposite direction, separating further particles. When the liquid level reaches the height of the overflow, it flows toward the outlet. So, it is similar to decanting wine. The decanter centrifuge can robustly process large volumes of suspension at g values of up to 5,000.

Of course, separation processes are not always perfect. Small particles can, for example, slip through the filter or leave the centrifuge with the effluent, with a concentration c_1. So, we need a measure for the concentration. The separation efficiency η_{sep} describes the effectiveness of a separation process as a ratio of the amount of material separated in the system $c_0 - c_1$ (filter or centrifuge) to the amount of material to be separated entering the system, c_0:

$$\eta_{sep} = 100 \cdot \frac{c_0 - c_1}{c_0} \tag{8.7}$$

The separation efficiency is usually – as in the formula – given in percent. This value must be small, especially if no particles are allowed to get into the discharge. This task is called clarification. This is the case, for example, in wastewater technology or in cases when no genetically modified microorganisms are allowed to get into the environment in the first place. A small number of particles sometimes slips through the filter membrane in the beginning of the process. The growing filter cake later acts as its own membrane and the process is improved. In these cases, the first particle peak in the effluent will be discharged separately. Another criterion in dewatering is the residual moisture in the filter cake in which the gusset water can collect. The water may be a wet film over the particles or in particle pores as well as water bridges in the gussets between the particles. However, the residual moisture does not always have to be small. As an example, the product should still remain a pumpable paste or it can be marketed immediately as a suspension. Residual water cannot be completely removed anyhow only by mechanical means; it can be removed solely by successive thermal drying.

Macromolecules, like proteins or polysaccharides, can also be separated by mechanical means, e.g., by ultrafiltration. However, selectivity, only based on diameter, may be low; also, energy demand is high. To harvest proteins, centrifugal forces of up to 200,000 g would be necessary, while industrial scale centrifuges end up at about 15,000 g. The same holds for smaller molecules like bioactive substances, down to ions, to be separated by nanofiltration. Although, this is state-of-the-art, e.g., in seawater desalination, it becomes more and more complicated to set up a scalable process.

8.4 The idea of phase transitions – system changing as a separation concept

To separate small "particles," e.g., in the range of macromolecules or even smaller, other properties of these molecules related to thermodynamics are used. Molecules can be separated according to whether they are hydrophilic or hydrophobic and/or following their surface charge, both connected with their solubility or their ability to attach to solid phases. In any case, these so-called thermal separation processes provide two different material phases, either to fluid phases or one solid and one fluid phase, or even a gaseous phase. The simplest case is initially that the water changes its aggregate state from the medium (see Fig. 8.6). To force the various phase transitions, appropriate conditions must be set in the process.

Characteristic for thermal separation processes is the inclusion of a phase transition of either the product or the solvent. The process step, following cell or macromolecular separation, is often drying, as a good example. In this process, the residual water in the slurry passes from the liquid state to the gaseous state over the vaporization curve as vapor or air humidity, i.e., it undergoes a phase transition. Ideally, the product is finally dry. To achieve sufficiently short drying times, the vapor pressure of the residual water must be increased relative to the air in the process chamber. This can be done by reducing the pressure in the air or by increasing the temperature of the slurry. This possibility is naturally limited to temperatures of, e.g., max. 60 °C for bioproducts. Further, the wet product slurry must be molded, so that it exhibits a high surface-to-volume ratio. This allows the water inside the paste to diffuse to the surface where it evaporates into the surrounding volume. Microalgae, for example, are pressed in spaghetti-like threads along which warm air flows. In addition, an infrared heater can be mounted over a plate with the slurry. For yeast flocs or alginate bead production, the suspension is broken down into small droplets by spray heads, from which the water evaporates into the warm air flowing up. This process, being similar to a fluidized bed,

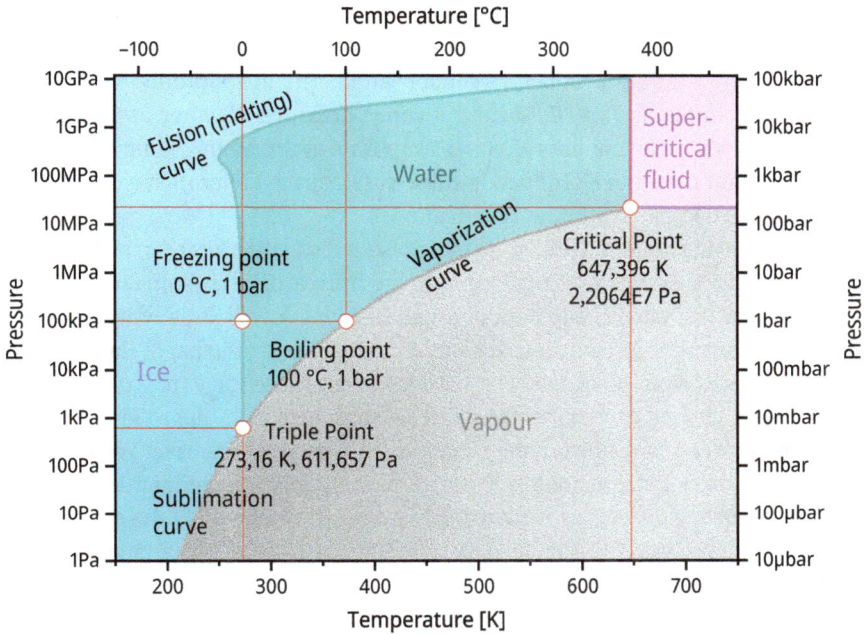

Fig. 8.6: Phase diagram of water; indicated are the names and conditions of phase transitions during thermal processing.

is called spray drying. However, the product sometimes suffers from heat damages on the surface of the granules formed and can experience loss in biological activity.

One problem during drying is that biochemical processes can continue during the process or can even by accelerated by warm temperatures and increasing concentration. Cells may, e.g., turn to anaerobic conditions inside the product or enzymes may start to work uncontrolled. Further, diluted salts may accumulate. To avoid such situations, freeze drying is suitable. Firstly, the initial material from the solid/liquid separation is cooled down below the freezing point at constant pressure. That stops any biochemical activity. Drying then takes place at lower pressures so that the water passes into the gas phase via the sublimation curve, resulting in the pure dry product. For small quantities in the laboratory, freeze drying is used, for example, to store small quantities of cells in strain banks. But there are many other examples that when drying is stopped and the outer water is removed, the intracellular water is kept to keep the cells alive.

In other cases, a product like ethanol or propanol, which is a liquid dissolved in another liquid, results in fermentation suspension. The distillation or method of separating liquid constituents requires heat application. Each liquid constituent has its own boiling point and volatility level. Increasing the temperature gradually brings the product first to its boiling point, converting it from liquid into vapor, which removes them from the original liquid mixture. The constituent with the lowest boiling point is the first and easiest one to be removed. In the second step, cooling of the gaseous phase leads to a liquid product by condensation. The evaporation temperature of ethanol, for example, is 78.4 °C. Since the vapor pressure of water at 78 °C is $p =$ 436 mbar (hPa), noticeable amounts of water evaporate at the boiling point of ethanol, so that one does not get a pure product. In these cases, special operations are applied, starting from repeated application to rectification.

By the way, production of distilled water is also a distillation process, as the name already says. Products being gases by nature, range from aromatic compounds – high-value product – to methane, cheap bulk product. Distillation was already used several thousand years ago and is already described by Aristotle. The extraction of flavorings or medicinal substances from plant material has also been in use for a long time. An entrainer, e.g. an oil, is used to dissolve the valuable substance from the solid matrix before the actual distillation. Unfortunately, distillation is energy-intensive. Would it not be a good idea to carry out ethanol fermentation with thermotolerant microorganisms? This would save cooling during cultivation and heating for distillation of the ethanol.

During crystallization, not the solvent (medium) but the product changes from the liquid to the solid phase, and can then be separated. Annealing (cooling) leads to finding the product molecule in a highly ordered crystal. This approach leads to the purest product, as contaminants are excluded by molecular forces from the growing crystal. Protein crystallization is one example, organic acids and penicillin are others.

The cells are also of small volume, containing a multicomponent system, which in a broader sense, can be regarded as a suspension with solid particles and a liquid matrix. In order to access the intracellular components, the cells must first be "disrupted". This means that the cell walls are disintegrated and the interior is extracted or further separated. Cell disruption is indeed an important step, which is often a bit tricky. Not always all cells are treated in a way to allow complete extraction. Also, for direct consumption of, e.g. Chlorella, the cell must be disrupted to get maximum bioavailability of the carotenoids. Some practical devices are shown in Fig. 8.7.

In the inlet of the high-pressure homogenizer, "French Press" ($p =$ 50–100 MPa), the suspension is initially exposed to a high pressure of $p =$ 50–100 MPa. This is not harmful in itself. However, when passing through the valve gap (micro-channels), the microorganisms are accelerated to 300 m/s within a very short time (approx. 250 ms) because they are forced through the 0.075 mm annular gap. The sudden drop in pressure causes turbulence and cavitation, which is the main cause of cell disruption. The elongational shear stress, in particular, tears the cells apart. About 90% of the energy is converted into heat. The device is usually used for milk homogenization and dispersion of the oil droplets.

Mechanical shear stress can be generated not only with two layers moving against each other, but also with two rotating cylindrical rings. This idea is realized in the so-called Ultra-Turrax. An internal rotor rotates against an external rigid stator. These are mounted on the end of a rod, which is inserted into the suspension. A velocity

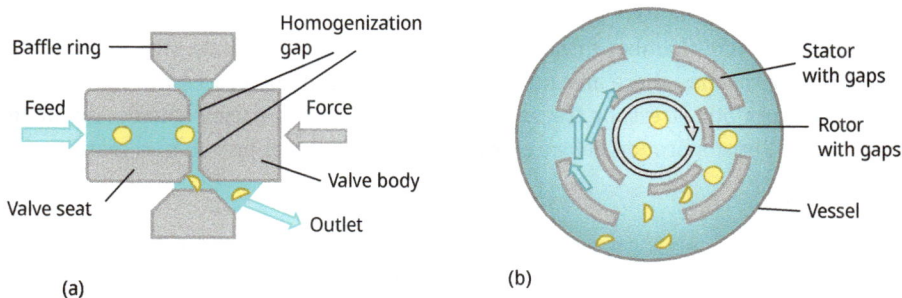

Fig. 8.7: Tools for cell disruption; a) high-pressure homogenizer b) ultra turrax. The arrows indicate local flows and velocities.

gradient is created in the slot between the two components, which leads to a shear force. The medium with the cells enters the system from below and is transported from the inside into the shear space through the gaps of the rotor. There, the cells break, and the cell debris leaves the shear chamber to the outside through the gaps of the stator. Typical rotation speeds are between 10,000 rpm and 30,000 rpm with a rotor diameter between 10 and 15 mm. The whole disperser works similar to a blender in the kitchen. This arrangement is particularly popular for small quantities in the laboratory in preparation for analysis.

An agitator bead mill is a cylindrical vessel with an agitator or a stirring rotor. The agitator itself can be made of rods or discs. About 80% of the vessel volume is filled with beads of less than 1 mm diameter. During milling, the grinding (cell) suspension is pumped into the chamber. When the rotor of the mill rotates, the beads move, leading to multiple impacts and friction between the beads. In this way, the suspended cells are crushed or dispersed by impact and shear forces between the beads. The disrupted cell suspension – the lysate – is drained from the mill. For this purpose, the mills have a sieve chuck to separate the beads from the slurry. Although the beads are made of special metal or ceramic, abrasion can occur, which affects the product quality. Bead mills can comminute cells down to the 100 nm range. Small-scale ball mills are also popular for sample preparation in cell analysis to obtain easily extractable material. Depending on the size of the cells and the thickness of the cell walls, different bead diameters and process parameters must carefully be selected. This also includes monitoring the degree of disruption, which should be at least at 90%.

In the ultrasonic homogenizer, the cell suspension is derived in front of the surface of a sonotrode. This emits ultrasound from 15 to 40 kHz. The resulting velocity/pressure waves, including cavitation, then rupture the cells. The process itself only takes place a few mm in front of the electrode head. Compare the ultrasonic spectacle cleaning device at the optician. Ultrasound is particularly recommended for the isolation of cell organelles. As with all cell disruption devices, a lot of energy is introduced that must be constantly dissipated. Even for lab samples, the power input can be up to

100 W or 1 kW/L. In the case of the ultrasonic device, the tip of the sonotrode is surrounded by a cooling chamber through which the suspension flows. In addition to these mechanical approaches, there are also chemical and enzymatic processes. Nitrogen decompression, osmotic stress, or freeze-thaw cycles can also be used stand-alone or in combinations. Electroporation before extraction is also a new approach in research.

8.5 Extraction – the standard approach to separate small particles or molecules

Solid extraction has been used by people as early as in the Bronze Age. The picture on Fig. 8.8 (left) shows a prehistoric pot that was initially puzzling. Only the collaboration with process engineers revealed the actual purpose, namely, to be an extraction device. The working principle is visualized on the right. Herbs were placed in the widened rim and extracted with the condensing hot oil or water. This looks very much alike a modern coffee machine. Extraction has lost none of its relevance since then. Ingredients from plants are still obtained by means of solid-liquid extraction. Extraction is, in general, any process, where the product is gained in one phase (solid, liquid, or gaseous) and is extracted by a second phase. This means that the product has to move from one phase to the other.

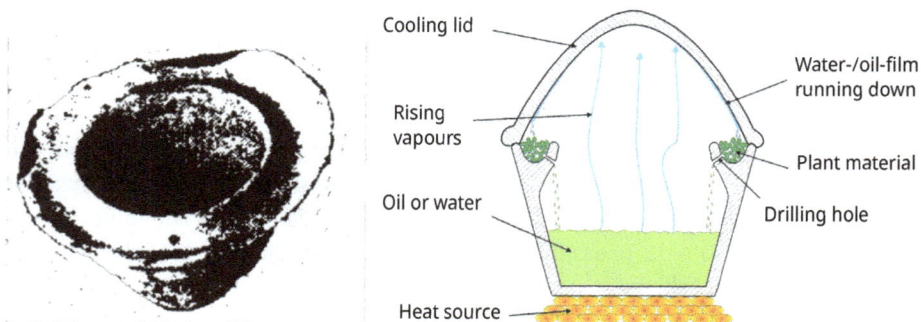

Fig. 8.8: Left side shows a prehistoric extraction pot. It is assumed that it worked according to the principle of solid extraction, right.

Liquid-liquid extraction is based on a different solubility of the product in the two liquids as the characteristic feature. It is better scalable in the process engineering sense and better suited to liquid raw materials such as fermentation suspensions. To do this, the suspension containing the product (feed) must come into contact with the receiving solvent to give the two binary liquids called the raffinate and the extract. Ideally, the product can then be located in the extract. In practice, however, a residue also remains in the raffinate. To better understand the situation, the situation is shown in a triangular diagram (see Fig. 8.9).

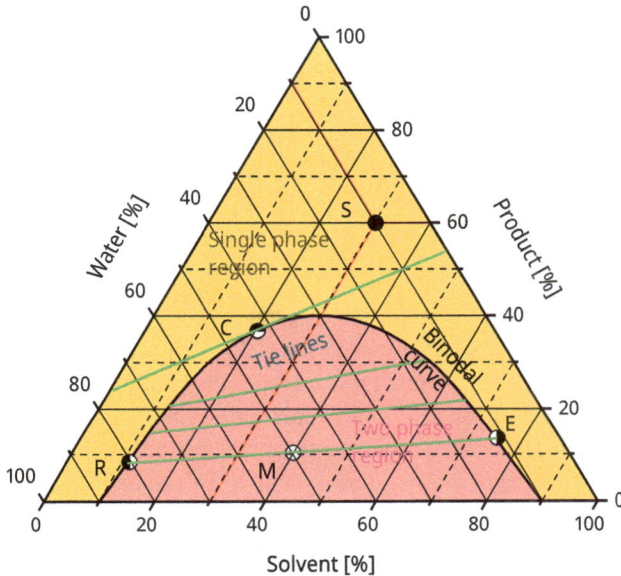

Fig. 8.9: The generic triangle (ternary) diagram shows the principal situation during liquid-liquid separation. The three components are mixed (point "M") in different proportions. The relative mass fractions can be read off the three axes, as shown for point "S". "R" means raffinate, "E" extract, and "C" the plait point.

The triangle diagram has three axes for the mass fractions x of the three compounds – solvent x_S, water x_W, and product x_P. Nevertheless, it is a two-dimensional sheet, the additional information comes from the definition that for all points, the sum of all x is 1% or 100%:

$$\sum_{i=1}^{3} x_i = 1 \tag{8.8}$$

To understand the application of the diagram, let us now imagine that feed and solvent are mixed, resulting in a mixture, represented by point M. These are then mass fractions for $x_S = 40\%$, water $x_W = 50\%$, $x_P = 10\%$. This would not change if the solvent was not chosen, so that it is almost insoluble in water. The liquid then breaks down into two fluid phases, one of which floats as a droplet in the other, which is the disperse phase. The two fluids then change further along the tie lines and end where these lines intersect – the binodal curve. These two points represent the thermodynamic equilibrium. The binodal (phase envelope) curve separates an area where all components are soluble in each other – from the area where phase separation occurs. This incompletely miscible state is referred to as a miscibility gap. The composition of raffinate and extract can be read off there. Tie lines and binodal curves have been determined experimentally at some point or can be calculated nowadays by molecular modeling. In the theoretical example chosen here, the concentration of the product in the extract is unfortunately only slightly higher than in the raffinate. The process must then be repeated several times in several stages. At least, half of the product is in the solvent. For further evaluation, additional coefficients and calculations are necessary. Of most importance is the distribution coefficient:

$$K_{\text{Solv,Water}} = \frac{c_{P,\text{Solv}}}{c_{P,\text{Water}}} \qquad (8.9)$$

It describes the enrichment of the product in the extract, compared to the raffinate at steady-state conditions. The octanol/water partition coefficient K_{OW} is particularly important as a standard for measuring the hydrophilicity/hydrophobicity of a substance. Pharmaceuticals with a high K_{OW} value, for example, accumulate in the lipid bilayer of cell membranes. The term partition coefficient means the same, but only for unionized chemicals. The total mass/volumes of the two phases can be calculated from the product balance. In the ideal case, it is the mass of water and solvent employed.

Two stages must therefore be passed through for the technical process implementation. The first is intensive mixing and the second is the separation of the two phases. This is done using mixer-settler systems. The mixer can be an agitated flask, for example, and the settler a separating funnel. The mixer can support the transport of product to the solvent phase, but the last mm come only by diffusion. Finally, the separation requires that raffinate and extract have different densities, which is another requirement, in addition to the solubility of the product. It is also too slow overall. The next step toward an intensive process is the use of extraction columns, where the lighter phase rises to the top and flows past the sinking phase. Screen plates mix the two phases. This apparatus variant is also the inspiration for the element in the flow-diagram but is also too slow. The reason for separating as quickly as possible is that product stability may not be guaranteed. In Fig. 8.10, typical partition coefficients for penicillin are shown. A value of 100 means that the concentration in the solvent is factor 100 higher than in the water phase. The positive effect is that less solvent than water is enough to accumulate the product. However, this can only be achieved if the extraction takes place at low pH values. Otherwise, the product can be damaged. The pH value must be quickly readjusted after extraction. Centrifugal forces must be superimposed for this purpose. This leads to machines like the disk stack separator, but then operated as an extractor.

In any case, the search for better extraction methods continues. This also involves improving sustainability, such as dispensing with organic solvents. One possibility is the PEG/dextran system. This produces an aqueous two-phase system (ATWS), as

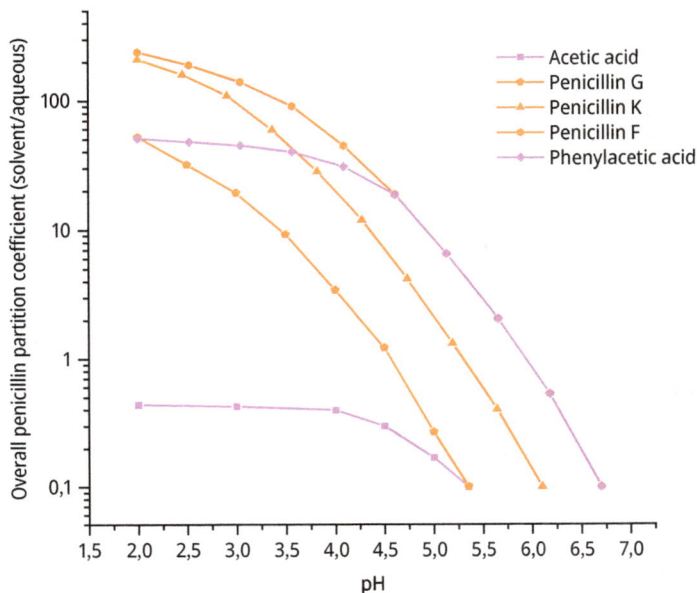

Fig. 8.10: Distribution (partition) coefficients in penicillin extraction; the dark orange lines stand for different penicillin species; the violet lines stand for intermediates to be recovered.

shown in Fig. 8.11. The two dissolved polymers do not mix freely in the water phase but separate to form microspheres or aggregates. It is a case of "like goes with like."

ATPS has attracted interest for its great potential in extraction of proteins, enzymes, or nucleic acids. It gives high recovery yield and is easy to be scaled up. It is also an environment friendly approach.

Then, there is the problem of recovering the solvent. This can involve energy-intensive evaporation and condensation, for example. A good idea is to compress gases, extract them, and then remove them from the product by releasing the pressure. At the very least, no harmful residues of the extraction agent can then remain in the product. One example recently coming more into focus is ammonia. An already well established method is extraction with supercritical CO_2. The phase diagram is represented in Fig. 8.12.

Supercritical gases are fluids, but the molecular interactions are different from a normal fluid as they show hardly any viscosity. Supercritical CO_2 is especially well suited for lipophilic extractions, which can also be solid-liquid separations drugs from medical plants, caffeine from coffee powder, or PUFAs from fish liver. But the big selling point is that the usable temperatures are in the physiological range (>30 °C) and the pressure (>7.4 MPa) manageable.

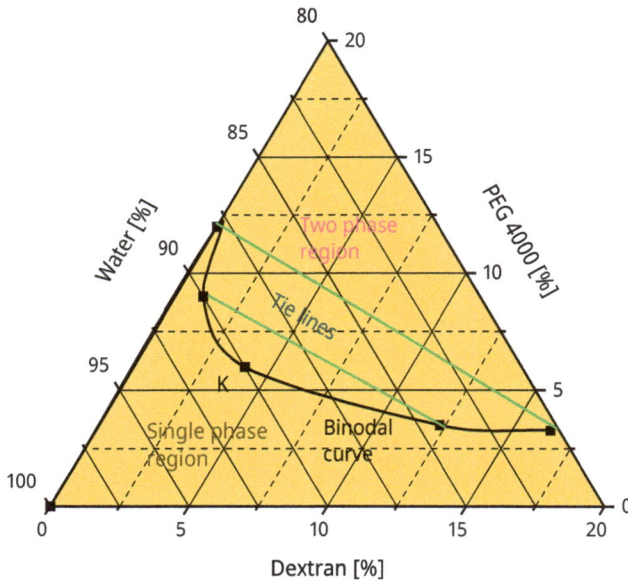

Fig. 8.11: Ternary diagram of an aqueous two-phase PEG/dextran system; PEG 4000 is polyethylene glycol with molecular weight of 4,000.

8.6 Running example yeast production – relationship between cell separation and cell physiology

Harvesting is, from the view of mechanical process engineering, a solid/liquid separation task to be accomplished by different centrifugation and filtration steps. In the following example, the solution for a real-life process is compiled.

In eqs. (8.3) and (8.4), the drag force and velocity during centrifugation have been noted. We can now discuss this equation with respect to cells. The first point is the radius of the particle. For bacteria, we can assume 1 µm and for eukaryotic cells, e.g., 5 µm; so low sedimentation velocities are expected compared to other technical particles. For bacteria, sedimentation velocity can even be a factor 25 lower as the velocity increases with the square of the radius. The second point to be considered is cell density. This can be roughly estimated to be 1.1 kg/L. This value is much lower than values, e.g., for mineral particles. Cell density may change with macromolecular cell composition. A high oil content can reduce the difference in density between the cell and the medium to nearly zero. As an example, the sedimentation velocity of a yeast cell yield for $r_{Cell} = 5$ µm, $\Delta\rho_{Cell} = 0.1$ kg/L, $\eta_{H2O} = 1$ mPa · s, and $g = 9.81$ m/s^2 is $v_{s,yeast} = 5.45 \times 10^{-6}$ m/s. So, we can wait for hours until a sedimentation can be observed in the measuring cylinder. As a third point, one has to be aware that media is not water and can have much higher viscosities – up to a factor 1,000 if polysaccharides are present.

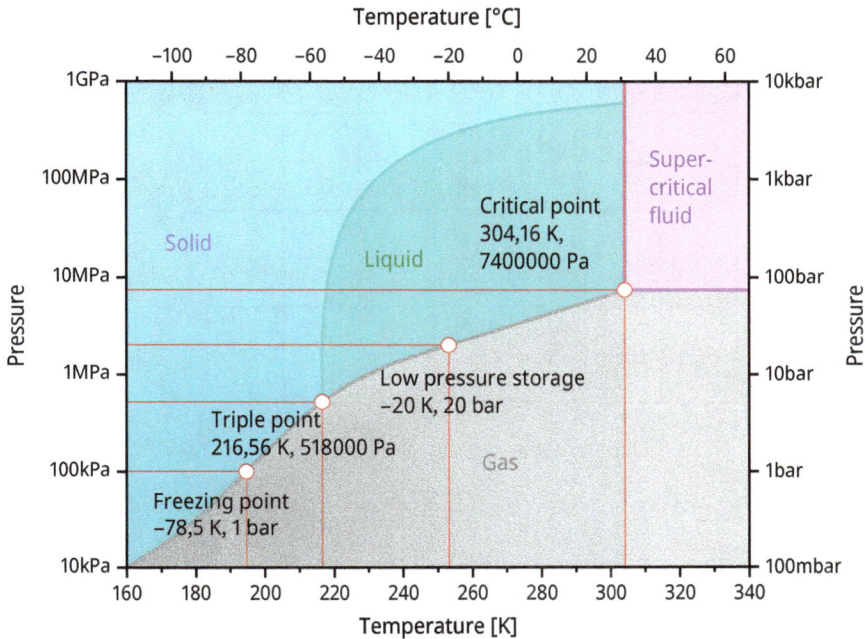

Fig. 8.12: Phase diagram of CO_2; interesting is the point where the gas behaves as supercritical fluid, which is suitable for organics. Dry ice used to produce disco fog is below the freezing point.

All three points make it obvious that specific measures must be taken. In a centrifugal field, sedimentation is faster by the factor *z*, the relative centrifugal force. This factor can reach values of up to 15,000 for technical centrifuges. Even then, centrifuges with very short sedimentation lengths are necessary. For yeasts, disk stack centrifuges – known from milk processing – are chosen, producing the yeast cream (180 g/L solids content), as indicated in the introduction Fig. 1.8. This can be sold as a product anyway.

For production of fresh yeast cubes, a second concentration step is necessary. Vacuum drum filter is the means of choice. Darcy's law describes the area specific flow of the filtrate through the filter cake, see eq. (8.1), assuming it as a porous medium. In our case, this means that despite Darcy's law, filtration resistance increases strongly over-proportionally, so a highly compressible filter cake behavior must be faced. Data are given in Fig. 8.13. Furthermore, budded cells show a higher resistance than unbudded cells. It could be, for example, that the buds fit better in the gussets between the large cells. This is another example of a direct functional chain from cell morphology's over feeding strategy to behavior in downstream. So, unbudded cells are preferred not only for high fermentative activity in the dough but also for better filterability.

Filter cake resistance turns out to be so high that only very thin filter cakes are obtained. Again, a structural change in the system must be brought about. This in-

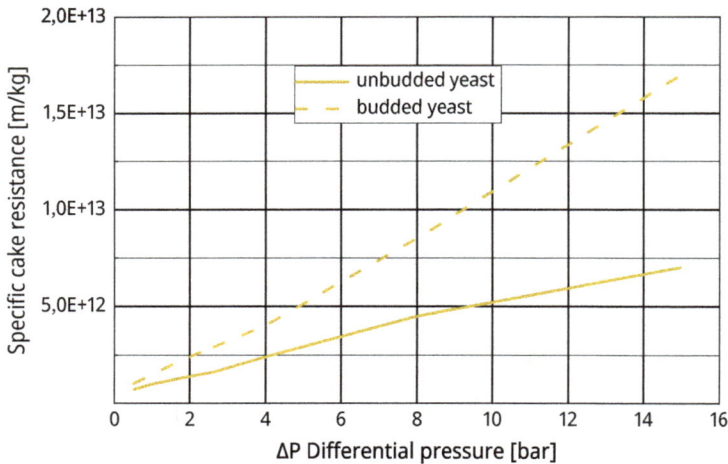

Fig. 8.13: Measurement data of specific cake resistance of unbudded (straight line) and budded cells (dashed).

cludes precoat filtration (Fig. 8.14), employing diatomaceous earth as filter aids. This material allows for better water permeation and better mechanical removal of the cake from the filter cloth. In the first step, the silica is floated onto the filter cloth in a layer several centimeters thick to build up the precoat. Then, the actual yeast filtration begins. Although only thin filter cakes are produced, they can be easily and continuously removed by the precoat.

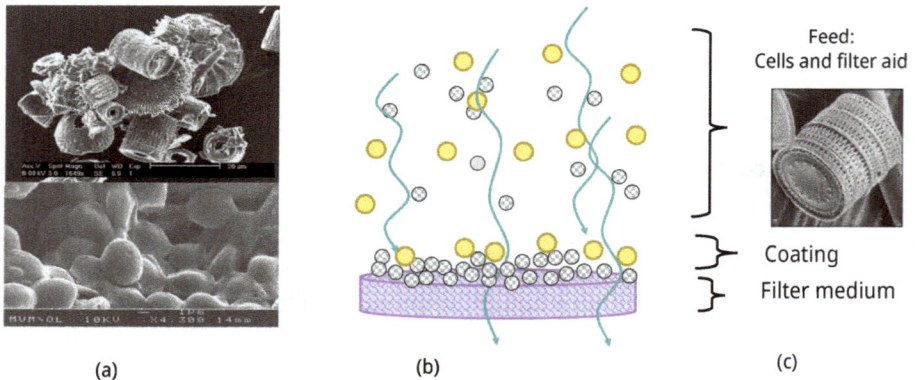

(a)   (b)   (c)

Feed:
Cells and filter aid

Coating
Filter medium

Fig. 8.14: Principle of filter-aid filtration with diatomaceous earth; the REM-pictures (small "tons") show the fossil frustules ("shells") of *Melosira*.

This approach is also used in other areas of food industry, namely fruit juice filtration. The juice, which is the actual product, is filtered, together with the diatomaceous earth particles, to remove turbidity and fibers. The good permeability of the resulting filter cake helps in this process. Interestingly, diatomaceous earth is a biological product formed by diatoms (microalgae) in the past geological eras. Artificial particles would hardly exhibit such microscale structures. Despite all our technical sophistication, the implication of the choice on humans should not be forgotten. The material is mined from fossils deposits. The dust is suspected to be a cancer promotor for the workers. Furthermore, it may contain heavy metals, which makes it necessary to bake the material, which increases costs. Finally, the used precoat material must be disposed, partly in landfills for special wastes. This is reason enough to look for new solutions to precoat filtration, in the best-case, biodegradable and biogenic particles, e.g., based on lignocellulosics.

The product from the filtration step is still a pasty and not a crumbly mass, as expected for yeast cubes (10^{10} cell/g; 30% solids content, mainly intracellular water). Here, another difference between cells and mineral particles comes into action. By washing steps, the osmotic environment of the filter cake is changed so that the yeast takes up the residual gusset water, leading to the desired consistency of the "fresh yeast" cake. This can be extruded to give the typical block in the supermarket. Granular "active dry yeast" or "instant yeast" is obtained by passing small amounts through fluidized bed drying, making the product more durable. Freeze-drying is standard for keeping strains in strain banks and collections.

Yeast production has long been the subject of social expectations and regulation, see Section 2.2.4. All ingredients used in yeast production must be of the appropriate quality. This is not the case with normal molasses because harmful chemicals are used in the sugar crystallization process. The technical ammonia is also unacceptable. Hydrolysates of vegetable proteins, for example, can be used as a substitute. The problems the gravel clay have already been discussed above. These are also issues that a process engineer must address.

8.7 Ideas for additional forces and structures – a toolbox with short user instructions

Up to now, we have been talking about specific weights of particles for centrifugation and steric hindrance for filtration. In both cases, the result is not optimal. So, it is a great idea to introduce additional forces and structures into the system.

8.7.1 Hydrodynamic forces

Since we are dealing with aqueous systems anyway, hydrodynamic forces are the obvious choice. Tangentially acting flows also cause a tangentially acting shear force, which contributes to the prevention of fouling. One application is cross-flow filtration (see Fig. 8.15). The suspension flows through a hollow fiber. The wall of the fiber consists of a filtering membrane. The fluid is discharged through the pores. If cells or other particles are deposited on the inside of the membrane and on the pores, they are removed by the tangential flow. Several fibers are usually combined to form a bundle, which is surrounded by a tube that drains the filtrate.

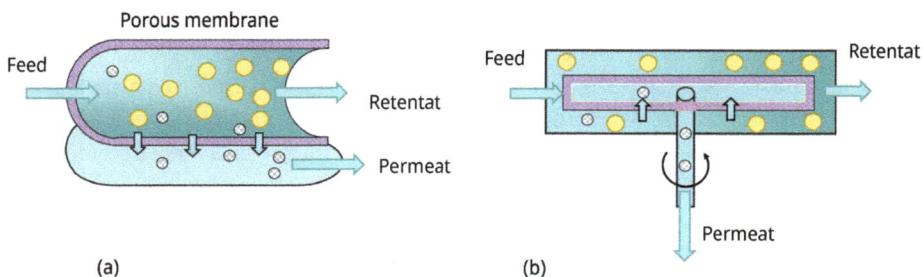

Fig. 8.15: Examples for tangential hydraulic forces: (a) cross-flow filtration, the inner hollow fiber has a membrane wall, the outer transports the filtrate; and (b) dynamic cross-flow filtration (DCF), the same principle, but with a rotating disk as the membrane element.

Hollow fibers can be used for cell recirculation or pre-concentration. A parabolic velocity profile of the fluid also occurs within the fiber. It flows fastest in the middle and slowest at the membrane. As a result, smaller particles accumulate on the outside and can be discharged through the pores. This free-flow fractionation can be used, for example, to separate extracellular proteins from cells. Disc-shaped ceramic diaphragms are also used to implement the principle. These rotate and thus generate a tangential flow over the surface. In this dynamic cross-flow filtration, the filtrate passes through the membrane and flows out of the system via the rotor shaft. This process is known as dynamic cross-flow filtration.

Other approaches are based on flow phenomena in which vortices are created on surfaces. These are known as Taylor vortices. These arise when a fluid flows axially through the space between two cylinders. Rotating tangential vortices form on the rotating inner cylinder, which is again designed as a membrane. Dean vortices are another option. These are created when a fluid flows through a coiled pipe. Standing ultrasound waves are particularly smart, but their effectiveness is controversial. These are created when two opposing sonotrodes emit sound waves. The cells then collect in the pressure minima. The method can be used, for example, for cell retention in slow-flowing outlets. One advantage is that it does not contain any elements that are difficult to clean.

8.7.2 Electrical forces

The charge balance of a volume of an aqueous liquid or of cells as a whole will always be equalized, i.e., the sum of all positive and negative charges is zero. However, this does not apply to smaller particles or surfaces. The most obvious ones are of course the ions (cations and anions) from the medium, but alsoother molecules can be charged. On the one hand, there are van der Waals forces that favor the accumulation of ions on (even hydrophobic) surfaces. On the other hand, dissociable groups and side chains have a greater effect, especially on organic particles such as cells or polymer molecules. Proteins are polymers of amino acids. Although these bind via an amino and a carboxyl group, some amino acids have a further dissociable group. While a second carboxyl group provides a negative charge by splitting off a proton, a second amino group can be positive by attaching a proton. Furthermore, the first and last amino acids, in particular, have at the amino-terminal end and at the carboxyl-terminal end, no partner on one side and therefore potentially carry a charge. Similar considerations apply to polysaccharides. In Fig. 8.16, some examples and the resulting interactions between particles are represented. Since cells also have organic molecules on their surface, they are also charged. There is even the assumption that there are specific charge patterns on the cell surface.

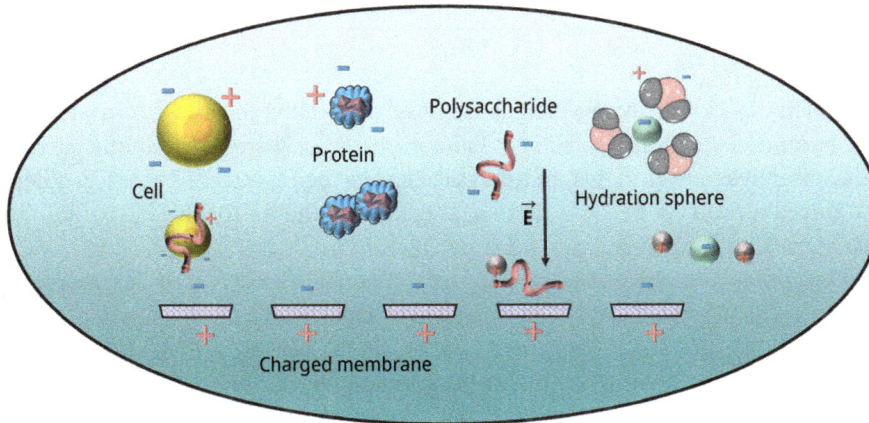

Fig. 8.16: Electric phenomena with impact on filtration in a watery suspension; details are given in the text.

It is well known that particles with opposite charges attract each other and those with the same charges repel each other. Mutual repulsion in the first place is the reason why proteins, for example, dissolve in water. The fact that water molecules arrange themselves around ions or charged polymers in a hydrate shell also contribute to solubility. The water dipoles are oriented with the negative side (O) toward cations and the positive side (H) toward the anions. The situation is further complicated by the fact that other ions are arranged in this hydrate shell. They then form an electrical double layer. Seen from the outside, the net charge then decreases. However, the overall aggregate has a larger diameter than the original particle, which has a negative effect on membrane filtration. By the way, movement of particles in an electric field means also movement of water.

Charged particles experience a force, the Coulomb force, in a constant electric field *E*:

$$F_C = q \cdot \vec{E} \tag{8.10}$$

This leads to a movement inside the solution. The velocity of the movement is then balanced by Stoke's law for the drag force (eq. (8.3)). This movement can then be taken to measure the charge of a particle, the so-called ζ-potential ("Zeta"). This considers that the electrical double layer is partially sheared off at the shear plane during the movement. This leaves a smaller hydrodynamic diameter, but a larger net charge. Nevertheless, the true diameter and the actual surface charge are not directly measured. Typical values for the zeta-potential are between −100 and +100 mV.

However, each charge also generates a field, which exerts a force on the other ions. The force effect of the mutual attraction of two particles carrying charges q_1 and q_2 then results in another representation of the Coulomb's law:

$$F_{C1,2} = \frac{1}{4 \cdot \pi \cdot \varepsilon_0 \cdot \varepsilon_r} \cdot \frac{q_1 \cdot q_2}{d_{1,2}^2} \tag{8.11}$$

The parameter ε_0 is the electric field constant or also called dielectric constant, the relative permittivity ε_r in the vacuum is 1 and about 80 for water. This mutual attraction also often leads to proteins and polysaccharides being attached to each other. This is indeed a problem because these aggregates are then difficult to separate for product extraction or analysis.

The processes described above are strongly dependent on the material matrix in the medium. The pH value and salinity play a particularly important role here. Additional ions (e.g. H_3^+) attach themselves to particles and thereby reduce the effective surface charge. It is known that proteins have an amphoteric character. That means that at a certain pH value, the zeta-potential approaches zero. In this state, the protein molecules can attach to each other and precipitate. The same applies if the salt concentration is increased. On the positive side, however, this is a method of protein separation called "salting out." At the same time, the diameter of the hydration shell is also reduced, which is also an advantage for filtration. Another example of the role of ions in moderating polymer interactions is electro-viscosity, which changes the viscosity of polysaccharides. Charged membranes are another mean, to influence attachment on filter membranes, intentionally (transmembrane transport) or unintentionally (fouling). This is an analogous behavior to electrophoresis in the fractionation of polymer mixtures.

One technical application of electrical forces is press-electro-filtration. In this process, an electric field is superimposed on normal press filtration, in which the suspension is located between two "cloths" or membranes. The plate electrodes are located behind the membrane, as shown in Fig. 8.17a.

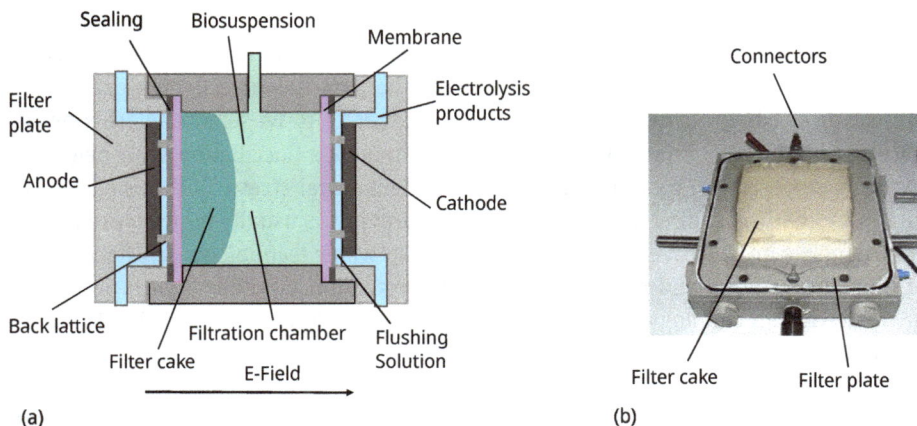

Fig. 8.17: (a) Sketch of how a system works; the electrodes are shown in dark gray, the membranes in violet. (b) Dismantled filter plate (anode) with a puncture-proof firm xanthan filter cake.

The normal pressure is often not sufficient to generate a filtrate flow through the highly compressible filter cake even when it is only slightly thick. The electric field now causes the normally negatively charged polymer molecules (proteins or polysaccharides) to move toward the anode. This creates a filter cake on the anode side. The cathode remains free of polymers and the filtrate can flow out. A second effect is electroosmosis. The negative ions are also moved in the direction of the membrane and pass through the membrane, together with their hydrate shell, further thickening the filter cake. During electrofiltration, the product must always be kept away from direct contact with the electrode. Otherwise, the molecules would decompose within the double layer in front of the electrode. This also happens with water, which leads to the well-known electrolysis. During electrofiltration, such high voltages (> 10 V) must be used that electrolysis will occur in any case. The electrolysis products must be removed using a flashing solution, together with the ions that have passed the membrane.

Figure 8.17b shows a Xanthan filter-cake in an electrofiltration plate. The filter cake, which is approx. 3 cm high, is firm and could not practically be achieved in this form with any other process. The classic method of separating polysaccharides is to precipitate them with alcohol or other solvents. The reason for the effectiveness is the fact that such hydrophobic substances reduce the dielectric constant and thus the electric field and the mutual repulsion. The disadvantage is the necessity to recover the whole solvent and remove it completely out of the product.

8.7.3 Ideas for changing the structure – learning from nature

As the previous chapter has shown, there is still room for improvement in bioseparation. So, it is time for a paradigm shift. Here, some examples are given and some approaches for solutions. In both filtration and centrifugation, the radius of the particles appears in the corresponding formulas. Typically, it makes life difficult for us by increasing the filter-cake resistance or decreasing the sedimentation rate. While we cannot change the size of the cells, we can make the particles or cells aggregate, thus increasing the hydrodynamic diameter of the particles to be separated. This procedure is called flocculation. Some cells show spontaneous flocculation with neighboring cells, upon stress, e.g., pH fluctuations or mechanical stress. Both measures are difficult to realize; salting out like for technical proteins is usually no measure for microorganisms or sensitive products at all. In wastewater technology, self-flocculation of filamentous fungi and bacteria supports cell retention. Another typical procedure in wastewater technology is flocculation with divalent ions. However, this is not an option for value creation in the life sciences sector. Biodegradable polymers with positive charges like cationic polysaccharides are sometimes an option. They are added to the suspension, where they then hold the cells together. However, it is not comfortable to have another substance in the product, which has to be removed later.

As bioengineers, we start looking for a solution in biology and find that many microorganisms are able to show flocculation naturally. In fact, most microorganisms do not live freely suspended in nature, but rather attached to surfaces or to conspecifics. In many industrial applications, like production of bioethanol, *Saccharomyces* show flocculation at desired points during the process, which is an important characteristic of a good production strain. Yeast flocculation depends on the expression of specific flocculation genes (FLO1 to FLO11). Expression of the flocculation genes is influenced by the nutritional status of the yeast cells as well as other environmental factors (e.g., T, pH, pO_2, Zn). The proteins being auto-associative (fit and bind to each other) then show up on the cell surface and can therefore be controlled by process design, facilitating later separation. Coupling the genes to other promoters leads to the possibility of "flocculation on demand". The situation in brewing is similar, where yeast flocculation makes the difference in different kinds of beer. Centuries of domestication have developed special properties in cells. These concern not only the production capabilities as such, but also the behavior in production and processing. This is standard in agricultural production, e.g., the grains in an ear of corn all ripen at the same time to make harvesting easier, or for apples to be stored for a longer period. For new production strains, this is often still in its infancy – one more reason to use tried-and-tested production platforms for various products. Some people even say that if recombinant proteins had only been invented today, E. coli would no longer be used as a production organism.

A kind of flocculant are also air bubbles. An easy-to-implement step for concentration is flotation. In this process, the suspension is gassed with small bubbles. The cells settle on their surface and float upward with them. The resulting foam can be harvested. The process is used for small cell concentrations and low added value, such as microalgae. There, foam is fed without further processing, e.g., in aquaculture. The characteristic property of the particles is the wettability (hydrophobicity). Sometimes, surface attachment of macromolecules or whole cells to surfaces offer an advantage. One application is cell immobilization. The cells are enclosed in a matrix or attached to the surface of a porous material. Shaping the material into spheres in the mm range then allows continuous process control with cell retention. In addition, possible shear force influences are minimized, but the exchange of substances may also be hindered.

A special method made possible by cell attachment is magnetic separation (Fig. 8.18). In the vast majority of cases, microorganisms are not naturally magnetic, but attachment to an iron particle would be a starting point for magnetic forces. In practice, iron oxide (Fe_2O_3) nanoparticles are used. These are super-paramagnetic. The "super" means that the magnetic forces disappear without remanence when the magnetic field is switched off, which makes it much easier to later separate the product from the carrier. For the magnetic separation itself, the suspension is passed through a strongly inhomogeneous magnetic field. The two poles are connected by a metal mesh, for example. Beside proportional to the particle volume, the magnetic force is proportional to the gradient of the magnetic field. The highest field strengths to which the particles bind are then the surfaces of the mesh fibers. The magnet can be an electromagnet or a permanent magnet. Remember from Chapter 10 that oxygen is also a super-paramagnetic material. An apology goes to the magnetic bacterium, *Magnetospirillum* (microbe of the year 2019), which also contains super-paramagnetic iron oxide crystals that are strung together like a string of pearls. They live in the oxic/anoxic transition zone. The lacking gradients would make it impossible to distinguish between top and bottom. The magnetotaxis helps them to make the difference. Has anyone tried out what happens if you dig up these bacteria in Europe and then bury them in Australia? There, the magnetic lines run in reverse. Many ideas for sophisticated employment in biotechnology or medicine are published.

The task that remains now is to look for better alternatives, apart from the strong but unspecific interactions between the product molecule and the matrix. Here again, it helps to make use of biological principles (see Fig. 8.18d–f). In order to utilize the interaction between the enzyme and its substrate, the substrate or an analogue is bound to the matrix. In order to guarantee steric accessibility, the chemical bond is made via a chain molecule. The target enzyme then binds with its active side to the ligand and thus also to the matrix. The idea is therefore quite analogous to the situation in measurement technology (see Fig. 10.15). Complementary oligonucleotides can also be used to achieve the precipitation of DNA as a product. In molecular imprinting, a kind of "imprint" of the target molecule is created in the matrix material. The molecule then binds to it, again similar to the situation with the substrate/enzyme interaction. Such approaches are called affinity separation. The applications are mostly in the small-scale or analytical area.

In a living cell, every bioactive molecule knows where it belongs and is transported there in a targeted manner. Cells are therefore masters of bioseparation. This idea is already followed in the production of extracellular proteins. What could be more obvi-

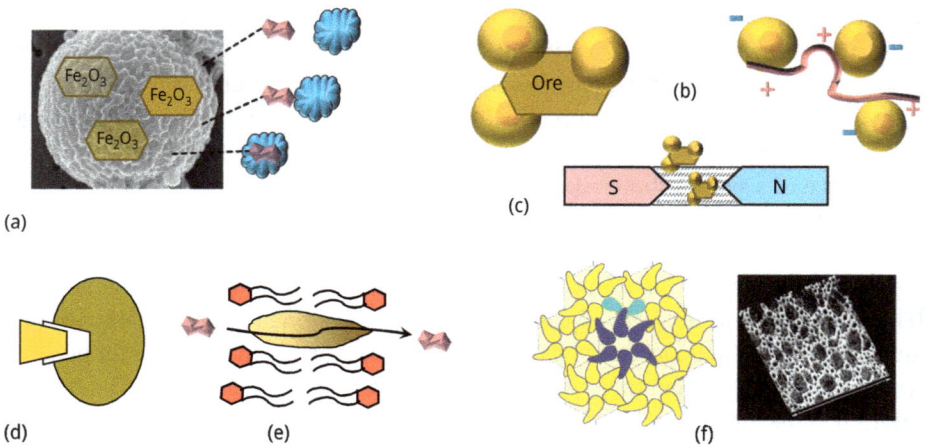

Fig. 8.18: Collection of structures suitable for bioseparation: (a) beads (mm-range) containing super-paramagnetic nanoparticles, on the surface are linked ligands to catch their corresponding enzyme; (b) flocculation with mineral particles and cationic polymers; (c) magnetic particles are collected on the surface of iron wires between the poles of a magnet; (d) molecular imprinting; (e) section of a phospholipid-bilayer. Incorporated is a channel (transport, pore) enzyme; (f) sketch of the hexagonal pattern of S-layer proteins of *Haloferax* ssp. and its electron microscope picture as affinity pore protein on carbon surface (© Bossmann, EBI, 1999).

ous than imitating cellular structures? Two examples are shown in Fig. 8.18. Such structures are often the result of self-organization. Phospholipids assemble into bilayers in the same way as cell membranes. These can be regarded as self-cleaning filtration membranes or as hydrophobic compartment during formulation of hydrophobic drugs. Equipped with enzymes, usually in the form of liposomes, completely new, highly specific separation processes can be developed. Proteins can also assemble into ordered structures. Such structures are found in many archaea, such as the S-layer protein ("S" for surface) of *Haloferax*. This can be used to create coatings with a defined pore structure usable as nanosieves, for example.

Filtration has also been invented by nature, where some animals – filter feeders – practice their ingestion of food by filtration with specialized morphological structures. Some examples are shown in Fig. 8.19. Sponges, sessile multicellular organisms at a low level of organization, allow ambient water to pass through pores on the outside into their hollow interior. There, scavenger cells absorb organic particles and digest them. The water leaves the sponge through the osculum. This is overflowed by local water movements, which leads to low pressure as driver for the water transport. Crinoids ("sea lilies") belong to the echinoderms. They transport food particles from the plankton with movable feet along the tentacles to the mouth opening, so they have developed a kind of intermediate between a filter and a conveyor belt. Mussels are filterers par excellence. The curtain of cilia at the edge of the opening ensures a flow of water from which plankton is filtered out. A single mussel can clean a 10-L aquarium full of suspended microalgae within a few hours. The ciliates also work with cilia that bring prey cells to the mouth opening, but two orders of magnitude smaller. Algae cells are phagocytized and can then be seen for a while inside the cell. In the realm of microorganisms, foraminifers, and sun animalcules (heliozoans) are unicellular protists of around 200–500 µm diameter. They invented sticky mucus filtration. Small plankton particles adhere to the filaments that are Incorporated into their shells. This has not really found a technical equivalence yet. It is best compared to a depth filter for dust removal. A characteristic feature is that the pores can be significantly larger than the diameter of the particles. Krill, which belong to the crustaceans, have one of the most highly developed filter systems. It uses a special structure on its front legs, the so-called filter basket, with which it can catch marine microalgae. These assemble on the Arctic and Antarctic ice ablation, where they profit from high concentration of mineral nutrition. The Krill follows them and propagate on the spot. It is just sad that the whales have the same idea. Humpback whales swim thousands of km to Arctic oceans to feed on krill or swarms of herring. They swim, mouths agape, swallowing thousands of crabs, or fish in one gulp. Pleated grooves in the whale's mouth allow them to drain the water, filtering out the prey. This animal can build up pressures of several bars to filter their prey as effectively as possible. So, this has great similarity to pressure filtration with filter plates. As in technical filter processes, a preconcentration

Fig. 8.19: This wonderful illustration by Marie Jamroszczyk is a homage to the creatures in the sea that catch their prey by means of filtration. The almost visible swaying movements in this watery environment cannot hide the fact that these filtering organisms are in a predator-prey relationship that ultimately leads to the world-wide food web, which is also vital for our survival. For details, see text. (© M. Jamroszczyk).

step is helpful, therefore several whales produce a bubble net, which confines the space for the prey (bubble net feeding).

Marie Jamroszczyk is a scientist in the field of biodiversity and ecotoxicology and a scientific illustrator. In both areas, she has specialized in the study of microalgae. She is also active in nature protection and in social media (#_algaeverse_ and #scientific.designs on Instagram), contributing to public mediation.

Her motto: "There is so much to explore!"

8.8 Questions and suggestions

1. One single elementary charge, which means one dissociated group in a macromolecule, can make a voltage on the surface of about 0.18 mV. Typical values for the potential are between −100 and +100 mV:

$$E_a(r) = \frac{q_{part}}{4 \cdot \pi \cdot \varepsilon_0 \cdot \varepsilon_r \cdot r_{part}^2} ; U_a(r) = \frac{q_{part}}{4 \cdot \pi \cdot \varepsilon_0 \cdot \varepsilon_r \cdot r_{part}}$$

Calculate the surface charge and voltage of a protein molecule carrying one elementary charge.

2. Look through the process flow charts of different processes in this book or elsewhere and find examples of separation steps and see for yourself how time- and money-consuming downstream can be.

3. Is it only by chance that centrifugation is not commonly visible in nature?

4. What requirements must a liquid fulfill for extraction? Make a list of the properties.

Chapter 9
Continuously operating bioprocesses – production under steady-state conditions

From the exercise in Chapter 7 on the fed-batch process, it became clear that the highest productivity can be reached by mini-harvesting only a small amount out of the reactor and filling up with fresh medium. The same holds for batch processes. The idea is now to go to infinitesimally small harvesting amounts but very frequently taken. Going to the mathematical limit finally leads to simultaneous feeding and harvesting with the same rate to keep the volume constant. The reactor configuration, as shown in Fig. 9.1, gives the basic variables. Assuming an ideally mixed reactor, the concentrations inside the reaction volume are of course the same as in the outlet. As there is no defined time of harvesting and the culture is active for a long time, this type of process is called continuous process or also (but not strictly correct) continuous cultivation.

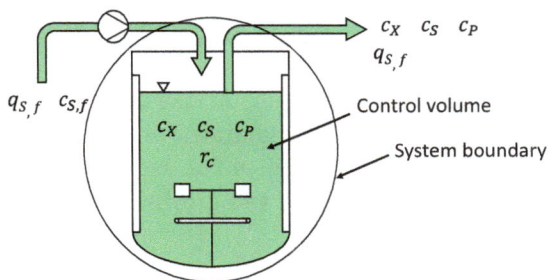

Fig. 9.1: Reactor configuration for a continuous bioprocess; usually, the outlet tube is mounted in such a way to suck surplus medium from the surface.

In this chapter, firstly the balance equations that are derived and discussed. This is done assuming steady-state conditions, which is one of the features of continuous processes. Another feature is that large continuously available amounts of medium can be handled. That is the case in wastewater treatment but also with other large-scale processes. Consequently, most of the processes are from these areas of application. Some structural modifications are possible for further process intensification. In chemical engineering, most of the processes are continuously operated. This is basically not the case in bioprocess engineering despite all the potential advantages. At the end of the chapter, some aspects of the current debate about the reasons for this observation are reflected.

Thermodynamically speaking, the continuously operated reactor is an "open" system, receiving energy and material from outside and giving used material with typi-

https://doi.org/10.1515/9783110773354-009

cally less energy, back to the environment. Thereby, the system can keep a state of low entropy at the cost of the environment. In this general sense, lakes, ecosystems, and also each living entity can somehow be regarded as a continuous process.

9.1 Setting up stationary balance equations – a good start for understanding process behavior

To get an adequate description of the process, we draw a balance around the system volume, counting the amount of all compounds going in or out through the system boundary. For dynamic reaction systems, a general scheme for material balances is

---

! Accumulation = input – output +/– reaction

For single compounds, substrate and biomass, the material mass balances read:

$$\frac{dm_S(t)}{dt} = q_{S,f}(t) \cdot c_{S,f} - q_{S,f}(t) \cdot c_S(t) - r_S \cdot m_X(t) \tag{9.1}$$

$$\frac{dm_X(t)}{dt} = -q_{S,f}(t) \cdot c_X(t) + r_X \cdot m_X(t) \tag{9.2}$$

The empty space in these equations is a hint to the missing input. Like in a bank account, we do not have to pay money in, but can make a withdrawal and get an increasing current balance, as long as the account status is high and so are the interest rates; here, the equivalent parameters are concentration and growth rate of the microorganisms.

While masses are conserved quantities, we are more interested in concentrations. As the working volume, V_R is constant, it is possible to divide the left and right side by V_R, getting:

$$\frac{dc_S(t)}{dt} = D(t) \cdot (c_{S,f} - c_S(t)) - r_S \cdot c_X(t) \tag{9.3}$$

$$\frac{dc_X(t)}{dt} = -D(t) \cdot c_X(t) + r_X \cdot c_X(t) \tag{9.4}$$

! The "dilution rate" D (h^{-1}) is the defined as the volume flow in to and out of a reactor per reactor volume. In the case, when nothing reacts in the reactor, an indicator substance inside like a color pigment would be transported out and "diluted" by freshly fed medium. The observable concentration would decline exponentially following the time constant $\tau_R = 1/D$, called the mean residence time:

$$D = q_f/V_R; \quad \tau = D^{-1} \tag{9.5}$$

The primary intention of dealing with the continuous process as such was to hope for steady-state conditions. So far, it is assumed that the process is indeed in steady state and the concentrations do not change over time. This will of course only be the case if D is constant for a long time as well. Furthermore, we are interested in the dependency of substrate and biomass concentrations on the dilution rate. For constant conditions, there is no accumulation and the balance equations read:

$$D \cdot (c_{S,f} - c_S) - r_S(c_S) \cdot c_X = 0 \tag{9.6}$$

$$r_X(c_S) \cdot c_X - D \cdot c_X = !0 \tag{9.7}$$

This is a set of two nonlinear (product of variables) algebraic equations for the two unknown variables c_X and c_S. The exclamation mark reminds us that the equality is not given explicitly following a mathematical deduction of the left side of the equation, but that it is for the moment only our own demand and we have to check under which conditions it will be reality. A closer inspection or explicitly solving the equations brings up unexpected results:

Looking at the stationary biomass balance (9.7), it can be noticed that it has two formal solutions, namely:

$$c_X = 0 \text{ and } r_X(c_S) - D = 0 \tag{9.8a, b}$$

Putting the first solutions into the substrate balance (9.6), we get the corresponding value for the substrate concentration:

$$c_S = c_{S,f} \tag{9.9}$$

Firstly, we try to understand this trivial solution $c_X = 0$, $c_S = c_{S,f}$. This case is indeed not only mathematically possible but happens in reality if we forget to inoculate the reactor or if the dilution rate exceeds the maximum possible specific growth rate. Biomass then grows more slowly than it is withdrawn from the reactor and vanishes completely after some time. The situation is therefore called the "washout case." From where does "the reactor know" which one of the two solutions is true? This depends on the starting conditions, a piece of information that got lost the moment we set the derivatives to zero. For application, the history of cultivation has to be checked.

We now come to the second and obviously more relevant and interesting operative solution. Most important in understanding continuous bioprocesses is that the specific growth rate is equal to the dilution rate:

$$\mu = r_X(c_S) = D \tag{9.10}$$

This is not only something that happens, but by changing the dilution rate, we can directly manipulate the specific growth rate.

This is a unique feature, as we can keep μ for a very long time on values of interest. For biological investigations, this is great as adaptation is completed and there are only minor physiological changes. Of course, now the microorganisms have a say. How can "the microorganisms know" that they have to keep μ at D? Substrate uptake rate and specific growth rate depend, of course, on the substrate concentration. A minimum requisite to maintain a constant μ is that the substrate concentration stabilizes on a level allowing the cells a specific substrate rate, in accordance with the required μ. For further calculations, we need to specify kinetics as a link between the substrate and the biomass. For now, we take the standard set of equations (4.43) and (4.44).

Without maintenance, the second solution reads:

$$r_X(c_S) - D = 0 \rightarrow y_{X,S} \cdot r_{S,max} \cdot \frac{c_S}{c_S + k_{SS}} - D = 0 \rightarrow c_S(D) = \frac{D \cdot k_S}{y_{X,S} \cdot r_{S,max} - D} \tag{9.11}$$

This equation looks different for different kinetics.

From the substrate balance, we finally get:

$$c_X(D) = y_{X,S} \cdot (c_{S,f} - c_S(D)) \tag{9.12}$$

This looks reasonable because only the used-up substrate is converted into biomass. The solutions are plotted in Fig. 9.2.

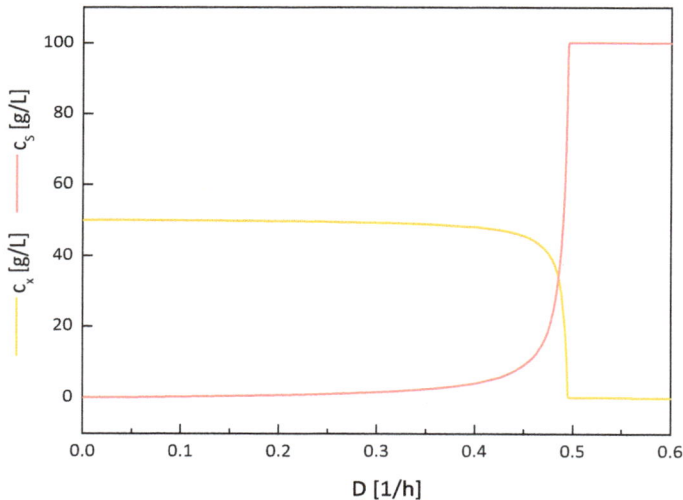

Fig. 9.2: Stationary states of glucose and biomass concentration over dilution rate. $r_{S,max} = 1$ g/g $\cdot$ h; $k_S = 1$ g/L; $y_{X,S} = 0.5$ g/g; $c_{S,f} = 1$ g/L.

Indeed, substrate concentration increases with increasing D. The higher D already is, the higher the necessary increase of c_S. What is actually to be seen is the Michaelis-

Menten kinetics for substrate uptake, but here, not as $r_S = f(c_S)$ but the inverse function $c_S = f(D = \mu)$. This is the important "second" solution, assuming Michaelis-Menten for substrate uptake and Pirt equation without maintenance. Turn the plot and hold a mirror vertically on the plot. You will then recognize the kinetics!

One detail deserves attention: when the dilution rate D approaches the theoretical maximum specific growth rate $\mu_{max} = y_{X,S} \cdot r_{S,max}$ the denominator approaches zero and the necessary substrate concentration, infinity. As the maximum possible substrate concentration of $c_{S,f}$, μ_{max} can be reached only approximately. We get the wash out case already for $D < \mu_{max}$.

For low substrate concentrations and dilution rates, the biomass concentration is nearly constant – close to the maximum possible value $y_{X,S} \cdot c_{S,f}$. For higher dilution rates, a higher substrate concentration is necessary (as explained above). This substrate is no longer available for growth and the biomass concentration drops.

The volumetric productivity for biomass is generally defined as the produced amount of biomass per volume and time:

$$P_{V,X} = \frac{\int_0^{\Delta t} q \cdot c_X \cdot dt}{V_R \cdot \Delta t} = \frac{q \cdot c_X \cdot \Delta t}{V_R \cdot \Delta t} = D \cdot c_X \tag{9.13}$$

The volumetric productivity is high for high dilution rates and for high biomass concentrations. But for now, we have to accept that both values do not reach their respective maximum at the same working point.

Like most bioprocesses, the continuous process is a transport/reaction system. The transport term for substrate $T_S = D \cdot (c_{S,f} - c_S)$ includes an inflow and an outflow component. The reaction term $R_S = r_S \cdot c_X$ embraces the biomass concentration and the biomass activity as a physiological component. Both terms (r_S, including substrate inhibition) are plotted in Fig. 9.3 versus a virtually given substrate concentration.

Possible stationary operating conditions require that the same amount of substrate is transported as used up by the cells, hence $T_S = T_R$, given by the points of intersection of the two curves. Firstly, we look at the left working point for low substrate concentrations. We now imagine that substrate concentration is lower than the concentration necessary to maintain the wanted operation point. The cells will grow with $\mu < D$, and the biomass concentration slowly drops below the precalculated value. This leads to decreased substrate consumption, and the transport term is higher than the reaction term, leading to an increase of the substrate concentration. Finally, the system will again approach the point of equilibrium. Accordingly, the same happens for substrate concentrations – arbitrarily higher than the value of equilibrium. This qualitative stability analysis shows that the culture, even when disturbed and not at equilibrium will autonomously find its way back to steady state without further control action from our side. This answers the question, what makes the substrate concentration stay at the required value. Such a stable operation is called a "chemostat" as all compounds involved are in steady state. A graphical representation of this feedback loop is shown in the sim-

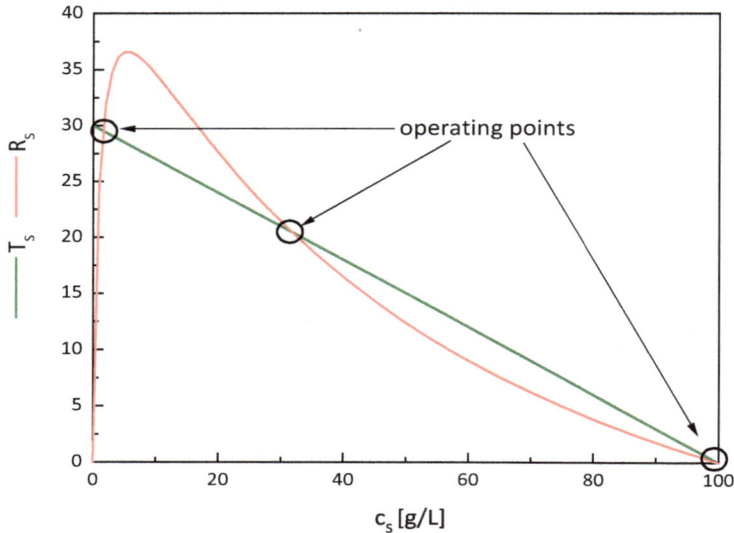

Fig. 9.3: Volumetric transport T_S (g/L· h) and reaction rates R_S (g/L· h) over substrate concentration for a process with substrate inhibition kinetics, parameters as above, $D = 0.3$ h^{-1}.

ulation example. The washout case at high substrate concentrations is stable as well. In cases of substrate inhibition, a third theoretical working point exists, which is unstable.

But how long does it take to reach the equilibrium, especially after the initial inoculation or after changing the dilution rate? To answer this question, a dynamic simulation helps, which is shown in Fig. 9.4.

The initial starting condition after the inoculation has been done is really close to steady state. Filling the reactor in the beginning with full medium would be counterproductive. Only in cases where not enough fresh culture for inoculation is available, can the continuous process be preceded by an initial batch ($D = 0$). Luckily, for most of the dilution rates, steady-state conditions are really reached, which is not self-evident, looking only at the stationary balance equations. A rule of thumb says that waiting time is around five residence time $1/D$. That is fortunately not true as can be seen from the derivatives, but the real time constants can be calculated, see below. One must be careful, as adaption could last for several cell generations. Further, we observe that the time to reach steady state increases with higher dilution rates. To understand this behavior, we linearize the system equations for different substrate concentrations and extract the dominant time constant; for details, see the calculation in the supporting material. In fact, transport and reaction contribute to transitional system dynamics. At low dilution rates and low substrate concentrations, reaction dominates and leads to fast time constants ($\tau \approx -r_{S,max}/k_S \cdot y_{X,S} \cdot c_{S,f}$), while at high dilution rates and substrate concentrations, the hydrodynamic time constant $\tau = 1/D$ dominates. In the special case of $D = \mu(c_{S,f})$, growth is very insensitive to substrate concentration and the time constant of biomass

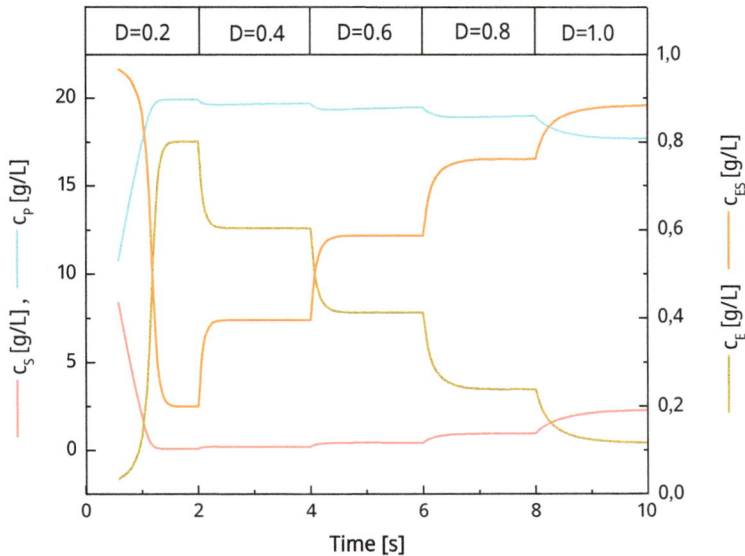

Fig. 9.4: Dynamic simulation with the same parameters as Fig. 9.2. The derivatives are a measure of how close the system is to equilibrium.

goes to infinity. It can indeed be tormenting to wait to see whether a culture is stabilizing or slowly being washed out.

9.2 Ethanol production in a continuous process – the window of operation

As a first data example, we investigate ethanol production. It is often carried out in batch processes, as we have already seen. As biofuels are produced in large amounts, the process is not too complicated, and downstream may be carried out continuously as well. It seems to be a good idea to apply continuous cultivation. An X/D diagram for ethanol production with the bacterium *Zymomonas mobilis* is shown in Fig. 9.5. This bacterium, known from "cider sickness," has been proposed as an alternative ethanol producer. Although yield and productivity were very good, the attempts failed because the bacterium could not use fructose properly and grow in a neutral pH range, abetting contamination with other bacteria. This is an example of how secondary aspects can also be a reason to prevent a "go" for new processes. It is a good example here to discuss chemostat cultivation.

The experiments were carried out with $c_{S,f} = 100$ g/L. The biomass concentration reaches about 3 g/L, which corresponds to a yield of 0.03 g/g. The maximum ethanol concentration is around $c_P = 50$ g/L, which also fits to a typical anaerobic fermentation

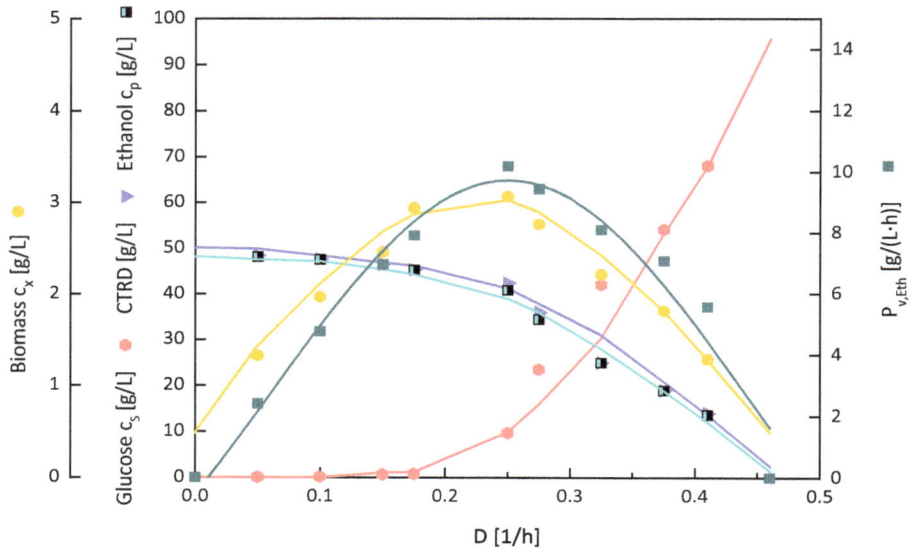

Fig. 9.5: X/D diagram of a continuous cultivation of *Zymomonas mobilis*. Beside biomass, substrate, and product, the volumetric productivity is also shown.

scheme. The data designated as CTR/D is a virtual variable assuming that the produced CO_2 would behave like a liquid product. This is a trick to visualize the relation of the two products – ethanol and CO_2. Both products are formed in the stoichiometric relation 1:1, so we expect a mass relation of $MG_{Eth}/MG_{CO2} = 46/44$. This is approximately true; maybe a part of the ethanol evaporates, together with the CO_2, into the off gas. But a further inspection of the data shows several differences to the standard curve we deduced in the last paragraphs. At low dilution rates, c_X is not constant but decreases for decreasing D. Obviously, the cells take up glucose but do not form much biomass from it. This reminds us of the maintenance term in the Pirt equation. A simulation (Fig. 9.6) shows that a maintenance term indeed leads to this typical curve.

The second salience is the slow decrease of c_X for increasing D at the range of higher dilution rates. In the standard simulation, this drop of biomass concentration was sharper. It is indeed not plausible that an increase of c_S from 10 to 50 g/L helps the cells to increase the specific growth rate from 0.25 to 0.35 h^{-1}. At such high substrate concentrations, we can exclude substrate limitation. Also, c_P drops in this range. The reason for this observed behavior of slow biomass decrease has been identified as diminishing ethanol inhibition. The volumetric productivity $P_{V,P} = D \cdot c_P$ decreases as well. As a "lesson learned," it can be stated that, in principle, ethanol can be produced in a continuous process. But for inhibiting products, high product concentrations and high volumetric productivities cannot be reached simultaneously.

The optimum working point would be somewhere between the highest productivity and the highest product concentration, depending on the reactor costs and costs

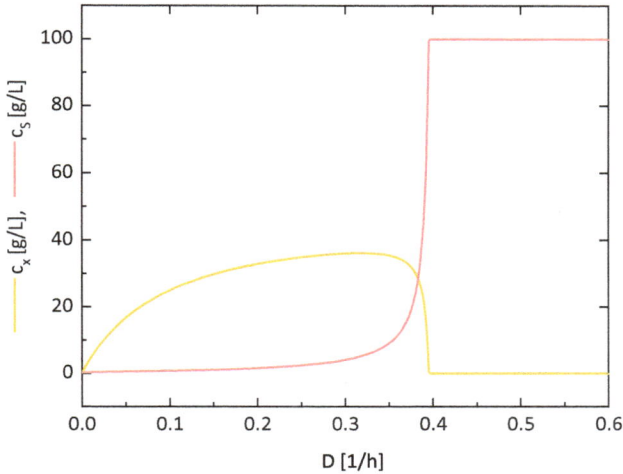

Fig. 9.6: Simulation of a chemostat with microorganisms exhibiting remarkable maintenance $\mu_m = 0.1\ \text{h}^{-1}$, other parameters as above.

for rectification. This leads to lower limits for both ethanol concentration and productivity. Other limits are given by the sugar concentration in the available substrate or the total amount to be produced in a given time ordered by a customer. These limits form a "frame" of sensible working points, visualized as an "operating window." In aerobic processes, this could be, for example, the highest possible oxygen transfer rate (OTR), or the highest possible biomass concentration (e.g., 100 g/L). Also, the shift of the intracellular product concentration to a specific point could be a limit if it is possible only at the cost of overall productivity. In other industrial areas, the integrity operating window (IOW) is a set of limits used to determine different variables that could affect the integrity and reliability of a process unit. This standard was set up by the American Petroleum Institute.

What could be a way out of such a window? As long as no changes in the microbial kinetics are possible, e.g., by strain development, structural changes in the process have to be foreseen. One strategy could be to decompose otherwise closely interlocked parts of the system like stoichiometry or other kinetically coupled physiological states. An example was the fed-batch process where different phases of the process – growth and production phase – were decoupled. For the current example of continuous ethanol production, not a timely but a spatially decoupled process structure could be envisaged. This means, in practice, the employment of a second reactor as shown in Fig. 9.7.

In the first continuously operated reactor, optimum growth conditions are adjusted. This means low or medium ethanol and high substrate concentrations. Biomass and unused substrate go into the second reactor, where additional substrate could be fed. Here, high product concentrations could be achieved at low growth

q_z

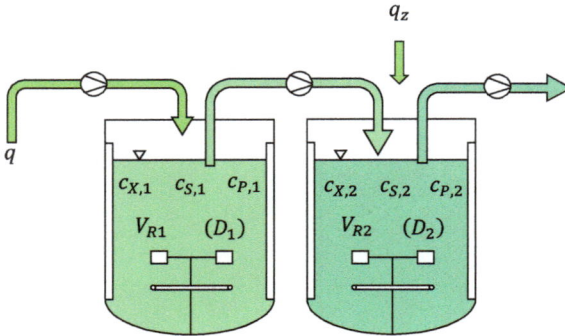

Fig. 9.7: Two reactors coupled in series to increase ethanol concentration and productivity.

rates. As biomass comes in from the first reactor, $\mu = D$ does not hold, and the dilution rate can be higher than the specific growth rate. Even zero growth could be an option, where the ethanol concentration may reach values inhibiting growth completely. Substrate turnover is ensured only by maintenance. This can even be an advantage as more carbon is allotted to the product.

9.3 Enzymatic processes – a simple but effective example for continuous bioprocess operation

Directly from the equations above, we see that one disadvantage of continuous processes is the permanent discharge of biomass. Especially in cases where product formation is at the cost of metabolic energy and therefore low specific growth rates, this is not acceptable. Extracellularly produced enzymes are an example. Here, the specific advantage of easy harvesting during the process would additionally be lost. In cases where resting cells or enzymes are used as biocatalysts, continuous production in the form described would be impossible. A way forward is changing the system to keep the catalysts inside the reactor independent of the medium and hydrodynamic retention time. The resulting reactor structure is given in Fig. 9.8. Such a measure is called either cell recycle, in cases of an external separation step, or cell retention, in cases where the cells are kept inside the reactor by immobilization or by submersed membranes. Basically, we have now established a perfusion system. This is well known from chemical engineering where the catalyst has to be kept in the reactor, in any case.

A flow of fresh medium ($q_{S,f}$) enters the reactor where the substrate is used by the growing cells. The reactor suspension is discharged via a solid–liquid separation step, where the "solids," namely the cells are fed back into the reactor in highly concentrated flow q_r. The permeate flow q_p contains mainly the water from the input and a possible product. A "bleeding" stream has to be foreseen to simulate an unavoidable

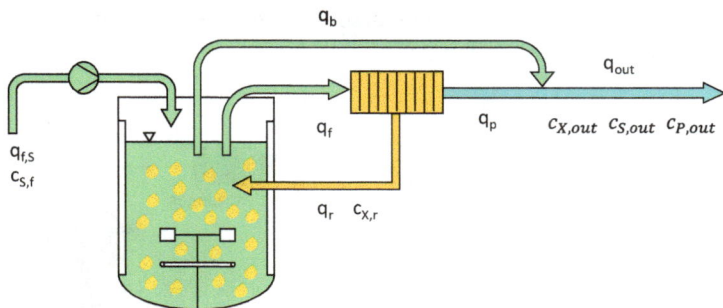

Fig. 9.8: Structure of a bioreactor with cell recycle or cell retention; q_b, bleeding flow; q_f, filter feed flow; q_p, cell-free permeate flow; q_r, retentate flow.

loss of cells or to discharge cells in excess. The total flow out, $q_{out} = q_{S,f} = q$. We can set now up the mass balance equations around the control volumes "reactor" (dynamic) and "filter" (static). The complete deduction is given in the supplementary material. Here, we select only two relationships.

Firstly, a recycle ratio R is defined:

$$R = \frac{q_p}{q_{out}} = \frac{q_p}{q_b + q_p} \tag{9.14}$$

$R = 0$ means no recycle and $R = 1$ means that there is absolutely no biomass in the effluent.

For biomass, we then get:

$$\frac{dc_X(t)}{dt} = r_X \cdot c_X - D \cdot (1 - R) = 0 \rightarrow r_X = D \cdot (1 - R) \tag{9.15}$$

A plausibility check for $R = 0$ leads to $\mu = D$ as in the regular continuous process. For $R = 1$, the result is $\mu = 0$ and the biomass theoretically goes to infinity. In practice, recycle ratios around 0.9 are commonly applied. In cases where product formation is directly coupled to growth, R should be even smaller.

Organisms and enzymes have the unique ability to perform enantiospecific reactions, which are not easily performed in chemical synthesis. Stereospecificity is indispensable in biochemical production, as only the L- or D-form of a substance is biologically active, while the other one is in the worst-case, toxic. Amino acids have at least one stereocenter and are a good example for a closer look.

Examples of continuous production processes employing immobilized enzymes or resting bacteria include the production of amino acids. L-Aspartic acid can be produced by enantioselective addition of ammonia to fumaric acid, a substance easily produced chemically at low cost. Racemic amino-amino caprolactam (ACL) can be hydrolyzed to give L-lysine, by an immobilized L-ACL-hydrase; R-ACL is racemized to replace S-ACL by a racemase until the chemical precursor is completely used up. A general scheme of such a process is shown in Fig. 9.9.

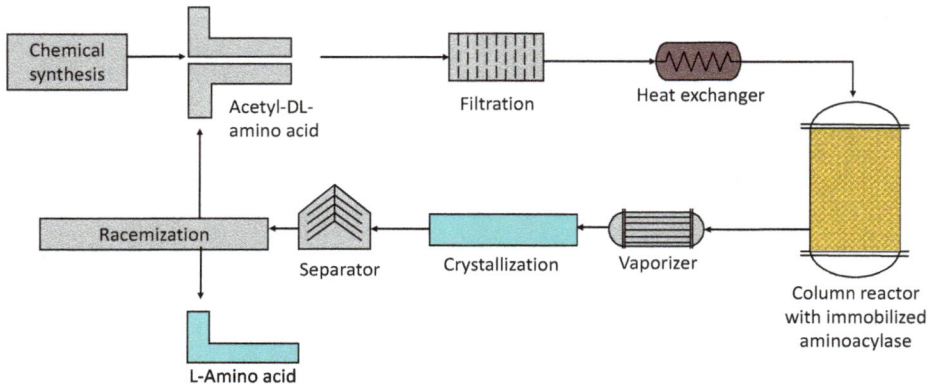

Fig. 9.9: Flow sheet to obtain L-amino acids from a D-L-precursor; here, with deacetylation and crystallization as the separation step.

Racemic mixtures of amino acids could, in principle, be produced chemically. The main problem is the separation of the two forms. Here, the stereospecificity of enzymes comes into action. The idea is to produce a precursor chemically and perform only the last step enzymatically. The L-amino acid and the remaining D-precursor can be separated, e.g., by crystallization, as they are two substances with different physical properties. This approach is an example of integration on the process level, as a bioprocess is coupled to a chemical process. Enzyme reactors can also be handled in an industrial environment, where no know-how or competence of microbial cultivation is available.

9.4 Biogas production via anaerobic digestion

Fermentative biogas generation via anaerobic digestion (AD) is a naturally occurring process that is readily observed when organic matter decomposes in anoxic milieu, e.g., in natural wetlands and rice fields, as well as the intestinal tract of ruminants and termites. Human beings have been using anaerobic digestion processes for centuries; however, the first documented digestion plant was constructed in Bombay, India in 1859. The first usage of biogas from a digester plant for street lighting was reported in 1895 in Exeter, England. Nowadays, such processes are regarded not only as techniques for treatment of sewage biosolids, livestock manure, and concentrated wastes from the food industry, but also as a potentially significant source of renewable fuel.

Due to the large waste streams, AD is carried out as batch, where the substrate is partially solid.

Organic matter is usually composed of complex polymeric macromolecules (often, in particulate or colloidal form), such as proteins, carbohydrates, and lipids. The anaerobic digestion process converts organic matter to the final products (methane and carbon dioxide), new biomass, and inorganic residue. Several groups of microorganisms (anaerobic bacteria and archaea) are involved in organic substrate transformation to CH_4, CO_2, and H_2O, and the overall process comprises multiple conversion stages with many intermediate products. Commonly, the conversion is simplified to four successive steps running in parallel in the process: (I) hydrolysis; (II) fermentation or acidogenesis; (III) acetogenesis; and (IV) methanogenesis (Fig. 9.10).

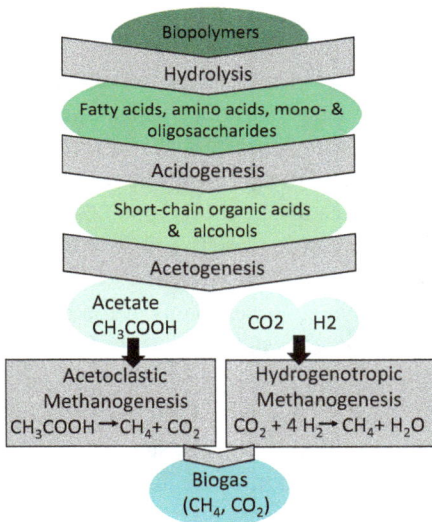

Fig. 9.10: Flow diagram of the anaerobic digestion process of organic matter; ellipsoids indicate substances with lighter color meaning smaller (volatile) molecules; rectangles are summarized conversion steps.

Biological biogas formation is a sequential process in the sense of consecutive biochemical steps and microbial consortia involved. During anaerobic digestion in biogas plants (Fig. 9.11), organic biopolymers such as lipids, polysaccharides, and proteins are converted by anaerobic hydrolytic bacteria into less complex compounds, which can then be further used by other microorganisms. According to microbiological examinations, hydrolytic species are found in a broad range of bacteria phyla and many of these bacterial have developed cell-bound multienzyme complexes, known as the cellulosomes, for the decomposition of cellulose- and hemicellulose-containing substrates. Many bacterial species do not have cellulosomes; however, they are able to secrete free hydrolases, containing multiple catalytic domains or they produce many other enzymes, such as glucanases, hemicellulases, xylanases, amylases, lipases, and proteases for efficient biomass hydrolysis. The abundance of every hydrolytic bacterial species is dependent on the inoculum type of the digester and substrate applied. Thus in biogas plants, the members of the phyla Firmicutes and Bacteroidetes are the most commonly found, while others belonging to Fibrobacteres, Spirochaetes or Thermotogae are less abundant. Thereby, the members of the genus *Clostridium* (Firmicutes) are described as usually dominating the bacterial community in the biogas plant.

In the next steps, acidogenesis, fermentative bacteria convert the breakdown products of hydrolysis to simple carbonic acids (e.g., propionate, butyrate, acetate, formate, succinate, and lactate), alcohols (e.g., ethanol, propanol, and butanol), and other compounds (e.g., H_2, CO_2, VFAs, and ketones). Some of these products (e.g., fatty acids longer than two carbon atoms, alcohols longer than one carbon atom, and aromatic fatty acids) are then used by acetogenic or syntrophic bacteria within the acetogenesis step for the conversion into acetate and C-1 compounds. Hydrogen-producing bacteria, like the homoacetogenic bacteria *Acetobacterium woodii* and *Clostridium aceticum* are usually described as performing the acetogenesis, resulting in the generation of acetate, CO_2, and H_2. The final step, methanogenesis, is performed exclusively by methanogenic archaea, whereby two functional groups are involved, namely, acetoclastic (utilizing acetate) or hydrogenotrophic (utilizing H_2, CO_2, or formate). Only few species are acetoclastic methanogens and thus are able to degrade acetate into CH_4 and CO_2. They belong to the order *Methanosarcinales* (e.g., *Methanosarcina barkeri* and *Methanotrix soehngenii*) and *Methanococcales* (e.g., *Methanonococcus mazei*), whereas all methanogenic archaea are able to use hydrogen to form methane. Methanogens of the orders *Methanosarcinales*, *Methanomicrobiales*, and *Methanobacteriales* are usually the most abundant within the archaeal subcommunity.

Overall digestion speed depends on the interplay of all involved microorganisms; the slowest step (often hydrolytic phase) is often also the rate-limiting step for the entire process. Despite the complexity of anaerobic fermentation by the involvement of the different microorganisms and degradation steps, the process is surprisingly frequently successfully used for diverse applications (methane production from biomass, water remediation, etc.). The main reason for this fact might be the age of the process (few billion years), and correspondingly, concurrent evolution of the interplay of the involved microorganisms and their metabolic self-regulation mechanisms. The optimal microbial community for maximal efficient degradation (under these anaerobic conditions) is permanently selected in these, for the most part, open (to the environment) systems. Nevertheless, this process possesses, like others, some limitations, which have to be taken into account for each application. An efficient AD process demands that both substrate degradation and methanogenesis are balanced. This can be accomplished by slow adaptation of the community to the desired substrate (stepwise adaptation duration of half-a-year to a full year). This is sometimes not respected for different reasons and the fermentation process collapses.

In the following, some scenarios will be discussed that can lead to process imbalances. Foremost, if the first process steps of hydrolysis, acidogenesis, and/or acetogenesis are too fast, the fermentation can fail due to intermediate product inhibition (scenario I). This can happen by the application of easily degradable substrates (e.g., glucose) since proliferation rates of methanogenic archaea are slower than those of the other bacteria in this process. The volatile fatty acid (VFA) concentration (intermediate product) rises within the digester, and the pH drops below the optimal range (pH 6.5–8.5) for methanogenic archaea, which can lead to decreased methane formation rates and subsequently, further accumulation of VFA, which lowers the pH until the process is completely inhibited. If the methanogenesis runs too fast, methane production is limited by the hydrolytic stage, however the process remains stable. Thus, the rate-limiting step depends heavily on the particular substrate used for biogas production. Another inhibition scenario occurs when there is an imbalance of nitrogen-to-carbon ratio (C/N ratio) of the substrate. An ideal substrate for anaerobic fermentation should have a C/N ratio in the range of 15–30. When the C/N ratio is higher than 30, microbial growths can be limited, and the substrate is not digested completely (scenario II). If the C/N ratio is lower than 15 (scenario III), the concentration of ammonia could increase during the continuous fermentation process, over the inhibitory levels of ~ 1,700 mg/L total ammonia nitrogen (TAN), where it can have an inhibitory effect, especially on methanogenic archaea. In fact, the inhibitory effect is caused by free ammonia nitrogen (FAN), which is part of TAN alongside NH_4^+, according to the dissociation constant that is dependent on pH and T. The amount of FAN (NH_3-N) is dependent on TAN concentration, pH, and temperature (eq. (9.14)) according to NH_3 – $NH_4^+ + OH^-$ equilibrium:

$$FAN = \frac{TAN \cdot 10^{pH}}{e^{\frac{63.44}{273.15+T}} + 10^{pH}} \qquad (9.16)$$

The temperature is usually regulated to a constant level (mesophilic 35–45 °C or thermophilic 45–60 °C) within the digester, so this factor is not relevant for this scenario. However, the pH is influenced by the NH_4-N concentration since its accumulation can strongly increase the pH in low buffered solution. Higher pH in turn leads to higher dissociation of ammonium (NH_4) to highly toxic ammonia (NH_3), which inhibits microorganisms within the digester (first of all, the methanogens) already in low concentrations (50 mg NH_3-N/L). Interestingly, the inhibition of methanogenesis leads to the accumulation of intermediate products (VFA) as described above (scenario I), and the pH drops alongside, decreasing the concentration of highly toxic FAN. Unfortunately, this self-regulation mechanism is very inert and, in some cases, the inhibitory potential of individual factors (TAN osmotic stress, VFA acidification) is combined, and results in a complete inhibition of the fermentation process (at pH below 6.4). Despite this biochemical self-regulation, ammonia inhibition is one of the common reasons for misbalance in industrial applications, since the adaption time for this more indolent biological system is not always applied in a sufficient manner.

Fig. 9.11: Arial view of a biogas plant in North Germany, maize silage clamps on the right upper side and digester/fermenter tanks in the center (© Martina Nolte).

Other process management parameters for an efficient operating AD process are represented by organic loading rate OLR ($kg/m^3 \cdot day$), hydraulic retention time HRT (day^{-1}), and solids retention time, SRT (day^{-1}), whereby, in common biogas plants operated as continuous stirred tank reactor (CSTR), HRT is equal to SRT since no substrate/water separation is carried out. For instance, the rapid increase of ORL, especially of easily digestible substrate, would cause fast acid formation and pH drop like in the above-mentioned scenario, I. The HRT determines the volume and capital cost for an AD system and might be helpful in avoiding scenario II by increasing the HRT (avoiding of washout microbial biomass and TAN) or scenario III, by lowering HRT (facilitate washout of inhibitory compounds TAN). SRT affects the volatile solids (VS) reduction degree and thus the methane yield from biomass.

Furthermore, for stable maintenance of the microbial community within the reactor and thus a stable fermentation process per se, other macro- and microelements (P, S, Co, Cr, Fe, Mn, Mo, Ni, Se, and W) can be crucial, especially if industrial byproducts (with unbalanced elemental composition) are used.

Particle size of the substrate has a direct influence on the biogas yield, since reduction of size improves the surface/volume ratio and correspondingly increases the rate-limiting step of hydrolysis. Increased reaction speed favors more complete substrate degradation rate by the given SRT and has a positive impact on the methane yield. Nevertheless, a balance between energy input (substrate size reduction) and output (additional biogas) is often negative or neutral, and does not pay off.

Despite the difficulties mentioned above, huge experimental and applied knowledge of the process has been gained in recent years. Efficiency by the conversion of chemically bound energy (biomass) into gaseous fuel (methane) is high. Theoretically, most of the energy (up to 90%) can be converted into methane, whereby the remaining energy is used for maintenance, metabolic activity, and de novo synthesis of microbial community. This high efficiency of energy conversion and comparably low

expenditure on equipment complexity by the application led to a proliferating industry in recent decades. The degree of application is growing worldwide, not least because of the need for alternative energy sources as replacement for fossil fuels.

Tab. 9.1: Characteristic biogas composition and theoretical CH_4 yields of different substrates.

Substrate	CH_4% in biogas	CO_2% in biogas	Theoretical/maximal methane yield L_N/kg organic dry weight
Carbohydrate	50	50	415
Proteins	60	40	446
Fats	72	28	1,014

Biogas (Tables 9.1 and 9.2) presents a suitable energy source, which can be used for
electricity generation via CHP (combined heat and power), car fuel (established technology), or launched into the gas grid. For direct use in CHP motors, the biogas must
be dry and free (<0.15% v/v) of H_2S gas in order to minimize corrosion effects. Purification of biogas from H_2S is usually carried out by a natural biological desulfurization
process, which takes place when oxygen is supplied. For the supply into the natural
gas network, the methane content must be increased to min. 89% v/v (whereby, max.
6% CO_2 and 5 H_2 are tolerable). For the separation of contaminate gases (mainly CO_2),
different methods are successfully applied (e.g. pressure swing adsorption, chemical
absorption [CO_2 scrubber], pressure water wash, and membrane separation processes). This CO_2 can be used as well, may be in the beverage industry or for feeding
of microalgae.

Tab. 9.2: Characteristic biogas and CH_4 yields of commonly used energy crops.

Substrate	Biogas yield L_N/kg organic dry weight	CH_4% in biogas	Methane yield L_N/kg organic dry weight
Maize silage	450–700	52	234–364
Maize cob	620–850	54	335–459
Sugar beet	800–860	53	424–456
Grass silage	560–620	54	302–335
Sunflower	420–540	55	231–297
Wheat grain	700–750	53	371–398
Rye grain	560–780	53	297–413
Red clover	530–620	56	297–347
Fodder beet	750–800	53	398–424

The versatility of applications of biomethane led to the creation of specialized facilities (biogas plants), which not only digest residual biomass but grow extra plant material (energy crops) for biogas generation (Tab. 9.3).

Tab. 9.3: Process parameter of a full-scale biogas plant fed with energy crop silage and manure at two different organic loading rates.

Organic loading rate (kg/m³/day)	HRT/SRT (day)	Biogas productivity (m_N^3/m³/day)	Biogas yield (m_N^3/kg VS)	Methane yield (Nm³ CH_4/kg VS)	VS degradation rate (%)
2.11	130	1.5	0.73	0.40	88
4.25	75	2.91	0.69	0.36	83

Typical process parameters of a full-scale biogas plant are shown in Tab. 9.3, whereby two different organic loading rates are applied. Higher OLR leads to much higher volumetric biogas productivity (almost two-fold). However, higher OLR has a direct decreasing impact on HRT/SRT, which leads to a slight decrease in biogas and methane yields on a VS basis, which, for its part, is also reflected in a lower degradation rate.

In some countries, the application degree is already reaching the natural limitation regarding substrate reinforcements. Here, intensive research is ongoing in order to find alternative regenerative substrates for anaerobic biogas generation. A very promising alternative might be the use of microalgae biomass, which has higher areal productivity than plant material (see Chapter 8 on microalgae). Nevertheless, high investment costs in algae cultivation plants and their natural ability to resist (partially) anaerobic fermentation processes prevent this technology from large-scale applications nowadays.

9.5 From minerals and particles – technical concepts for using chemolithotrophic microorganisms

In Chapter 3, we learned that culture media consist of components for biomass and organic product formation. For the most part, this includes a set of standard elements (C, O, H, N, P, and S), which are balanced with the stoichiometry of the organism. Independent of the species to culture, these elements always make up a substantial part in the biomass. For redox reactions, gaining energy to drive the metabolism, substrates are consumed in excess and products others than biomass are formed. Both contain mainly C, O, and H. Aerobic heterotrophs use organic compounds as substrates to be oxidized and oxygen as end-electron acceptors, on the one hand, and CO_2 and H_2O as products, on the other.

One particular group of microorganisms uses inorganic substrates in redox reactions for energy generation: the so-called chemolithotrophs (see Chapter. 2). These organisms are taxonomically very diverse and have developed manifold strategies to survive in habitats with low organic nutrient concentrations. Using the energy gained from the redox reactions, many groups are able to use CO_2 as carbon source, and are then called chemolithoautotrophs. This ability allows chemolithotrophs to even populate terrestrial sub surfaces 3 km below sea level. Chemolithotrophs play a crucial role in nature since they participate in the biogeochemical cycling of nitrogen and sulfur, and the formation of soil from inorganic material. Consequently, technical applications are already in use or under investigation (Tab. 9.4).

The capability of chemolithotrophs to oxidize inorganic material has been known for a long time. It is therefore not surprising that they are industrially utilized. Relevant bioprocesses, according to the redox reactions in this list, are briefly outlined in the next paragraphs.

Ammonia oxidation is an important step in the natural N- cycle, enabling plants to take up nitrate as the N- source. Furthermore, it is a crucial step in wastewater treatment. Ammonia, e.g., from municipal sewage is converted to nitrate. This process step is called nitrification. Wastewater treatment, based on chemolithotrophs, is therefore one of the most important bioprocesses in terms of material conversion. The next step in wastewater plants, the so-called denitrification, is the microbially facilitated reduction of nitrate to molecular nitrogen, which evaporates into the atmosphere. Otherwise, nitrogen would contribute to eutrophication of the environment. The nitrogen in wastewater mainly originates from artificial fertilizers. Plants incorporate the nitrogen and are processed into human food. The nitrogen ultimately finds its way into wastewaters in the form of urea and protein. Considering the huge amount of energy for ammonia production from atmospheric nitrogen by the Haber-Bosch process (1% of world energy consumption), denitrification is a huge waste of resources. A much more reasonable way of nitrogen disposal could be to use wastewater-derived nitrate directly as fertilizer. An obligatory prerequisite for this purpose is the efficient separation of nitrogen from other wastewater pollutants. One option to do so is the cultivation of microalgae. Microalgae are one of few organisms able to incorporate diluted nitrogen and phosphorous compounds in their biomass.

Syngas (short for "synthesis gas") is a gas mixture consisting of H_2, CO, and CO_2. It is produced, e.g., from renewable hydrocarbon feedstocks by steam reforming (Fischer-Tropsch process). In chemical engineering, it is used for the synthesis of large organic molecules or as fuel for combustion engines. Looking at Tab. 9.4, we learn that microorganisms are able to perform synthesis steps on syngas as well. This gains more and more relevance, as H_2 will be available in the future from electrolysis, where electricity is generated from sun or wind energy. However, fluid fuels will still be necessary, e.g., for aviation. Due to this strong demand, current research focuses on producing fuels or bulk chemicals by syngas fermentation. The metabolic pathways in biological conversion show a relatively high product specificity, e.g., for alcohols like ethanol or butanol

Tab. 9.4: Most important examples of inorganic redox reactions in microorganisms and technical applications.

Reaction	Reaction pattern	Group/genera microorganism	Technical process	Remark
Nitrification	$2NH_3 + 3O_2 \rightarrow 2NO_2^- + 2H^+ + 2H_2O$ $2NO_2^- + 2O_2 + 2H^+ \rightarrow 2NO_3^- + 2H^+$	*Nitrosomonas* *Nitrobacter*	Wastewater	Anaerobic
Denitrification	$2NO_3^- + 10e^- + 12H^+ \rightarrow N_2 + 6H_2O$ Intermediates NO_2^-, NO, N_2O	*Pseudomonas* *Paracoccus denitrificans*	Wastewater	Anaerobic
Sulfur oxidation	$H_2S \rightarrow HS^- \rightarrow S^0 \rightarrow S_2O_3^{2-}$	*Thiobacillus*	Bioleaching	Mainly aerobic
Sulfate reduction	$SO_4^{2-} \rightarrow S^0 \rightarrow H_2S$ Organics or H_2 as electron donor	*Desulfuromonas* *Desulfobulbus* *Desulfobacter*	Side-reaction in anaerobic digestion	Anaerobic
Iron oxidation	$Fe^{3-} \rightarrow Fe^{2-} + e^{1-}$	*Thiobacillus* *Gallionella* *Leptothrix*		Also other metals
Hydrogen oxidation	$H_2 + O_2 \rightarrow H_2O + e^-$ $H_2 + CO_2 + CO \rightarrow H_2O + e^-$	*Hydrogenobacter*	Power to fuel	Microaerophilic

or organic acids like acetate. In this regard, it is remarkable how an efficient chemical process and a biological continuous process based on a gaseous substrate are coupled. Nevertheless, the low solubility of H_2 reduces volumetric productivity, which can be counteracted by higher pressure. Reported values are up to $P_V = 50$ mg/L · h.

Chemolithotrophic oxidation contributes significantly to geochemical iron and sulfur cycling and the responsible organisms inhabit many different habitats. However, the ability to oxidize sulfide minerals was already industrially tapped in the early industrial ages. This is important because, unless oxidized, heavy metal sulfide minerals are insoluble in water or acid solutions. There is historical evidence that bacterial leaching was used for copper recovery from Rio Tinto in the eighteenth century. Today, bioleaching is a standard technique in hydrometallurgy, where chemolithotrophic bacteria are used for the extraction of copper, zinc, cobalt, nickel, and uranium from their ores. One of the major advantages over traditional mining is that bioleaching works with relatively low concentrated ores, with possible extraction yields of over 90%. Bioleaching is generally simpler and cheaper to operate. The operation costs for cathode copper, for example, range between US $0.18 and US $0.22 per pound, and can compete with traditional smelting.

There are three different bioleaching techniques, depending on the resources to be processed. These are "dump leaching" for low-grade ores and waste rock (Fig. 9.12), "agitated leaching" for high-grade chalcopyrite concentrates, and "heap leaching". Heap leaching is mainly used for newly mined run-of-the-mine ores, which contain intermediate grade oxides and secondary sulfides. The material can be leached during mining or crushed and acidified before deposition on the heap. Traditional mining can also only be partially replaced by bioleaching, for example, to replace energy-demanding and therefore cost-intensive crushing and grinding steps. This can be seen as a continuous process, where the solid substrate and the attached cells are kept in the reaction volume, which is indeed a very large volume, even if not exactly enclosed.

The main principle of leaching is to transform the insoluble sulfides into soluble salts, which can be collected. This involves acidophilic metal sulfide-oxidizing bacteria and archaea that actually oxidize Fe^{2+} and/or sulfur compounds. This reaction provides Fe^{3+} and protons, which subsequently attack the metal sulfides:

$$2Fe^{2+} + 0.5O_2 + 2H^+ \rightarrow 2Fe^{3+} + H_2O \tag{9.17}$$

The bioleaching process of copper is schematically depicted in Fig. 9.12. The most important copper minerals are sulfides, for example, chalcopyrite ($CuFeS_2$), bornite (Cu_5FeS_4), chalcocite (Cu_2S), and covellite (CuS). The process involves two bacteria, *Thiobacillus ferrooxidans*, which catalyzes the oxidation of iron ($Fe^{2+} \rightarrow Fe^{3+}$) and *Thiobacillus thiooxidans*, which catalyzes the oxidation of sulfur.

These sulfur-oxidizing bacteria oxidize sulfides to sulfates and protons, which keep the pH low and therefor support the solution of Fe ions and the solution of metal sulfides:

$$Cu_2S + O_2 \rightarrow CuS + Cu^{2+}$$

$$CuS + O_2 \rightarrow Cu^{2+} + SO_4^{2-}$$

$$CuS + 8Fe^{3+} + 4H_2O \rightarrow Cu^{2+} + 8Fe^{2+} + SO_4^{2-} + 8H^+ \qquad (9.18)$$

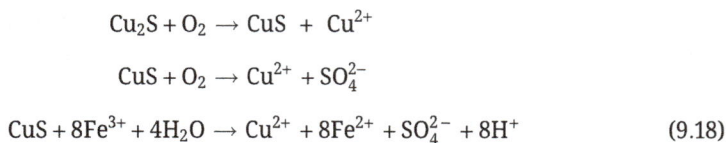

The dissolved copper ions (Cu^{2+}) are then extracted, for example, by ligand-exchange solvent extraction.

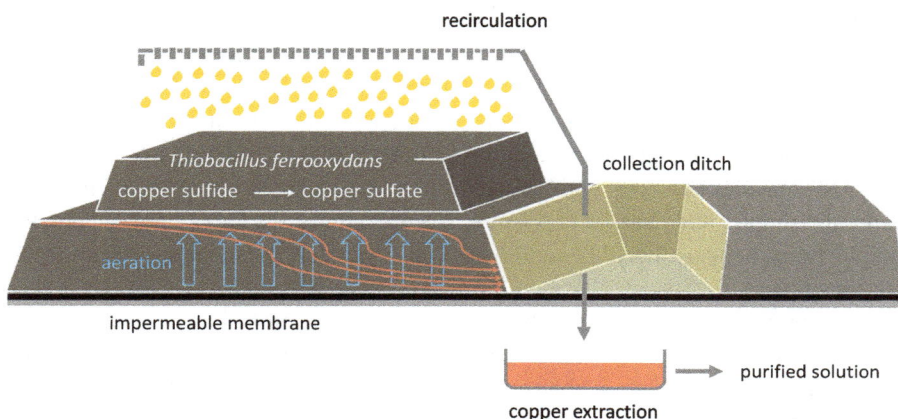

Fig. 9.12: Dump and peripheral devices for bioleaching of copper. Microorganisms oxidize weak soluble metal sulfides for mining. Fe^{3+} is regenerated by *Thiobacillus thiooxidans* in the collection ditch and recirculated onto the heap.

In the beginning of industrial bioleaching, only microorganisms were used, which naturally occurred on the leaching site. With rising success and acceptance of this technology, attempts were made to optimize the performance of the microbial population involved in the process, for example, toward acid and temperature tolerance. Some of the hyperthermophilic organisms used today were extracted from natural sources like hot springs and volcanic lava, and can catalyze reactions at 65 °C and higher. Although a large proportion of approximately 30 naturally occurring microbial strains that can be found on a bioleaching site are chemolithotrophic bacteria, most extremophiles belong to the archaea. Leaching sites are often inoculated with industrial strains of microorganisms in order to ensure a high rate of microbial activity prior to the start of the bioleaching operation.

Today, scientists are thinking one step further and numerous applications for microbial utilization of inorganic material are currently under investigation. These include extraction of rare-earths from industrial slags or waste electronics, heavy metal resorption from wastewaters, or even biological rust removal.

Beside redox reactions for the delivery of redox equivalents and energy production, some microorganisms perform inorganic reactions for other purposes. Among these are detoxification processes, or building up skeleton elements, or functional structures for sensorial tasks. In this regard, we should take a look at well-known examples of inorganic particle formation. Nanoparticles of silver and gold are reported, while less noble metals like Ni, Fe, Cu, Zn, and others are precipitated as insoluble salts, including sulfides or carbonates. Nanoparticles have very interesting qualities, which attracted the attention of material sciences in these processes. A typical "nano-" attribute is that the electron's mean free path exceeds the dimensions of the particle. This leads to specific qualities, different from the bulk material. An interesting example is magnetite nanocrystals formed by *Aquaspirillum* (Fig. 9.13).

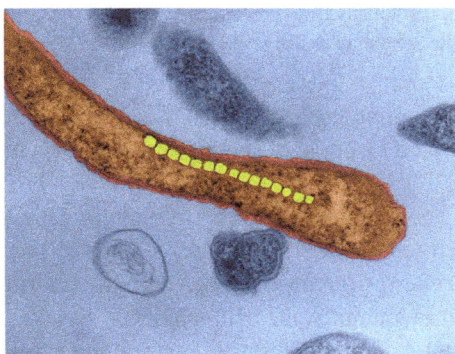

Fig. 9.13: Electron microscopic picture of *Aquaspirillum magnetotacticum*. The cell uses magnetite nanoparticles (artificially colored in yellow) for orientation along the Earth's magnetic field (© Science Photo).

While magnetite Fe_3O_4, as such, is nonmagnetic, these particles are superparamagnetic, allowing the cells to detect the direction of the terrestrial magnetic field. *Magnetospirillum* cells inhabit the microaerobic zone in the soil. In the case of anoxia, they use this orientation signal to move in an upward direction in order to find better conditions. What happens if a soil sample from the northern hemisphere is buried in New Zealand? Please do not try this out as it is forbidden. Magnetite nanoparticles (nowadays chemically produced) are of some technical interest, as they are applied in magnet separation (see Chapter 8) in biotechnology.

Another example of nanoparticles used in biotechnology are CdS crystals for their strong fluorescence abilities. Cd is a highly toxic element for all organisms. Yeasts living in contaminated soils can take up Cd and precipitate it as CdS in a detoxification reaction. The particles are coated by a protein layer to prevent Cd diffusion inside the cell (Fig. 9.14). Such yeasts can survive in heavily contaminated environments and can accumulate up to 20 mg Cd/g CDM, which exceeds the lethal dose for other cells by orders of magnitude.

The mechanism works in a way that phytochelatine molecules can bind to a stoichiometrically fixed number of Cd^+ ions (one or two). They then arrange themselves in a defined three-dimensional order to cover the CdS. The technically interesting re-

Peptide

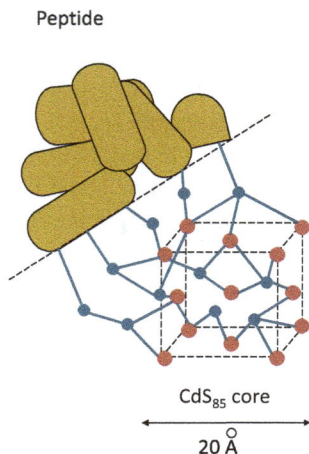

CdS$_{85}$ core

20 Å

Fig. 9.14: Structure of CdS particles in yeast as the result of self-organization on the molecular level.

sults are very (probably to one atom accurate) monodisperse particles, which cannot be produced in any other manner. This is also an example of spatial self-organization as a basic and important principle in nature.

Besides inorganic particle formation for detoxification, there are popular examples for microorganisms, which build up inorganic matter to form exoskeletons that cover the cell surface. One of the most impressive organisms is the coccolithophorid alga *Emiliania huxleyi*. This alga produces large amounts of calcite platelets, so-called coccoliths, which are 3–5 μm in diameter and composed of several crystalline subunits (Fig. 9.15). This allows for an exceptionally sophisticated three-dimensional structure, while shape and size parameters are very narrowly distributed – a property that is exclusively delivered by nature. Until now, there is no possibility to replicate coccoliths or similar particles synthetically. Another interesting difference to industrial calcite particles is the presence of organic components. These components form an organic matrix, which is embedded in the calcite structure, and which is known to control the crystal formation during the biomineralization process.

Coccoliths represent an innovative particle species, potentially suitable for numerous industrial applications. Current suggestions include all sectors of conventional calcite applications, such as bulk products like carrier material for paints and lacquers, fillers for tablets, adhesives, and even cements. High-tech applications such as semiconductors, lasers, optics, liquid displays, ultrafine surface modification, high-quality photo papers, and self-cleaning surfaces are also conceivable possibilities. The production of coccoliths demands an efficient cultivation process with high calcite productivities. This means that the alga needs large amounts of initial substrates to form equivalent amounts of calcite. Unlike other media recipes, cultivation media for coccolith production needs to contain large amounts of calcium salts like calcium chloride or even calcium carbonate. These substrates demonstrate a completely new challenge in media design, since calcium chloride inhibits growth in large concentra-

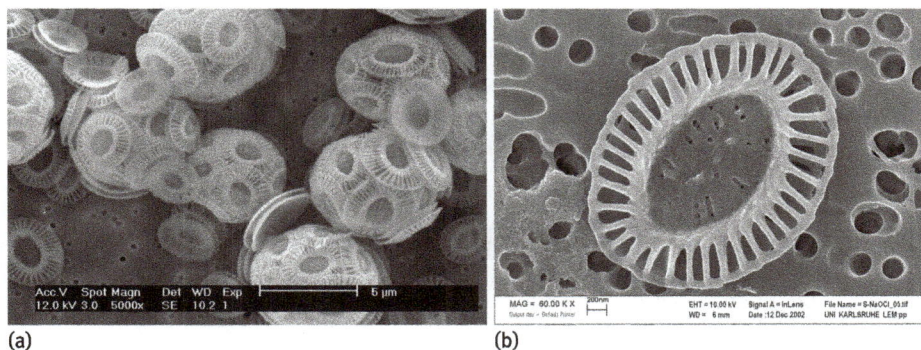

Fig. 9.15: Left: ESEM picture of a single coccolith of the coccolithophorid alga *Emiliania huxleyi*. Right: 10–15 single coccoliths build up a coccosphere, which surrounds the cell. The function of this particle envelope is still unknown (© KIT).

tions and calcium carbonate is hardly soluble in water. Besides coccoliths, there are other phytoplankton particles, for example, the silica shells of diatoms or the enormously diverse skeletons of Radiolaria, and their potential in industrial application is yet to be discovered.

9.6 Complex molecular structures – how they influence the process

Most products produced by biotechnological means are basically compounds with natural occurrence in microorganisms, plants, and animals, and are therefore preferably employed in food or as drugs. The other way round, production is not only possible by biotechnological production with microorganisms, but also by chemical synthesis or by extraction from higher organisms, especially plants. The decision for a biotechnological process depends on several aspects, of which the structure of the molecule is one. From the structure of penicillin, it can easily be understood that a complex molecule can be more easily produced by a microorganism than chemically. This also holds for the last step in production, the enzymatic cleavage of the residual. Here, at a very specific site of the molecule, a substitution is introduced. Only enzymes or cells can be so specific; this characteristic feature is called "regioselectivity." Where this is the case, the term "regiospecifity" is also used very strictly. A bit further, we will have a look at the production of vitamin C, which is an example to follow in this aspect.

It is since the beginning of the twentieth century that scientists understood the role of specific substances in food as protective agents against different kinds of deficiency diseases (Casimir Funk, 1912 and later). The seaborne disease, scurvy, e.g., could be effectively combated by vitamin C. These "vitamins" were later identified, described, and synthesized. The impact of these findings is so high that 12 Nobel Prizes for 20 laureates were awarded. In the beginning, these trace nutrients were extracted from plant material. In the case of vitamin C, the human need is about 75 mg/day, usually completely covered by fruit or vegetables. Note that paprika contains 100 times more vitamin C than citrus fruits, which was first detected. Production on a technical scale was introduced in 1934 by Tadeus Reichstein, 20 years after its discovery, a milestone in the history of nutrition. This included even a biotechnological step. Figure 9.16 gives an overview over the so-called "Reichstein-Guessner-process."

1. In the first step, D-sorbitol is formed by the hydrogenation of glucose over a nickel alloy.
2. By microbial oxidation with *Gluconobacter oxydans*, it is possible to oxidize the OH group at C 5 of D-sorbitol specifically, and thus generate L-sorbose. The dehydrogenation reaction with *Gluconobacter oxydans* is regiospecific and follows the empirical Bertrand-Hudson rule. Due to this rule, polyols with cis arrangement of two secondary hydroxyl groups in D-configuration to the adjacent primary alcohol group (D-erythro configuration) are oxidized to their corresponding ketoses.
3. The open-chained form of L-sorbose is in equilibrium with the cyclic hemiketals, L-sorbofuranose (α- and β-forms) and L-sorbopyranose (α and ß forms).
4. In an acid-catalyzed reaction of α-L-sorbofuranose with acetone, 2,3,4,6-di-ispropylidene-α-L-sorbofuranose (diacetone-L-sorbofuranose) is formed. This insertion of the protective groups is necessary to enable the selective oxidation of the terminal OH group of L-sorbose. Otherwise, all hydroxyl groups are available for oxidation.
5. Diacetone-L-sorbofuranose is then oxidized over a Pt or Pd catalyst to di-*O*-isopropylidene-2-ketogulonic acid.
6. The acetal protecting groups are removed by treatment with acid.
7. The keto form of the resulting 2-ketogulonic is due to keto-enol tautomerism, in equilibrium with the endiol form. It rearranges in acidic conditions to L-ascorbic acid. In an intramolecular esterification, the OH group in position 4 reacts with the carboxyl group spontaneously under elimination of water to a γ-lacton.

The important step is the regioselective transformation from sorbitol to sorbose. This partial oxidation is interesting insofar as sorbitol is a nearly symmetric molecule. A chemical catalyst could hardly distinguish between the two sides and would deliver many different isomers. So, a chemical synthesis would be much more complicated, as can be seen in the following steps where protective groups have to be introduced. Here, high regioselectivity of biosystems shows a clear advantage over chemical synthesis. This process has been standard for many years and is in operation even today, although the final yield is only 50% and accompanied by the need for solvent recycling, high temperatures, and high pressures in some of the steps.

The story shows how fast biotechnological realization can be, after scientific recognition is achieved. Politics also play a role. In 1994, the World Bank published a paper on the importance of vitamin supplementation of food for overcoming vitamin malnu-

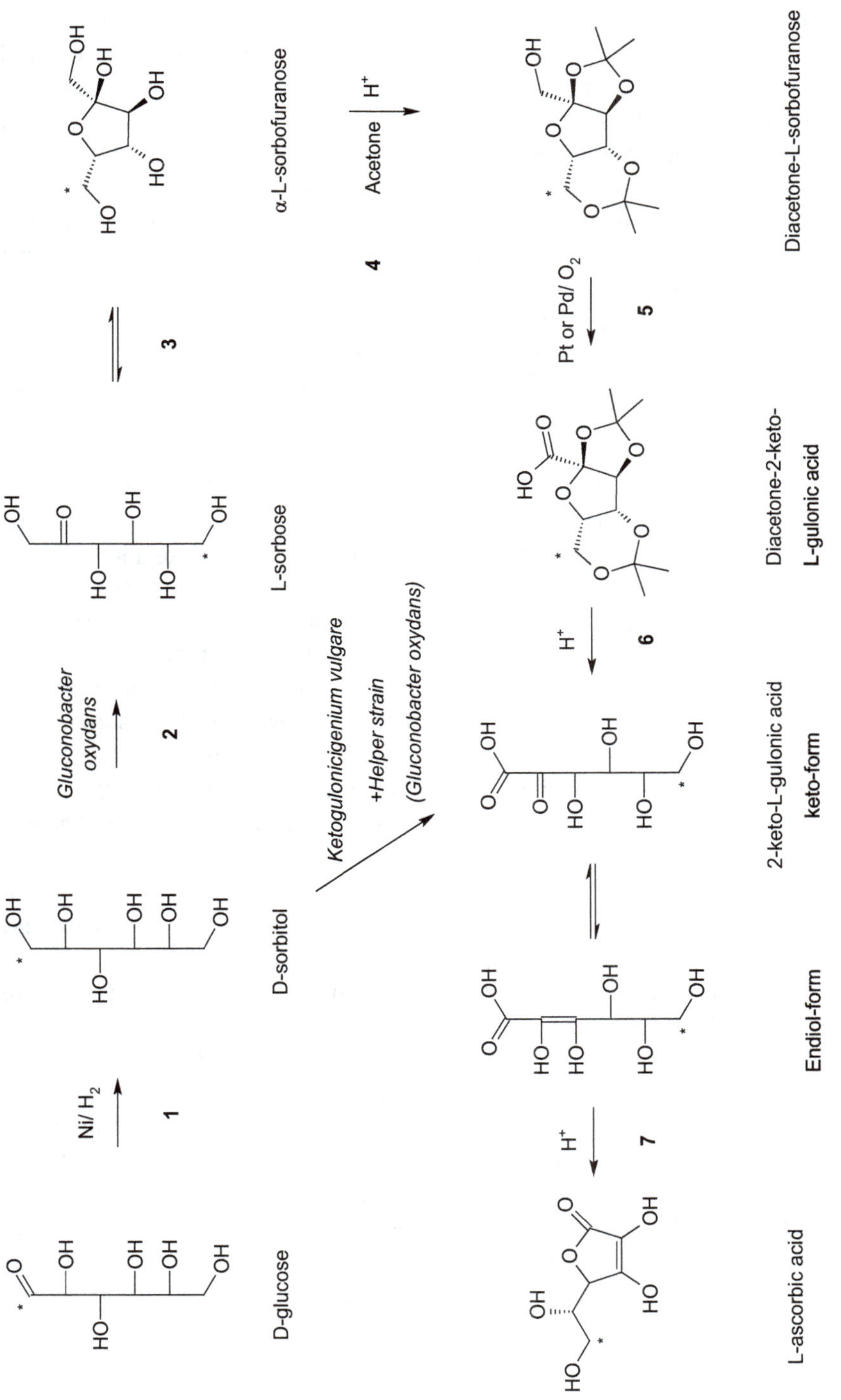

Fig. 9.16: Production steps of the Reichstein-Güssner process employing protective groups and modern shortcuts by microorganisms.

trition. The Chinese government declared vitamin production a key technology. This encouraged further research and led to today's dominance of Chinese research and production of vitamin C. The target of new processes is of course, as becomes clear from the sketch in Fig. 9.16, the substitution of the protective group steps by a direct partial oxidation sorbose, directly to 2-KLGA, with another microorganism, *Ketoguloronicigenium vulgare*. This only works together with a "helper strain," e.g., *Bacillus megaterium*, for not completely understood reasons. This modern two-step process delivers up to 130 g/L vitamin C, with a yield of 80%, related to sorbitol. Further improvement by different process variations and other genetically engineered microorganisms is in progress. In 2014, the world production amounts to 110,000 tons per year, mainly in China, and is sold in a competitive market for approximately US $6–8/kg. Western producers serve a premium market segment, protected by quality brands. Around 50% goes into the pharmaceutical market, 25% is sold as antioxidant, while 15% is used as an additive in beverages.

All these achievements would not be possible without the ability of *Gluconobacter* and *Ketogulonicigenium* performing partial oxidation. So, a short look is advisable. The enzymes involved are mainly located in the periplasm, so the substrate does not have to enter the cytoplasm and the product leaves back to the environment. Most of these dehydrogenases are characterized by a remarkable regiospecificity, but relatively low affinity, and by a broad substrate range. This makes a two-step process necessary to avoid direct contact of *K. vulgare* with sorbitol, possibly leading to unwanted byproducts. Both stages are batch processes with product titers above 200 g/L for sorbose and 100 g/L for 2-KGA. As oxygen transfer is comparatively low, airlift reactors are chosen. Substrate overshoot metabolism is not observed. In fact, this is reminiscent of an overshoot mechanism, as the TCC is incomplete and could be bridged by the NADPH gained from the uncomplete oxidation. To ease the growth of the cells, a rich medium with high titers of corn steep liquor and yeast extract is used. Nevertheless, careful control of metabolic turnover rates could finally lead to a one-step process, provided *G. oxidans* works faster, keeping the sorbitol titer low.

Besides further improvement of the bacterial route via genetic engineering, we can also think of alternatives, e.g., enzyme reactor with NADPH recycle, or complete synthesis by plants cell culture or microalgae. This is actually possible, but product titer of 2 g/L makes downstream processing, including cell disruption, too expensive, for the time being. Even genetically engineered higher plants cultivated in greenhouses are not excluded. Other examples for incomplete oxidation are vitamin B_2 (precursor D-ribose with *Bacillus*), vitamin B_{12}, biotin (possible) or regiospecific oxidations of progesterone by *Rhizopus nigricans*, leading to hydroxy-progesteron, and finally, hydrocortisone. Further and more general examples with some insight into the relation between the molecular structure and the process are collected in Tab. 9.5.

Tab. 9.5: Unique selling points, molecular structures, and processes.

Aspect	Substance class	Example products	Process
Enantioselectivity Stereospecificity		Glutamate	Direct synthesis by *Corynebacterium*
			Enzymatic
Regioselectivity		Penicillin	Direct synthesis Enzymatic substitution
		Vitamin C	Partial oxidation Extracellular
Macromolecules Sequential order	Proteins	Technical proteins	Direct synthesis, extracellular, *Bacillus*
		Hormones	Intracellular e.g., *E. coli*
		Antibodies	Animal cell culture
	Polysaccharides	Xanthan Schizophyllan	Direct bacterial synthesis extracellular
	Nucleic acids	RNA therapeutics DNA computer (?)	Direct synthesis Intracellular
	Fatty acids	Health food Lubricants	Direct synthesis
Autocatalysis	Whole cells	Baker's yeast	Fed-batch
		Starter cultures	Batch

9.7 Current discussion about continuous processes – motivation and obstacles

Continuous processes are, for the time being, applied for large amounts but small molecules; nevertheless, not in the extent as could be expected or envisaged when summing up the advantages and obstacles (Tab. 9.6).

Especially in the pharmaceutical industry, there are some reservations but with a positive trend for more applications. Here are statements from a representative from the industry:

Today's reality follows the "legend" that continuous bioprocesses are more for academics ("conti cult") pretty closely, although there are quite some approved processes that use continuous (perfusion) cultivation processes for an extended period of time. For those processes, genetic stability has been demonstrated during the entire perfusion process. For the downstream process, things seem to be moving now. In today's approved processes, all DSP operations are indeed batch-wise. New technologies have been developed that would allow transforming the DSP steps into an inte-

Tab. 9.6: Pros and cons of continuous processes.

Physiological aspects		Technical aspects	
Cells are fully adapted	+++	Reactor always full at high biomass concentration	+++
Insidious contamination	−−	No batch-wise quality control possible	−−
Creeping loss of plasmids	−−		++
High product concentration and productivity, not simultaneous	−	Can be compensated by cell recycle	+
No high dilution rates for product inhibition	−	Can be compensated by reactors in series	+
Substrate lost in effluent	−	Low reactor downtimes	+
		Continuous downstream desirable	+/−

grated continuous bio-manufacturing process. This includes single-pass tangential flow filtration and multicolumn chromatography. Although some of these technologies may require slightly more complex equipment, the process itself should be less complicated because (a) they are operating at a steady state, (b) they provide better segregation between different process fluids, and (c) the impact of dead volumes is far less significant in continuous processes than in batch process. The potential advantages of a downstream processing platform are improved productivity (producing more product in smaller process equipment, thereby enabling a fully disposable process), enhanced process control, and with that, better product quality control. This last point has also inspired regulators (e.g., FDA) to support the trends towards continuous processing in pharma and biopharma. So -in short- yes: the current opinion still follows the mentioned legend, but things are rapidly changing. I would hope that academic institutions, who educate tomorrow's process engineers and process scientists, would appreciate the current state of the art (batch) but also address the trends that are natural in a maturing industry (which includes implementation of continuous processing).

In scientific social networks, the topic is further discussed; links are given in the reference list.

9.8 Questions and suggestions

1. Is it possible to increase the maximum biomass productivity in the standard example by increasing substrate affinity, e.g., by genetic engineering?
2. Calculate the specific substrate uptake rates in the data example of continuous ethanol production in Fig. 9.5 and discuss the influence of ethanol concentration on kinetics.
3. You produce a recombinant protein with a genetically engineered cell line of $D = \mu_{eng} = 0.2\ h^{-1}$. By an occasional reverse mutation, 1% of the cells are wild strain cells in the reactor. They can growth faster without metabolic burden (higher yield) – $\mu_{wild} = 0.3/h$. Calculate the time until they are 10% and finally overgrow the mutant.
4. What other mechanisms can affect a long-term continuous bioprocess?

Chapter 10
Measuring principles – how to put an end to the blind flight

Nobody would enter an airplane knowing that the pilot can control the flight only by a view out of the cockpit window. Bioprocesses are very complex as well require detailed and continuous monitoring and technical specifications for the course of the process. This chapter gives an overview of the technical possibilities, potential benefits, and constraints with respect to employment of at least basic measurement devices. The direct physical principles generating the relevant signals will be outlined and followed by the measurement chain up to a level where the meaning for the process can be evaluated. This gives the basis for a generalized view of the structure of sensors for applications in complex environments.

10.1 Only a look through the keyhole – what measurement means for process understanding and optimization

One of the major advantages of bioreactors over shaking flasks or roller bottles is the possibility to measure and control many environmental parameters influencing the microbial growth process. Even slight deviations from optimal values can change the specific growth rate or specific production rate by several percent. Microorganisms change many of these parameters by their metabolic activity. The other way around, changes of pH or pO_2, for example, offer useful information to assess the current physiological state of the cells. Targeted changes in environmental parameters are a means for optimized process strategy. For the production of (bio) pharmaceutical products a transparent and comprehensive documentation of each process run is demanded by the authorities. Measurement data are required to prove that all of the important parameters are within a prescribed range to ensure product safety. The key phrase is "process analytical technology" (PAT), a set of rules for how these aims are to be achieved. Especially in technical environments there are some reservations against installation of sensors as they are suspected to increase contamination risks. Detractors speak of making the reactor into a "Swiss cheese." Furthermore, measured values are not rigorously evaluated to give hints for process enhancement. In bioprocess technology and more in environmental engineering, the term "monitoring" is frequently used.

https://doi.org/10.1515/9783110773354-010

10.2 State of the art – overview of measurement at a standard reactor

To observe and control the environment many measurement variables are important and usually measured. Others should be measured but no reliable standard sensor exists. Specific sensors for dedicated processes are in use only in rare cases. In Fig. 10.1, an overview of usual measurements on a bioreactor is shown (Tab. 10.1).

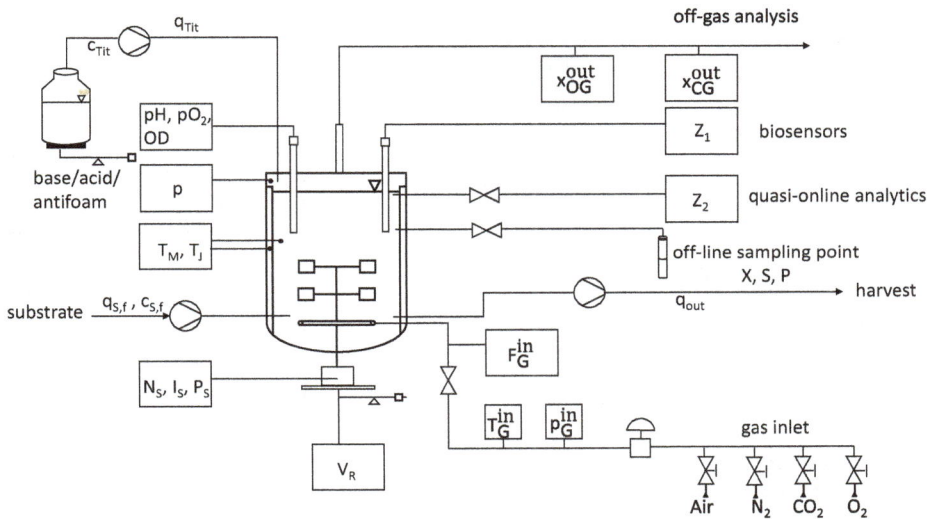

Fig. 10.1: A bioreactor equipped with on-line and at-line sensors for different parameters; the parameters to be measured are shown as rectangles; the related devices as usual flow sheet symbols.

Tab. 10.1: Labels, names, and some additional information on standard sensors for fermentation.

Label	Name	Physical principle	Meaning for process
T_M, T_J	Temperature (medium, jacket)	Change of electrical resistance, PT100	Dedicated optimum
P	Pressure	Mechanical force by pressure difference	Drives exhaust gas flow, control during sterilization
pH	pH value	Electrochemical cell	Dedicated optimum
pO_2	Oxygen partial pressure	Electrochemical cell	Minimum to be respected
OD	Optical density	Light scattering	Estimation of biomass concentration

Tab. 10.1 (continued)

Label	Name	Physical principle	Meaning for process
$N_S, I_S,$ P_S	Speed, electrical current, electrical power of stirrer	Electromagnetic	Power input for mixing and gas dispersion
V_R	Working volume	Weight via balance	
F_{liquid}	Mass flow, liquid	Differentiation of balance signal, conversion to volume by calibration	Medium supply, balancing
F_{gas}	Mass flow gas	Hot-wire anemometer, conversion to volume by calibration	Balancing gas phase
$x_{O2},$ x_{CO2}	Molar fraction gas	IR for CO_2 and paramagnetism for O_2	Gas transfer, physiological state
K, rH	Specific conductivity, Redox potential	Electrochemical sensor	Sum parameter for dissolved ions

> Sensors are classified according to the measurement principle but also according to the purpose of application. The definitions of some technical terms are given in the following paragraphs as background information for this chapter.

"In-line" measurements are based on sensors integrated into the reactor and usually have direct contact with the medium. They need specific flanges at the reactor and must survive high temperature during sterilization. "At-line" measurements are mounted outside the actual reaction space but in the gas phase or in a bypass (side stream, slip stream). Here size or technical design is not so limited, but measurement is more indirect. The terms "in-line" and "at-line" relate therefore to the spatial structure of the process. In cases where a sensor measures a compound but without direct contact with the medium or the gas phase the term "noninvasive" is used.

Another classification of terms connected with measurements is related to temporal resolution. Most in-line sensors deliver values in rapid succession. That holds also for some at-line sensors. The last value represents the current state of the measured variable. Such sensors are called "on-line" sensors. According to the Nyquist-Shannon sampling theorem the measuring frequency should be at least 10 times faster than the characteristic time constant of the process. The maximum frequency is limited by the establishment of the measured variable and the physical state of the sensor. The term "off-line" measurements in bioprocess engineering refer to sampling and analysis in the lab. The results are typically not available during the process. To overcome this problem some analytic devices can be connected to the reactor by an automated sampling system. These measurements are referred to as "quasi-online." While it takes several minutes to get a new value it is fast enough to make decisions while the process is running.

There are other terms that are in use when designating the physical working principle. "Optical" sensors use the interaction between light and the analyte, while "electrochemical" sensors are based on an electrochemical effect as described below. Here "potentiometric" (also "galvanometric") means an electric potential and "amperometric" (sometimes "polarographic") an electric current as the primary measuring signal. In contrast to "analog" sensors, "digital" sensors contain an analog/digital converter (AD converter) to reduce signal corruption by noise in rough environments. "Software sensors" combine different hardware sensor signals and process the data with computational methods to elucidate new information. They calculate a virtual measurement like the specific growth rate µ from real measurements like off-gas analysis. It makes process monitoring more convenient. An example is the integration of $dc_X/dt = \mu \cdot c_X = y_{X,O} \cdot$ OTR using the stoichiometry between cell growth and oxygen uptake. This gives estimation of biomass as

$$c_{X,est} = c_{X,0} + y_{X,O} \cdot \int OTR \, dt \tag{10.1}$$

and

$$\mu_{est} = \frac{OTR}{c_{X,est}} \tag{10.2}$$

A sensor's sensitivity indicates how much its output changes when the input quantity being measured changes.

10.3 Physical parameters – adaptation from other fields of process technology

Temperature is a central environmental parameter. In Chapter 2, it was already shown that a deviation from the temperature optimum by 1 °C can cause a loss in growth rate of several percent. Measurement is a standard done by a PT100, a platinum wire of 100 Ohm electrical resistance having linear resistance/temperature kinetics.

Pressure in the headspace is measured firstly for keeping 2 bar during sterilization and secondly during cultivation for driving the exhaust gas through the off-gas tubes. In pressure sensors a pressure difference leads to deformation of a membrane, which is secondarily transduced by, e.g., piezoelectric effects to an electric signal. In big reactors the pressure inside the medium, e.g., at the bottom, is of interest. The pressure is a measure of the fill level and influences the solubility of gases.

To measure the working volume of the reactor during cultivation, which is the amount of medium, is not simply done by a float. The signal would be corrupted by foam, the vortex induced by stirring, and last but not least by the changing gas hold up (bubble volume) in the reactor. The only solution is to put the whole reactor on a balance, which is precise enough to follow changes of weight during the process. Similar reasons hold for employing balances for measuring the amount of acid, base, and antifoam fed into the reactor. The related fluid mass flows are calculated from the balance signal by differentiation and pump calibration.

Gas mass flow measurement is foreseen only at the gas inlet. It is based on hot-wire anemometry. An electrically heated wire at constant temperature is overflowed by the gas. Heat transfer from the wire to the gas measured by electrical heating power depends on gas flow and molecular mass. Hence, calibration has to be done for the gas composition in use. Although mass flow controllers as the standard device to control gas mass flow are quite expensive one should give precedence to them over the classic rotameters.

Off-gas composition is basically a chemical parameter but based on physical effects. Standalone measurement devices are adapted from other industrial applications, as are most of the physical sensors. A specific feature of CO_2 is the strong light absorption in the infrared (IR) range. This is commonly known from the global greenhouse effect but employed here for measurement purposes. IR light is absorbed by the gas sample, which is consequently heated up. The pressure difference to a comparison sample is the final physical parameter to be calibrated to x_{CO2} and displayed. This is an example of a measurement chain where the signal carrying the measurement information is several times transformed before it reaches the data acquisition system.

A unique characteristic of oxygen is its paramagnetic property. A gas sample experiences a force in a strongly inhomogeneous magnetic field into the direction of increasing field strength. This induces a measurable deviation of the gas flow or the torsion of a dumbbell-shaped element carrying a comparison sample at the other end.

10.4 Chemical parameters – employment of electrochemical effects

Chemical parameters can be measured in the suspension but also in the gas phase. As a first target we have a look at dissolved ions, which are important as medium compounds and influence the pH value. To understand the physical basis of related sensors an excursion to electrochemical effects generating measurement signals is useful. Three basic arrangements are shown in Fig. 10.2.

In the first arrangement (Fig. 10.2a) a solid metal electrode is in contact with water or a salt solution, which could be the medium of a bioprocess. Metal ions will diffuse into the medium. Thereby an electrical potential is built up, which exerts an electrostatic force (Coulomb force) on the ion in the opposite direction to diffusion until equilibrium is reached:

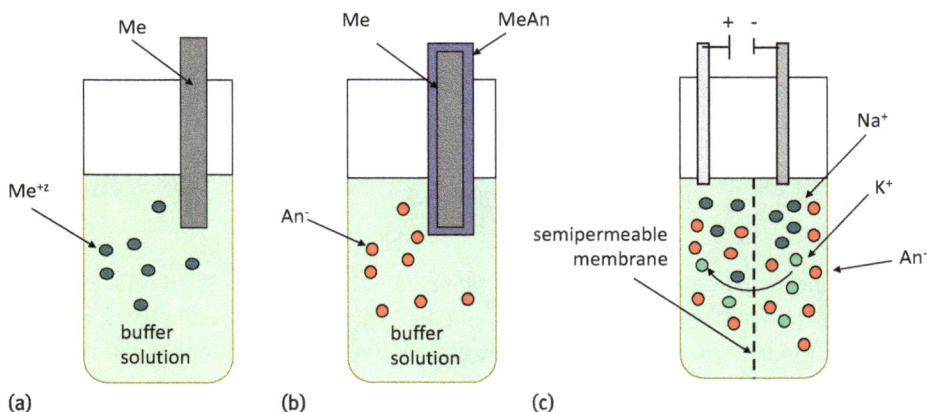

Fig. 10.2: Basic principle of electrochemical effects; the potential at the electrodes depends on the ion concentration: a) metal electrode; b) metal-oxide electrode. Semipermeable membranes c) build up a transmembrane voltage.

$$Me^{+z}(\text{solute}) + z \cdot e^-(\text{solid}) \rightleftharpoons Me(\text{solid}) \qquad (10.3)$$

The resulting electrical potential is given as

$$\phi = \phi^\circ + \frac{R \cdot T}{z \cdot F} \cdot \ln a_{Me+z} \qquad (10.4)$$

This equation is called the Nernst equation according to Walther Nernst (Nobel Prize Chemistry in 1920). The activity is defined as an activity constant γ – for simplicity here set to 1 for low ion concentration multiplied by c_{Ion}/c^0, where c^0 is the unit 1 mol/kg. Summarizing, this arrangement acts as an ion-sensitive electrode and enables us to measure cation concentrations. See a calculation example under questions.

In the second arrangement (Fig. 10.2b) the electrode is covered by a layer of a metal salt, e.g., metal oxide or metal chloride. At the inner interface a dissociation process generates electrons e^-:

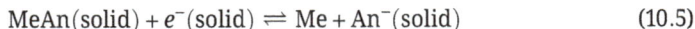

$$MeAn(\text{solid}) + e^-(\text{solid}) \rightleftharpoons Me + An^-(\text{solid}) \qquad (10.5)$$

At the outer interface again diffusion and electrostatic forces come to equilibrium:

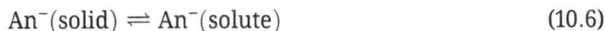

$$An^-(\text{solid}) \rightleftharpoons An^-(\text{solute}) \qquad (10.6)$$

In this case, the Nernst equation reads:

$$\phi = \phi^\circ - \frac{R \cdot T}{z \cdot F} \cdot \ln a_{An-z} \qquad (10.7)$$

By this ion-sensitive arrangement obviously anion concentrations can be measured.

In the third arrangement a semipermeable membrane separates two compartments filled with salt solutions, e.g., KCl in compartment I (left) and NaCl in compartment II (right). Assuming that only K^+ can diffuse through the membrane, some K^+ ions will follow the concentration gradient until they are stopped by the self-induced electric field. In this case, the Nernst potential becomes

$$\phi = \frac{RT}{z \cdot F} \cdot \ln \frac{a_{K+I}}{a_{K+II}} \tag{10.8}$$

This membrane potential plays an important role in biology. Nearly all cells exhibit a membrane potential over the cell membrane. Biological receptors often work by modulation of the membrane potential. In the case of neurons, the potential can be more than 100 mV. Also, where no membrane is present, an initial gradient in ion concentration can lead to an electric potential, because different ions diffuse with different velocities. In cases where they have different charge, a diffusion potential is built up.

As an example, for the application of electrochemical sensors, we will have a closer look at pH-value, which is one of the most important environmental parameters to control in a bioreactor. Starting with a correct pH value at the beginning of the process the hydrogen ion concentration (correctly speaking, oxonium H_3O^+) is subject to change due to activity of the microorganisms. They may produce acids or take up acid or basic salts. One candidate is NH_4^+ where the uptake as ammonia leaves a proton in the medium. In most cases, the pH value drops during a cultivation making titration by a base necessary. This could again be ammonia solution. As long as the biomass contains a constant quota of nitrogen and does not produce acids, ammonia dosage by pH control is proportional to biomass growth. Here we have another example of a popular software sensor, where the growth is estimated via ammonia dosage.

To measure pH, in principle a hydrogen gas electrode according to arrangement Fig. 10.2a would be necessary. However, that would be very impractical. At this point an observation, made firstly in the early nineteenth century, helps. It says that protons can diffuse into glass material generating in this way a potential between the glass surface and the surrounding medium. This effect is made possible by the small size of protons and the ion exchange properties of glass, where Na^+ ions, or Li^+ in modern versions, can compete with H^+ ions for the negatively charged SiO^- compounds in the glass. "Glass electrodes" using this effect are probably the most widespread electrochemical sensors. A commercial pH sensor additionally provides all necessary parts to form a galvanic element as a closed electrical circuit, where the voltage generated by the protons can be measured. A schematic drawing of the tip of such an electrode is shown in Fig. 10.3.

The central part is the bulb shaped glass "membrane" (about 0.5–1 mm thick) carrying inside and outside the pH sensitive layer of a hydrated "gel" (about 10–100 nm thick). Increasing the surface of the membrane is necessary to reduce the electric resistance. In between the two sensitive layers Li^+ ions (Na^+ in older versions) take over the charge transport. In principle, the electrical potential ϕ on both sides of the membrane

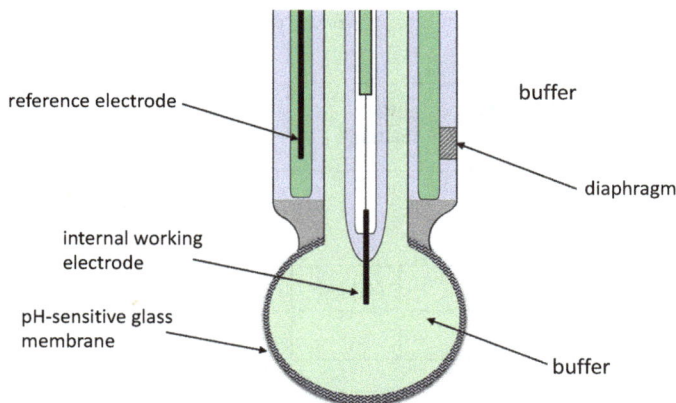

Fig. 10.3: Tip of an electrochemical pH-electrode; the two compartments contain buffer; the working and the reference electrode are visible as well as the glass membrane (sensing gel layer) and the diaphragm.

must be measured. The difference $U_{\text{meas}} = \Delta\phi$ is the voltage as the primary signal. The cylindrically shaped electrode shaft consists of two separate annular cavities forming two electrochemical half-cells corresponding to these two potentials. In the inner volume of the probe an $Ag^+/AgCl$ electrode – the "working electrode" – is in contact with the inner surface of the membrane via a KCl solution (0.1–1.0 molar). The potential at the Ag^+ wire is therefore the sum of the outer membrane potential, the cc potential, the inner membrane potential, and the potential at the buffer/AgCl interface.

The potential outside of the glass membrane is held on ground potential as the reactor and the medium are earthed. However, this is not defined precisely enough and must be measured inside the medium as close as possible to the place of the glass membrane. This is the task of the outer reference electrode. In fact, the medium is in electric contact with the outer annular volume of the probe via a perforated ceramic plate, the so-called diaphragm, or via small pores (gel electrolyte). The reference electrode, again $Ag^+/AgCl$, measures the potential in the outer annular volume delivering in this way a measure for the potential at the outer gel layer in contact with the medium. The outer volume is filled with a KCl solution of the same concentration as the inner volume. This salt is chosen as K^+ and Cl^- exhibit similar mobility thus avoiding a diffusion potential in the diaphragm. Finally, an amplifier closes the circuit. As the probe consists of two single electrodes, it is called a "combination electrode." The course of the potential along the circuit is given in Fig. 10.4.

The potential of the outer gel layer gives a logarithmic measure of the pH value according to the Nernst equation. As pH itself is a logarithmically defined quantity, a linear relation between voltage and pH of approximately 60 mV/pH is provided, this value being the sensitivity of the pH electrode. All other potentials along the measurement chain are constant but require a zero-point adjustment and a calibration prior to use. Cross-sensitivity especially with Na^+ or K^+ in the medium may happen prefera-

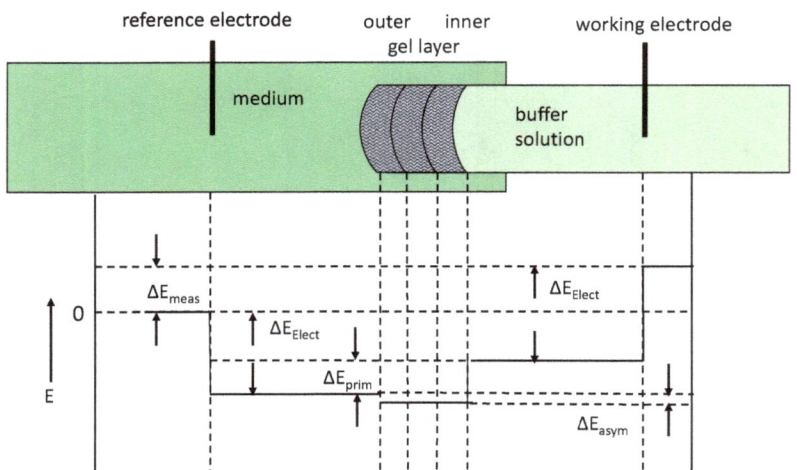

Fig. 10.4: Course of potential E along the measurement chain in a pH electrode; E_{prim} primary measuring effect (approximately 60 mV/pH), and E_{meas} measured value (appr. 60 mV/pH), E_{elect} electrode potential, E_{asym} asymmetry potential (a few mV) also containing the diffusion potential through diaphragm (approximately 2 mV).

bly in the alkaline range. Temperature sensitivity (see Nernst equation) is actively compensated by the amplifier based on a Pt-100 temperature sensor integrated into the pH probe as well.

The importance of knowing the oxygen partial pressure pO_2 during cultivation has already been discussed in Chapter 6. While pH measurement is an example of a potentiometric (galvanic) sensor measurement of pO_2 is based on an amperometric principle. Therefore, a voltage has to be actively applied to the electrode inducing several electrochemical reactions. The principle is shown in Fig. 10.5.

At the cathode – the working electrode (Pt or Au) -O_2 is reduced to OH^-:

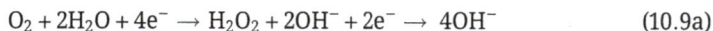

$$O_2 + 2H_2O + 4e^- \rightarrow H_2O_2 + 2OH^- + 2e^- \rightarrow 4OH^- \tag{10.9a}$$

At the anode – the reference electrode (Ag) – electrons are produced at the cost of the charge of Cl^-:

$$4Ag + 4Cl^- \rightarrow 4Ag^+ + 4e^- + 4Cl^- \rightarrow 4AgCl + 4e^- \tag{10.9b}$$

The electrons close the charge balance via the electrodes and the external amplifier. Consequently, the current I_{meas} is directly proportional to the number of oxygen molecules being oxidized at the cathode. Employment of the two electrodes in direct contact with the medium would induce many other electrochemical reactions. So, we need an additional element that is selectively permeable for oxygen and prevents ions from diffusing into the space inside the sensor. As oxygen is quite hydrophobic such a separating membrane can be selected out of a range of hydrophobic materials;,

Teflon is a usual choice. Remember that also for bubble-free aeration of animal cell culture Teflon membranes are employed. The electrode is called a "Clark electrode" in honor of its inventor Leland C. Clark (development around 1956–1962).

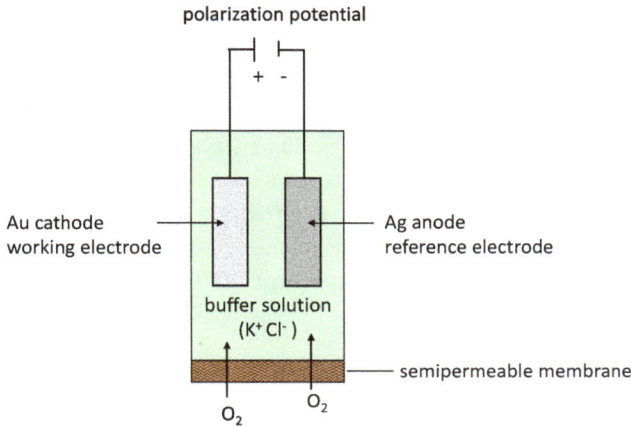

Fig. 10.5: Measuring principle of a Clark electrode.

After switching on the sensor, oxygen inside the probe is used up, while fresh oxygen diffuses through the membrane. Accordingly, a current is observed increasing with increasing voltage. pO_2 sensors are designed in a way that at around 0.6–0.8 V – the typical working voltage – the oxygen concentration in the probe drops to zero and only fresh oxygen diffusing through the membrane is reduced at the cathode making the current proportional to the oxygen concentration in the medium:

$$I_{meas} \infty \ Q_{O_2} = k_{Diff,O_2} \cdot \left(c_{O_2,medium} - c_{O_2,sensor} \right) \approx k_{Diff,O_2} \cdot c_{O_2,medium} \tag{10.10}$$

Higher voltage (>1.6 V) would lead to hydrolysis of water inside the probe. As the diffusion coefficient $k_{Diff,O2}$ (m^2/s) may be subject to change the sensor has to be calibrated against a solution with known c_{O2}. That can be provided by saturation of a medium sample with air, nitrogen (zero value), or other gases with known composition. While the physical measurement principle is based on concentration usually calibration is done based on partial pressures of the calibration gases. This has to be considered in cases where oxygen solubility changes, e.g., by changing ion concentrations or pressure changes during cultivation. The Teflon membrane has to be changed from time to time as proteins can precipitate on it especially during sterilization. Furthermore, it has to be noted that the ions are not balanced, and the electrolyte has to be changed after a number of hours of operation.

Electrochemical sensors have some disadvantages as they have to be calibrated and cleaned regularly to avoid aging and drift. The response time is typically in the range of up to a minute. The time constant is given by the product of the diffusion coefficient k_L for the analyte through the membrane and the storage capacity of analyte in the sensor volume. So, the membrane surface has to be as large as possible (sensor diameter, spherical shape) and the volume as small as possible. Nevertheless, there is a need for more robust sensors with shorter response times. An interesting property of dissolved oxygen has been found in fact that it is soluble in a polymer membrane containing a luminescent ruthenium(II) dye complex. There it acts as a luminescence quencher. The effect can be measured by irrigating the membrane with blue light and measuring the red luminescence light (Fig. 10.6). Oxygen sensors working according to this principle belong to the photometrical sensors and increasingly replace the classic electrochemical oxygen electrodes. Even with a response time of around 40 s (membrane volume) the advantages are robustness, manageability, high sensitivity especially at low oxygen partial pressure, accuracy (±0.1 mg/L), and low cross-sensitivity.

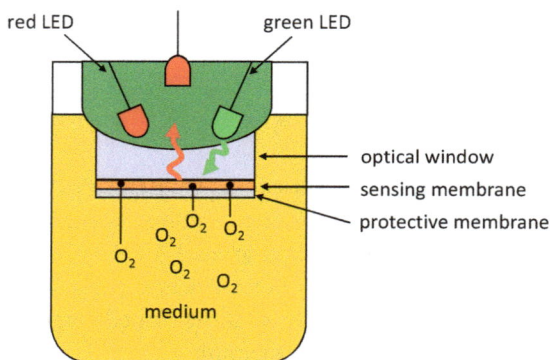

Fig. 10.6: Principles of optical sensing principles: the optical pO$_2$-sensor sends excitation light (green LED) to the dye, and emitted light (red) is detected by a photodiode. The red LED is used for calibration purposes.

In the previous paragraph we learned about a first example of an optical sensor, also called "optode." A color reaction is also the classic litmus test for pH measurement. This lab test gave the idea of looking to dye reactions for use in online sensors. In modern versions dye patches are employed allowing for optical online measurement, see Fig. 10.7. The dyes are applied as a coating on the tip of a glass fiber through which a light beam can read out the color.

Such fiber optical sensors are available for pH, pO$_2$, and pCO$_2$. Constraints for the dyes are that they are not toxic, can be sterilized, and that they are stable in the long term. The measurement principle allows for fabricating "microsensors," which are quite small so that they can be used to resolve spatial resolution of the measured parameters. The response time can be less than a few seconds. Pasting the pads inside a

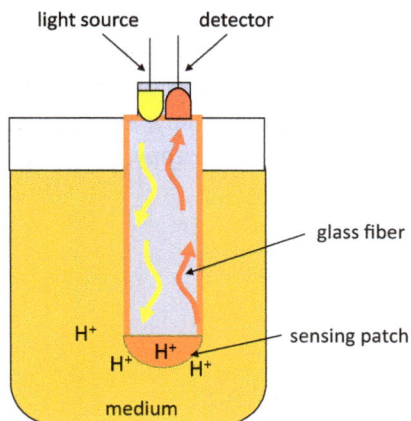

Fig. 10.7: Principles of optical sensing principles: in fiber optical sensors a patch containing a dye, e.g., with a pH-dependent color, is read out by light emitted by a diode the reflected light contains the information of the pH value in the medium.

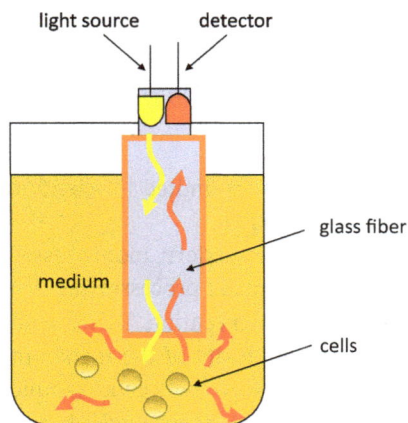

Fig. 10.8: Principles of optical sensing principles: Fluorescence sensor to measure "biomass"; excitation light is provided by an LED or by an adjustable light source delivering the whole spectrum of interest; emission light spreads in all directions, then only a small part reaches the glass fiber and finally the detector (photodiode).

glass reactor or shaking flask is also possible so that they can be read out from outside. Such noninvasive techniques have their advantages for lab work including microtiter plates and for disposable plastic reactors.

Not all analytes we are interested in do us the favor of showing a convenient color or electrochemical reaction. Here we must refer to lab methods from laborious wet chemical procedures up to employment of costly measurement devices. Often, results are available only the next day. However, it has succeeded in bringing these procedures from the lab closer to the process. A sample is automatically taken, e.g., by a small bypass stream and led through an automated device to handle the sample and give a measurement value. These quasi-online measurements can happen within a couple of minutes: fast enough and with high repetition rate to make use of the result for process control. The idea of at-line sensing techniques is shown for flow injection analysis (FIA) as an example in Fig. 10.9. A droplet of the bio-suspension is injected into a stream of a carrier fluid to be transported to several other injection points, here to add a dye for a color reaction, which can be read out by a detector. This scheme can be more complicated according to the related lab protocol. Other options for at-line measurements are online high-performance liquid chromatography (HPLC), spectroscopy, microscopy, flow cytometry, or even nuclear magnetic resonance (NMR) spectroscopy.

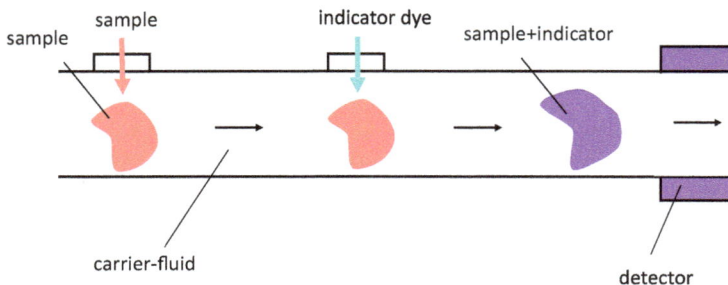

Fig. 10.9: Working scheme of flow injection analysis (FIA) as an example of an at-line measurement. The green sample is mixed with a dye injected exactly at the spot of the sample droplet; then color reaction has some time to occur before it is detected.

As setting up at-line measurements is not always easy, the use is limited to scientific projects and select industrial processes with high-quality demands.

Nevertheless, quantitative measurement is essential for modern bioprocesses being used in pharmaceutical and food production. These complex processes require fast and high-performing analytics for effective control and intervention. This should be specific enough to separate overlapping signals. The complexity of the matrix in most bioprocesses has so far prevented their standard use. However, good sample preparation can have a benefit in the off-line area. The most convenient methods are, as already mentioned, optical sensors using an interaction between an electromagnetic wave and a particle or a molecule. To solve current problems with specificity and sensitivity, Raman spectroscopy is repeatedly discussed. The wavelengths are fixed by being tied to molecular properties, e.g., 20–200 µm. If two molecules have overlapping signals those can hardly be shifted away from each other experimentally. The contributions of individual intracellular and extracellular components that make up a mixed-up signal can nevertheless often be separated mathematically. To cope with the great number of possible measurement situation, recently machine learning (AI) is being employed.

10.5 Measuring biomass – the great unknown "X"

Throughout this book and generally in the bioengineering community we speak about "biomass" in terms of cell number or cell dry weight. Going through the catalogues of equipment suppliers we will not find an online sensor that gives a direct signal of these quantities. The cell is something like a diamond, which looks differently in different light and from different angles. From measurement of the physical and chemical variables (paragraphs above) it became clear that we have to define a specific property of cells characterizing biomass and make it different from other compounds in the suspension. In a second step a specific interaction with a test signal has to be found. This could be light, concentration differences, indicator molecules, electric fields or mechanical forces. Starting from a basic engineering view some aspects are defined and listed in Tab. 10.2. From a physical view the cell is a particle with a spatial structure. Chemical reaction technology may look at it as a heterogeneous catalyst. The view of the cell as an ensemble of macromolecules was already employed during media design. Each of these aspects can lead to different ideas for what could and should be measured.

Tab. 10.2: Characteristics of living cells.

Basic view on the cell	Characteristic	Specific interaction	Measurement technique
Particle	Size	Image	Microscopy
	Size, refraction index	Light scattering	Optical density
	Mass, density	Weight force	Sedimentation, cell dry mass
	Spatial structure, cell membrane	Electrical capacity	Alternating electric field
Catalyst	Chemical conversion	Analyte formation	Analytics + software
	Charge transport	Electric field	Electrodes
Macromolecule	Light absorbance	Light Absorbance	Spectroscopy
	Aromatic residues	Fluorescence	Fluorimetry

With this background in mind, different measuring technologies from other fields of physics, chemistry, and those already adapted for use in chemical engineering can be assessed and luckily new measurement effects are found and utilized for the specific purpose.

Optical methods are well-developed in particle technology. Here the interaction of particles with light namely light extinction by absorption and scattering caused by diffraction, refraction, and reflection are measured. This makes samples of suspensions look turbid in contrast to molecular solutions. Unlike in fluorescence, the scattered light has the same frequency as the incident light. The basic setup to measure scattering is shown in Fig. 10.10.

Scattering depends on the wavelength and the size of the particles. Rayleigh scattering occurs where particles much smaller than the wavelength of the light interact with the light beam, e.g., small molecules. Raman scattering is caused mainly by intramolec-

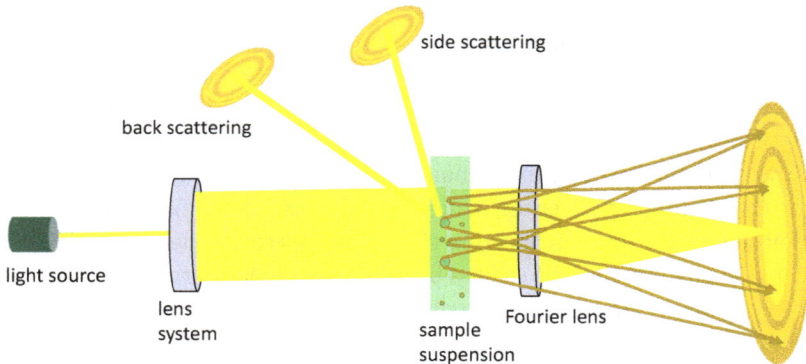

Fig. 10.10: According to Mie theory spheres produce light scattering patterns of concentric rings of light intensity; large particles scatter at small angles and vice versa. The Fourier lens focuses scattered light from different spots in the medium but with the same direction onto one ring.

ular vibrations and rotations. Raman spectroscopy is frequently used in biological and chemical labs. It can be applied also to complex fermentation suspension to measure, e.g., products but mainly in a scientific context. Of higher importance for measuring fermentation suspensions is Mie scattering describing especially the interaction of light with particles of similar size as the applied wavelength. The intensity of the scattered light is a function of the cell's optical properties and dimensions. In principle, it is possible to resolve size distribution of cells of a particular shape from the measured angular scattering intensity pattern. In practice, not all the information contained in the scattered signal is used. In particular, the scattering patterns are ignored or averaged because the detection of the transmitted signal is foreseen only at a single light detector mounted in the direction of the incident light. This reduces the approach to a turbidity measurement. The basic application is the so-called optical density, which is the standard approach for off-line determination of biomass in a spectrometer. Assuming a light beam with a known intensity $I_{Light,0}$ travels through a suspension of biomass along an optical path with length d_{Path} an attenuated light intensity $I_{Light}(c_X)$ can be measured. Usually, the Lambert Beer law is employed for evaluation:

$$I = I_0 \cdot \exp\left(\varepsilon \cdot c_X \cdot d_{path}\right) \tag{10.11}$$

The parameter ε is called the extinction coefficient. Historically the absorbance $A = -\log(I/I_0)$ is defined as common logarithm and the optical density (OD) as

$$OD = \frac{A}{d_{path}} = -\frac{1}{d_{path}} \cdot \log\left(\frac{I}{I_0}\right) \tag{10.12}$$

Now some simplifications in this deduction can be mentioned. The Lambert Beer law was originally stated for molecularly dispersed systems at low concentrations. In the case of measuring cells multiple scattering events occur, making the real light extinction nonlinear with respect to light path length and biomass concentration. For these reasons the light path length is standardized to 1 cm but not always in online sensors where it is usually smaller. From Fig. 10.10, it becomes obvious that also the distance of the detector from the sample and the detector area have a direct impact on the numerical results. In principle, during a measuring campaign the same device is used always. A calibration curve can be carefully determined with a set of samples of known cell dry mass concentration.

The choice of the measuring wavelength has to ensure that no light absorption occurs. For bacteria with characteristic gauges of 1 µm usually 600 nm is chosen. For microalgae 750 nm is a good choice (no chlorophyll absorbance), and some online sensors even measure in the near-IR range at 880 nm. Eukaryotic cells with diameters of several µm are considerably larger than the measuring wavelength. Here scattering at the organelles or at surface structures will influence the results. This also means that during cultivation the relation between OD and CDM concentration will change following cell differentiation or adaptation. Table 10.3 shows a few results for different cell types.

Tab. 10.3: Measured values of biomass related variables for different cell types.

Cell type	Cell diameter (µm)	λ_{meas} (nm)	CDM (g/L) /n_{Cells} (10^8 m/L)	OD (-)/CDM (g/L)	OD (-)/n_{Cells} (10^8 m/L)	Remarks
Matured yeast	7	750	4.26	3.32	14.1	
Budding yeast	7	750	3.49	3.20	11.2	Buds (30%) counted in n_{Cells}
Chlorella vulgaris	7.5	750	7.55	5.34	40.3	Organelles contribute
Bacillus subtilis	2 * 4	550	0.34	3.62	1.23	

The basic properties of cell suspensions measure different aspects of biomass and can be relativized to each other. Optical density is considered to be linearly related to cell dry mass concentration as well as cell number concentration. Small cells contribute less to OD than the same number of larger cells (last column). However, the higher weight of large cells compensates this effect, so that the OD/CDM ratio is often in the same order of magnitude for different cells. The lower CDM/n_{Cell} values for bacteria against yeasts correspond to their smaller volume. Cells are not simply particles but are semitransparent and contain other structures like organelles in the size of the light wavelength. These contribute significantly to light scattering. As the size and structural parameters (organelles) may change during cultivation all relations are subject to change as well.

While measuring OD off-line there is the possibility to dilute the sample until it falls into the linear range of the calibration curve between CDM concentration and OD; this is not the case for online turbidity sensors. Change of OD over two orders of magnitude may happen during cultivation. Some sensor systems measure transmission, side- and backscattering (reflection) for low, middle, and high cell concentrations to offer a signal that can be evaluated. Others can increase the sensing light intensity according to the increasing biomass density. In every case careful calibration is necessary.

Even from a formal physical view the cell is more than a particle. It exhibits a structure and a chemical composition, which may be exploited for measurement purposes. From the principle of measuring cell size with a Coulter counter it is known that cells do not conduct electrical current. The reason is that the cell membrane is a chemical barrier against diffusion of ions and electrons, whereas the plasma is a good conductor. A Nernst potential builds up over nearly all cell membranes and is controlled by specific ion channels. Technically speaking the cell membrane behaves like a battery or an electrical condenser. The electrical capacitance should be measurable by its alternating current resistance, the impedance. A corresponding device is shown in Fig. 10.11. An alternating electric field is applied via two electrodes leading, like in capacitors, to a small charge displacement at the membrane. Here the cell behaves as an induced electrical dipole. Then the current can propagate inside the cells better than in the medium. The impedance spectrum between 0 Hz and 10 MHz delivers finally a signal depending on the fraction of cell volume in the medium and a shape factor of the cells. For higher frequencies the organelles also contribute to the measurement. The main field of application is consequently large cells like mammalian cells and yeasts as well as high cell density cultivations of bacteria. Dead cells (broken membrane) and other particles are not measured, which is why people speak of "viable cell density" measurement.

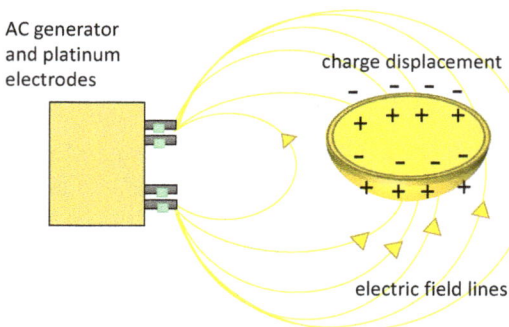

Fig. 10.11: Dielectric measurement of cells in cultivation; only one cell is shown. Charge displacement is indicated by + and –, while the membrane potential itself is not shown.

Among the molecules which cells consists of, two candidates are known to show fluorescence: aromatic amino acids in proteins and NADH. Fluorescence is therefore an option to measure biomass. This can be done using an in-line fluorescence probe as shown in Fig. 10.8 allowing us to record 2D fluorescence plots (Fig. 10.12), where emission is measured for a wavelength range of interest. As the protein content may vary during cultivation and the NADH/NAD relation depends on the cultivation conditions as well, fluorescence measures only a facet of the cells depending on cell composition and physiological state. Direct calibration is usually successful only in well-known processes, being a restriction for use. An advantage is that conclusions regarding the physiological state can be drawn. The possibilities are better with microalgae because the medium is clear and many of the molecules to be determined, such as chlorophylls or carotenoids, have characteristic fluorescence spectra. Time-resolved Chl-fluorescence can also be used to analyze the metabolic state of the cell in more detail (PAM).

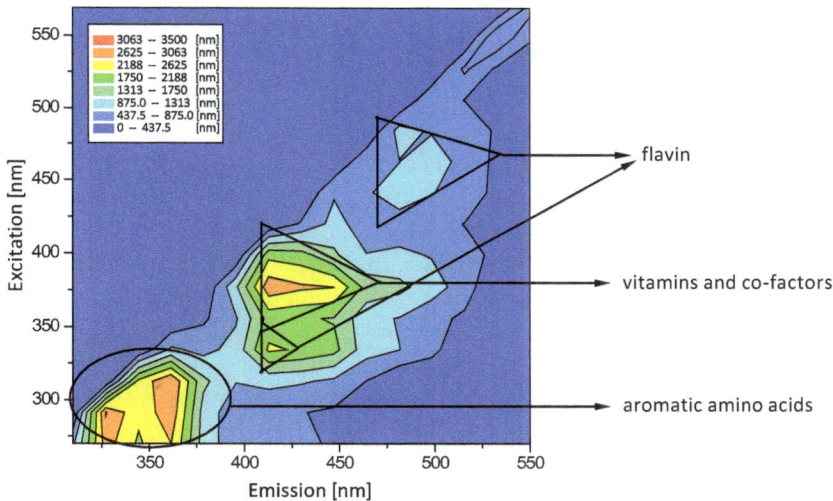

Fig. 10.12: Typical outcome of a 2D fluorescence measurement here for yeasts; the contribution of protein and NADH florescence can be distinguished for further physiological interpretation.

Multivariate measurements promise to overcome limitations of single approaches. However, this requires a lot of process knowledge and mathematical data evaluation. In Fig. 10.13 some different sensing signals are compared, all of them meant to measure "biomass." Data are from batch cultivation with a nitrogen limitation and subsequent accumulation of a nitrogen-free storage compound.

It is obvious that each of the different principles shows only one aspect of what biomass is. In any case a combination of cell mass and physiological state is detected. Furthermore, different physical sensing methods are corrupted by different distur-

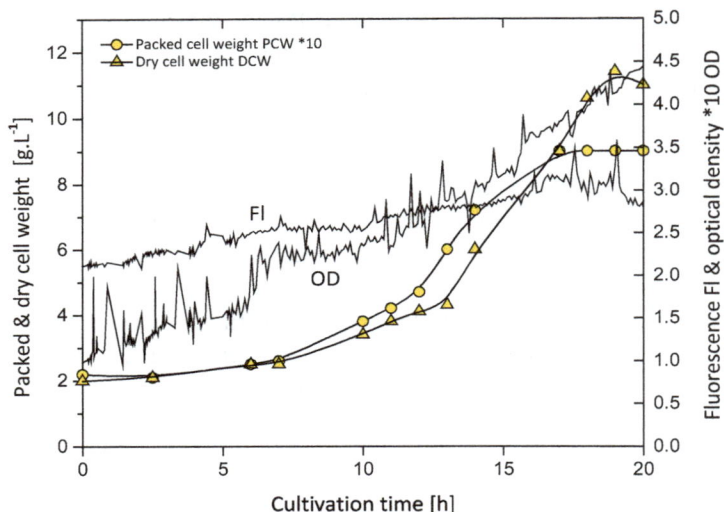

Fig. 10.13: Comparison of different on-line signals for biomass determination with an obvious deviation during the cultivation.

bances, which must be removed as well as possible by technical or data processing means. Then the information can be carefully interpreted. In this case cell wet mass and cell dry mass follow the growth process with a constant ratio of about 10, corresponding to a water content of the cells of 90%. In the late phase of the cultivation, cell dry mass overtakes cell wet mass due to accumulation of the storage compound. Starch or PHB (polyhydroxybutyrate) granules contain only a little water. This interpretation is supported by the increase of optical density and the constant fluorescence signal indicating constant protein content. Both optical measurements are very noisy due to air bubbles and show a strong background signal from the medium as can be seen by the high initial values.

10.6 Compiling a construction kit for sensors – a general approach for biological sensors

All sensors have a general structure in common, as depicted in Fig. 10.14. Firstly, the analyte interacts with a "receptor," which accepts only molecules with a characteristic property. The receptor must be as specific and selective as possible for this property. Information flow is always coupled to an energy and/or mass flow. This can be diffusion and/or chemical reaction energy. This primary energy flow to and in the receptor – the primary measuring effect – is very small. Therefore, it must be converted and amplified by a second element of the sensor: the "transducer." The transducer can be unspecific because further energy or material transformation and amplification happens in the protected space of the sensor. In the transducer a much higher energy flow, supported by auxiliary energy, is generated, and modulated by the primary receptor signal. Again, different kinds of physical or chemical effects can make up the specific manifestation of the modulation. Finally, the transducer delivers the measuring signal at a technically feasible energy level, commonly electrical energy. The signal is then further amplified and processed for monitoring, control, and data storage.

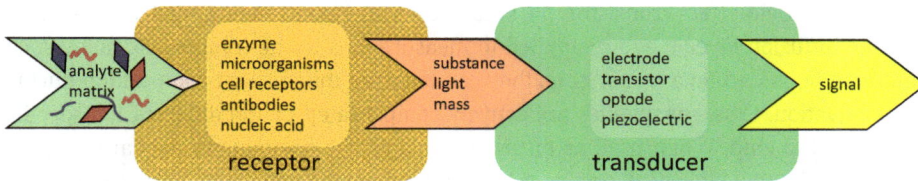

Fig. 10.14: Basic structure, in general, of all sensors and here particularly for biosensors; the two stages "receptor" and "transducer" are shown as rectangles; the corresponding material/energy flows as arrows. The pink hash at the analyte/receptor interface indicates that only red diamonds can bind but not the blue ones and no strings irrespective of the color. That underpins the concept of selectivity.

Examples of this structure can be found among the sensors in the previous paragraphs. For CO_2 the characteristic property was high absorbance of IR. For oxygen the semipermeable membrane is the receptor, while the active electrode is the transducer. In the case of pH the selective property of H^+-ions is the charge and the size of the specifically permeable gel layer is the receptor.

The hardest problem in measuring biological molecules is to find specific properties and a selective receptor for this property. Thinking of enzymes for example, size or surface charge are options. These parameters are used in chromatography. However, the real thing to be exploited is the specific interaction of the enzyme with its substrate. Otherwise, it will not be possible to distinguish a particular enzyme from other proteins during cultivation. The great idea now is to employ biological principles to solve technical problems. The receptor should include the chemical structure of a substrate to measure a related enzyme in this way guaranteeing the most specific interaction. The bioactive molecules for this purpose are called "ligands" like in other areas of specific biological interactions. This will also work vice versa. Furthermore, the receptor provides a signal, which is more or less unspecifically further amplified by the transducer. This is mounted of course directly behind the receptor. The first sensor based on this principle was a glucose sensor, where immobilized glucose oxidase is the receptor to react with glucose and oxygen. Oxygen consumption is then detected by a classic Clark electrode as described above acting as the transducer. In fact, two sensors are connected in series, where the output of the first one is further detected by the receptor (oxygen permeable membrane) of the second. This sensor, developed for use in medicine, was firstly described by Clark and Lyons in 1962. Sensors including a biological component and typically meant to measure biological analytes are called "biosensors." Specific interactions between molecules are characteristic of living systems. These include enzymes and their substrate or a substrate analog, antibodies and antigens, hormones, nucleic acids, or cell receptors. All these interactions are employed as receptors in biosensors.

Transducers themselves can be sensors, e.g., for the product of the reaction between an analyte and an enzyme as receptor. This could be electrochemical cells to detect, e.g., the pH value. Optodes are employed in cases where the substrate or product is a dye. This can be enforced by coupling a fluorophore to the reaction. The employment of an ion-specific field-effect transistor (FET) is shown in Fig. 10.15. Coupled with an ion-specific diffusion layer they are applied to measure ions in the solution. As many such units can be placed on a sensor chip, they can measure the nutrient composition during the cultivation. Cross-sensitivities are actively compensated by the different ion specific layers on the chip. When used as biosensor for organic compounds they are covered with an additional biologically active layer to transform the organic analyte to a measurable ion. Other possibilities are under investigation or in use. One example is the employment of piezoelectric crystals. The additional mass induced by binding of the analyte to a ligand changes the resonance frequency of the piezoelectric crystal.

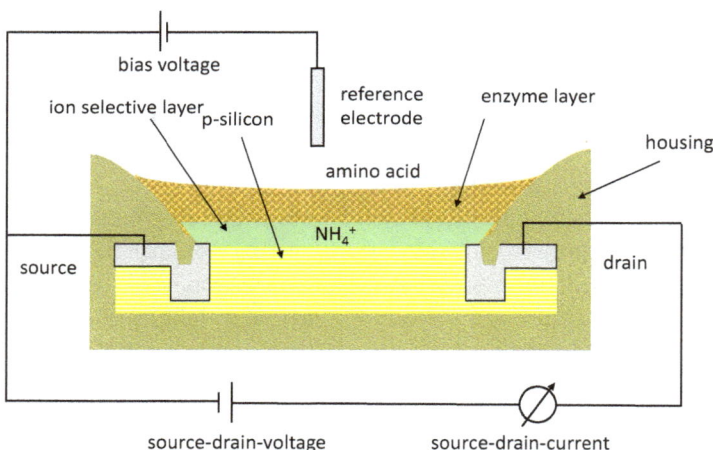

Fig. 10.15: Biosensor based on an enzymatic reaction and an FET. An amino acid ammonia lyase cleaves a dedicated amino acid. The charge of the produced ammonium ions is strong enough to act as input for the ion-sensitive gate, modulating the source to drain current being the output signal. The bias voltage has to be applied to adjust the working point.

A last example for a biosensor in this chapter is shown in Fig. 10.16. Here microorganisms have the job of measuring heavy metals in wastewater. The kernel of the idea is a recombinant promotor, which is inducible by heavy metals. Coupled to the promoter is the LacZ gene from *E. coli* enabling the microorganisms to produce galactosidase and therefore to metabolize lactose. Applying a wastewater sample together with lactose leads to cleavage of the lactose and respiration on glucose and galactose, but only if the wastewater contains heavy metals. The corresponding oxygen consumption is measured with a classic pO_2 electrode. Running this electrode requires control of the different measuring and flushing phases during a measurement cycle and evaluation of the dynamic responses.

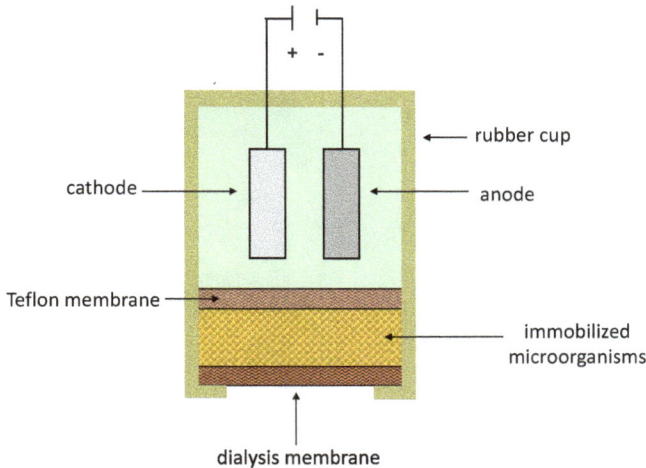

Fig. 10.16: Scheme of a biosensor with immobilized living microorganisms to measure heavy metals in wastewater.

Biosensors are, for many measuring tasks, the only possibility for on-line sensing. Despite this potential there are some drawbacks to be mentioned, which include lack of long-term stability or high temperature resistance during sterilization. Solutions to such problems can be found in avoiding direct contact to the cultivation suspension by a separating membrane or by mounting the sensor in a bypass similarly as in quasi-online devices. Typical fields of application are environmental biotechnology or food technology, e.g., as single-use disposable sensors. Modern applications of biosensor principles also target measuring in micro titer plates or even in the intracellular space. In fact, sensory organs in animals work also in accordance with this principle. Light for example triggers a photochemical reaction of rhodopsin – being the receptor – in the retina, then an ion channel opens triggering in the nerve membrane – being the transducer – an electric impulse, where the membrane charge is the auxiliary energy transducer.

The "cell as a factory" reminds us that it is full of molecular sensors and control circuits. This applies both intracellularly to stabilize biochemical and physical actions and extracellularly to react specifically to environmental stimuli. Biology has described many tropisms in this way. It is a pity that this aspect plays almost no role in bioprocess technology. Nonetheless, biohybrid sensors and actuators are being developed. One example is the "luminous algae" *Pyrocystis*. It reacts to mechanical movement stimuli with blue luminescence. This is intended to attract squid, which in turn pursue the microalgae's enemy, krill, as food. There are probably many such reactions to stimuli that we do not perceive so directly and prominently due to their chemical nature. In any case, there is the idea of developing *Pyrocystis* as a microscopic mechano-sensor without a power supply but with optical readout. Art in the form of "media art" or "biomedia" has also become aware of this microalga. Dr. Tim Otto Roth is a German conceptual artist, composer, and scholar. He deals with the artistic realization of wave phenomena such as light and acoustics in nature and technology, "a plea for the physics of art." An example for using microorganisms in art is shown in Fig. 10.17.

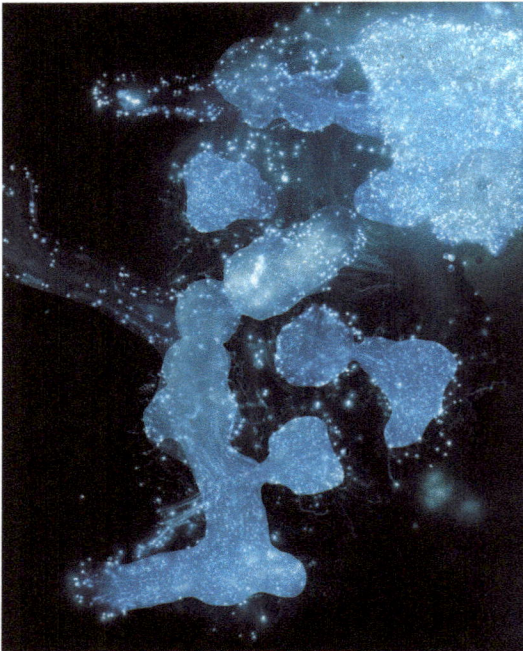

Fig. 10.17: Light art canopy: "Meeresleuchten" ("sea lights"), plankton "luminogram" of *Pyrocystis elegans* (dinoflagellates), enlarged lambdaprint on slidefilm, 78 × 100 cm, 2020. In absolute darkness, the seawater sample with algae was poured on a photographic sheet film. The stressful situation triggers the emission of blue light recorded immediately as light impression on the film (©Tim Otto Roth, imachination projects).

At first glance the light from distant celestial objects wandered through the depths of the universe seems to shine in "Meeresleuchten." Actually, the picture records the trace of a cell culture of the marine algae *Pyrocystis elegans*, which causes among other marine organisms the legendary marine bioluminescence. In this way, the discourse between art, design, biology, and engineering will lead to a new perspective in an application or research-oriented manner.

10.7 Questions and suggestions

1. In nature, many specific interactions occur at the molecular and cellular level and are essential for the functioning of the organism. Try to find examples where a direct interaction with the situation in the reactor could be of interest.

2. Cells react to the environment. Collect examples where mankind already depends on these reactions also with respect to current ecologic problems.

3. Collecting examples in technical applications is also sensible. The λ-sensor, for example, built in the exhaust pipes of cars, measures oxygen content of the exhaust gas. Find out what is the specific element of this sensor. Can it be used as off-gas analyzer for bioreactors?

Chapter 11
The practice of fermentation – a step-by-step guide through the workflow

In the previous chapters, the processes inside a reactor were highlighted and we described how they can be influenced by the inputs provided by the reactor. Nevertheless, it is another problem to stand in front of a real bioreactor and to consider what to do next. In this chapter, a step-by-step guide for the cultivation of microorganisms is given and broken down into single actions. The main aim is to become acquainted with fermentation systems and learn how to solve technical problems using concrete examples. Since the single steps for conducting fermentations in a bioreactor are, in general, quite similar, we encourage transferring the presented steps to other cultivations. By following the instructions of this guide, even inexperienced users should be able to run a standard bioprocess. The checklists and example protocols given in the particular paragraphs can easily be transferred to other fermentation systems or to different kinds of cultivations.

11.1 Structuring the reactor and the cultivation – zooming in from principle understanding to visibility of process details

Before beginning with the actual work, it is mandatory to comprehend the material flows in the reactor system and to identify the respective technical realization. A graphical representation is given in Fig. 11.1, showing basic reactor modules. This hand drawing may represent the impression of somebody having a first look onto an unknown bioreactor. In the next step, it is important to achieve a structured mind mapping regarding inputs and outputs of gas, steam, and liquid streams, and why they are required in the respective sections of the reactor. Follow all necessary media (gas, fluids, etc.) along the material flows! Standard procedures for how to set up the reactor vessel and how to prepare the pre-culture are also mandatory. Therefore, this guide is split into four parts: cultivation design, cultivation preparation, the cultivation itself and data evaluation, providing concrete and detailed solutions concerning the exemplary fermentation process. In the following paragraphs, we will move closer toward an engineering view to become more familiar with further details and the relevance of single-reactor parts. Handling of the reactor and timely structuring of the fermentation become clear by performing a real concrete fermentation. In this chapter, an aerobic fed-batch process is chosen, with *Bacillus amyloliquefaciens* producing phytase, an extracellular phosphatase, which is excreted into the cultivation suspension.

https://doi.org/10.1515/9783110773354-011

Fig. 11.1: A graphical sketch of a pilot reactor. Basic modules are visible; more details only open up by getting closer and running hands-on cultivations (© bioengineering).

11.2 The reactor – a complex assembly of parts and modules

The cultivation will be performed in pilot-scale using a stirred tank reactor (STR). This is the most commonly used reactor type for bioprocesses. Approaching the reactor, the first thing we will see is a stainless steel cylinder with a lot of periphery attached, next to a rack made out of more steel with little displays (Fig. 11.2). For clarification, each single element will be identified and classified, regarding material and energy flows.

Every organism needs nutrients. In the case of bacteria, these are oxygen for cell respiration and culture media, which are designed to provide all elements the cells need. To supply each single cell with enough nutrients, it is necessary to mix the culture broth and to disperse the gas bubbles. In stirred tank fermenters, this is accomplished by mechanical agitation. The pressurized inlet gas enters the reactor vessel from the bottom. The inlet gas, usually air, owing to its low cost, provides oxygen for aerobic cells during cultivation and removes gaseous byproducts. A reactor jacket allows heating and cooling of the culture broth. As a training example, a commercial pilot reactor with a total volume of 19 L and a work volume of 12.6 L is used. Table 11.1 gives an overview of the technical specifications. Such a table is usually given in the manual or has to be set up by the user.

The vessel itself consists of stainless and acid-resistant steel. It is fixed with clamps on a rack. The five lateral ports in the bottom of the vessel allow the arranging of sensors in the fermentation broth. During fermentation, the broth can be observed via a longitudinal window of security glass. With an operating temperature of 150 °C and an operat-

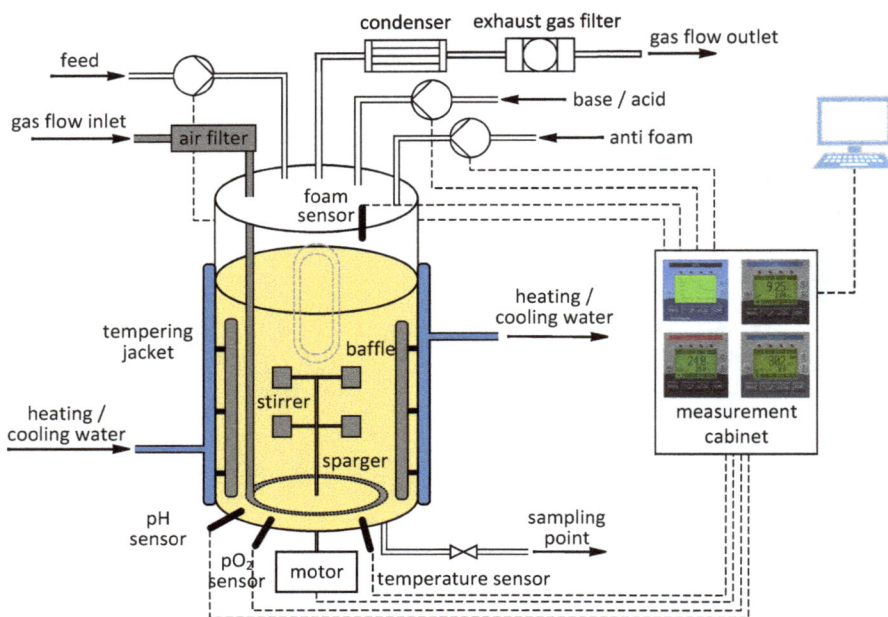

Fig. 11.2: Scheme of the pilot-scale ´fermenter´ for use in development labs; the symbols represent parts and devices, which have to be identified at the real reactor.

Tab. 11.1: Specifications of the bioreactor system.

Component	Technical data
Vessel	Stainless steel 316L
Heating circuit	Electrical heater; heat exchanger for cooling circuit
Mechanical seal	Silicon carbide; glycerol lubricated
Stirrer	Blade stirrer, baffles
Inlet air	Absolute filter; gas mixing station
Agitation	Belt driven from bottom; double mechanical seal
Aeration	Submerse; stainless steel ring sparger

ing pressure of 2.5 bar, sterilization by superheated steam is possible. The lid is fixed by tightening wing nuts on studs. It is sealed by an O-ring seal, made of a synthetic elastomer rubber (EPDM). Like the vessel, the lid is made of stainless and acid-resistant steel. The ports in the lid are round threads. Thus, they are hygienic and mechanically stable. Gas inlet and gas outlet are connected to the reactor via the lid. In our setup an auto-sterile filter is attached to the gas flow inlet. Medium, base, acid, and antifoam feed as well as the foam sensor are connected to the lid. The stirrer is belt-driven by an electrical engine (AC; 1.1 kW; max 1,500 min^{-1}) from the bottom. If necessary, a gear reduction up to 1:5 is possible. The power electronics for the engine is located in a separate cabinet.

To prevent leakage between the shaft and the vessel, a double mechanical seal is used as shaft bearing (see Chapter 5). A foam separator is integrated to eliminate foam and remove particles from the gas discharge. The resulting pressure drop between the vessel and the gas discharge enables the foam to be pushed against the rotor. The mechanical forces induced by the rotor separate the gases, which then can pass off via the degassing chamber and the exhaust gas line. The liquid particles are thrown back into the foam mass.

Process parameters are controlled by 19 inch plug-in boards integrated in a measurement cabinet and a higher-level process control, based on a graphically programmable process control system. Temperature regulation is necessary to change and control the heating of the bioreactor during the autoclaving process and to keep the temperature constant during the cultivation process. A Pt-100 temperature sensor is used to control the cultivation temperature. The cultivation broth is tempered by a double-wall vessel. An electrical heater, directly connected to the double-wall vessel, sets the temperature in the primary circuit. The electrical heater is connected to a heat exchanger, which is connected to an external cooling circuit.

11.3 Structuring the cultivation process – from planning to final data acquisition

This subsection presents a typical cultivation, an aerobic fed-batch cultivation of *Bacillus amyloliquefaciens*. A systematic approach along with the sequence of necessary steps is given to explain the typical procedure regarding planning, assembling, and commissioning of a standard fermentation. We take it for granted that the main connections for gas supply and cooling water are already available. Air is usually supplied by a compressor and should be free of oil and moisture. Pneumatic valves should be utilized to provide the process gas at 1–3 bar. The water used should be of a quality that meets the process requirements. We furthermore assume that all analytical devices and methods as well as the needed software, which may be necessary for sample processing and the evaluation of process data, are available and were tested before. The utilized process control system has to be attuned for the cultivation purpose by either self-programming or in consultation with a software manufacturer. In the case of a fed-batch process, a feed controller has to be implemented and parameterized in order to control the feed-pump flow rate.

Before starting to work with the reactor, some safety instructions must be internalized:
– Never heat an empty reactor; devices within the fermentation vessel could be destroyed.
– Always use a safety overpressure valve to prevent pressure increase in the case of a blocked output filter. In the case of exceeding the maximum pressure, the security glass inside the fermenter vessel must burst.

- Make sure to use constant gas pressure <0.2 MPa (2 bar). A security gas valve can reduce pressure fluctuations.
- During sterilization, the exhaust gas line of the reactor system has to be open to compensate possible overpressure during the sterilization process.
- Sterilize dangerous and corrosive liquids and storage bottles separately. Transfer the chemicals to the sterile storage bottles under sterile conditions.
- Wear goggles and gloves and follow the safety instructions while handling corrosives.

Prior to realizing the experimental setup, it is important to carefully consider the process strategy and data evaluation. Figure 11.3 shows a principal pattern of an experimental fermentation procedure. Summarized under the term ´upstream processing´ are the preparation of the medium, sterilization of the reactor components, and inoculum preparation.

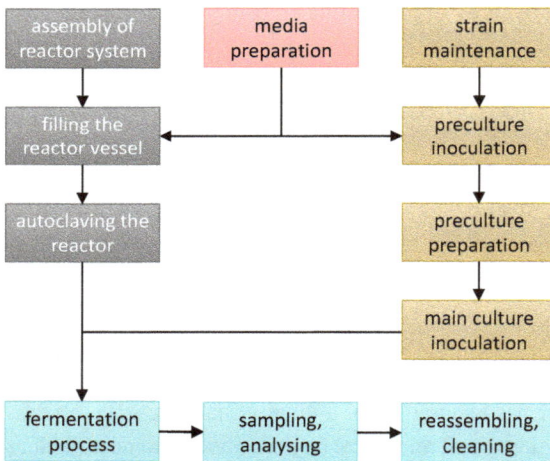

Fig. 11.3: Block diagram of an experimental fermentation procedure; the red, brown, gray, and cyan boxes mark the steps in the workflow. The colors represent related steps belonging to a respective work package.

At first, it is necessary to prepare the culture medium (red box) in sufficient quantities. Especially for fed-batch processes, calculation of the required amount is necessary since culture medium is needed in sufficient amounts for the preculture, batch phase, and fed-batch phase. The second work package (brown boxes) deals with preparing the inoculum from strain maintenance in the lab, up to a sufficient amount of preculture for inoculation of the reactor. Depending on the strain, medium, and the objective of the experiment, the amount of inoculum needed can vary. However, the inoculum volume used for this experiment was 5% in terms of the liquid reactor volume. Inoculation needs precise timing as a vital culture in the exponential growth phase is required. Thus, the lab part of this work package has to start first. Media preparation and preculture preparation are the typical upstream parts of the process. In the third work package (gray boxes), the bioreactor is set up by assembling the fermentation vessel and peripheral devices. Filling with medium and sterilization are the next steps. Inoculation finally completes this work package. The cultivation itself (cyan boxes) starts with transferring the axenic preculture into the sterilized reactor and activating the process control system, which controls the feeding rate and records all digital process parameters. In the next four sections, the working lists, according to the four work packages, are further elaborated.

11.4 Media preparation

The media composition used for the preculture and the main culture of the strain *Bacillus amyloliquefaciens* are given in Tables 11.3 and 11.4, respectively. An amount of 10 mL/ L trace elements solution has to be added to both solutions. The composition of the trace element solution is given in Tab. 11.5 while the stock solution – phosphate and glucose, which are used for the fed-batch process in the main culture, are listed in Tables 11.6 and 11.7. For the purpose of the exemplary cultivation, a modified synthetic medium was used; see also Section 3.3. Usually, the medium is sterilized in the preculture flask or in the fermentation vessel. However, sterilization in advance of its use is also possible. The most popular methods for media sterilization are filtration or thermal treatment.

Tab. 11.2: Components of the preculture medium; pH has to be adjusted to 7.5 and sterile glucose solution should be added under sterile conditions after autoclaving; further details are given in Section 3.3.

Components	Concentration (g/L)
$(NH_4)_2SO_4$	2.6
$MgSO_4 \cdot 7H_2O$	0.41
HEPES	12
Trisodium citrate dihydrate	1.14
KH_2PO_4	3.3
Glucose monohydrate	30

11.5 Preculture preparation

Preculture preparation has to provide sufficient amount of active cells to inoculate the bioreactor and is generally accomplished under sterile conditions, e.g., in a biological safety cabinet. Different cell banking systems are used to store the desired strain,

Tab. 11.3: Components of the main culture medium. Potassium dihydrogen phosphate and glucose should be added after the autoclavation process under sterile conditions, using feed bottles.

Components	Concentration (g/L)
$(NH_4)_2SO_4$	2.6
$MgSO_4\ 7H_2O$	0.41
Trisodium citrate dihydrate	1.14
KH_2PO_4, potassium dihydrogen phosphate	0.3
Glucose monohydrate	42

Tab. 11.4: Trace elements solution (100 fold).

Components	Concentration (g/L)
$CaCl_2$	0.291
Na_2MoO_4	0.024
$FeSO_4\ 7H_2O$	0.52
$ZnSO_4\ 7H_2O$	0.72
$MnCl_2\ 4H_2O$	1.00
$CuCl_2\ 2H_2O$	0.085

Tab. 11.5: Components of the phosphate feed solution.

Components	Concentration (g/L)
KH_2PO_4	5.0

Tab. 11.6: Components of the glucose feed solution.

Components	Concentration (g/L)
Glucose monohydrate	600

e.g., in liquid nitrogen (long-term storage), agar plates, or liquid cultures (mid- and short-term storage). In the current case, the preculture is present as an agar plate culture. To get liquid precultures of *Bacillus amyloliquefaciens*, sterile 250 mL Erlenmeyer flasks containing 20 mL sterile preculture medium (Tab. 11.2) are prepared. By a flamed inoculation loop, bacterial cell mass is transferred from the agar plate to the Erlenmeyer flasks. After sealing, the liquid precultures are incubated at 37 °C and rotated with 150 rpm on an orbital shaker. As soon as the liquid cultures enter the exponential growth phase, it is recommended to prepare a second charge of liquid precultures from this charge to ensure proper adaption of the microorganisms to the new environmental conditions. The second preculture charge can then be used as inoculum for the main cultivation process. It is essential to prepare sufficient inoculum with the appropriate cell density at the end of the exponential growth phase. Growth conditions should enable the generation of high cell densities within a short period. Preliminary examination of preculture growth rates is therefore helpful to determine an appropriate inoculation time, which should be aligned with the main cultivation. Tables 11.7 lists the single steps to take into account for the preparation of precultures.

Tab. 11.7: Checklist for preculture preparation.

Tasks for preculture preparation	Done
a) Prepare preculture medium for liquid shaking flask cultures.	
b) Prepare Erlenmeyer flasks (250 mL) with sterile closures for gas exchange.	
c) Fill each flask with 20 mL medium and sterilize these for 20 min at 121 °C.	
d) Inoculate liquid media with preculture from agar plate under sterile conditions.	
e) Incubate liquid precultures on an orbital shaker at 37 °C and 150 rpm.	
f) In advance, start the preparation of another preculture generation before the first culture enters the stationary phase.	
g) Autoclave older precultures if they are no longer required.	

11.6 Reactor installation and setup

All major bioreactor components have to be set up to operate the fermentation process. The medium has to be prepared in advance (see Fig. 11.4). Several openings in the reactor cover and the reactor hull allow the installation of the periphery. It is important to make sure that the sterile air inlet device is installed correctly (connection defined by tubular die) before the reactor cover devices are installed. Furthermore, it has to be taken into account that the ports for inoculation and sample taking are easily accessible. Septa have to be inserted in the proper direction. Do not connect the external flask to the reactor, prior to autoclaving. To verify the functionality of the setup after assembling, it is mandatory to track the complete pathway of the gas and the cooling water flow. Online sensors are connected to the system via the control unit. For a typical fed-batch cultivation, the necessary installation and setup steps will be described in detail in the following sections.

11.6.1 Fermentation vessel setup

For the set up of the fermentation vessel, the following parts have to be mounted: sparger, air input including sterile filter for gas inlet, agitator shaft, including six-

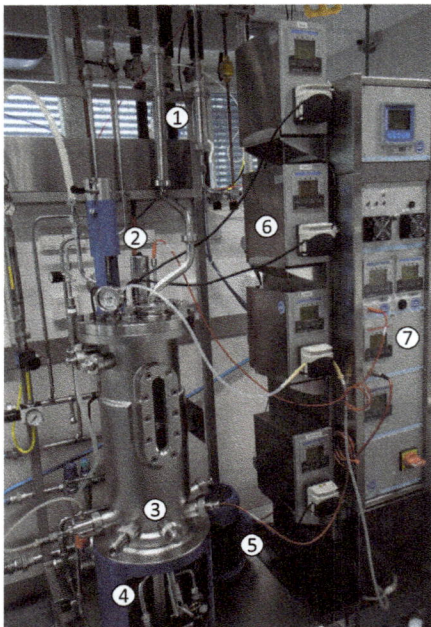

Fig. 11.4: Picture of the example reactor; the main parts are indicated by numbers: (1) exhaust gas cooling and filtration; (2) auto sterile filter with air inlet, ports for metering, illumination, antifoam sensor, exhaust gas port, and manometer; (3) lateral ports for pH, pO2, pCO2, temperature, and sample port; (4) big sample port; (5) motor; (6) pumps for acid/base/antifoam metering and feed; and (7) measurement cabinet (© bioengineering).

blade Rushton stirrer, baffle, dummy plugs, septa, and optionally, an illumination device to illuminate the inward reactor vessel. The sparger is needed to disperse oxygen in the liquid phase. It is usually installed below the impellers. Before entering the bioreactor sparge line, the airflow passes a sterile filter (pore size 0.2 μm) to minimize the risk of contamination. The supplied air is then released into the liquid phase as gas bubbles and dispersed by the agitation and baffling system. It is necessary to verify the functionality of the whole gas pipe system before the reactor set up is continued. After the set up of the sparger, the tubing for gas inlet and outlet, including sterile filters, and the agitation and the baffling system are mounted. The fermentation vessel has to be hermetically sealed when it is connected to the gas line. By gassing the sealed reactor system with air, it is possible to verify the functionality of the gas pipe system by routing the gas flow via a bypass into a water bottle and observing the bubbling.

11.6.2 Peripheral devices

The material flows can be driven by pressure or pumps, and have to be directed by and through additional peripheral devices. Some of these are connected to the inner reactor space via the reactor head (see Fig. 11.5).

Fig. 11.5: Head of the reactor; the parts to be mounted are indicated by numbers: (1) condenser; (2) valve; (3) autosterile filter; (4) illumination device; (5) antifoam sensor; (6) acid/base/antifoam metering; (7) manometer; (8) feed; (9) exhaust gas outlet; and (10) septum for inoculum (© bioengineering).

Overpressure security valve: Mount an overpressure relief valve for transient overpressure protection during cultivation or sterilization. Additionally, the installation of a manometer is recommended to monitor pressure history during the autoclaving process.

Sample unit and fittings: Installation of a sterile sampling unit and fittings for the addition of inoculum, medium, or stock solution feed as well as solutions for pH adjustment (acid/base) are mandatory. Liquid compounds are pumped into the fermentation vessel from sterile storage bottles.

Outgas condenser: To prevent condensation of water and the related risk of a blocked gas flow, and to protect the off-gas analyzer, the installation of an outgas condenser is necessary. The condensed water will flow back into the reactor vessel instead of clogging the outgas filter. It is common practice to use an intermediary bottle between the outgas condenser and the output gas filter. To improve the protection of the output gas filter against clogging, this bottle can contain a small amount of antifoam.

Gas outlet filter: To guarantee the preservation of a sterile reactor system after autoclaving, a gas outlet filter is needed. For this purpose, an autoclavable sterile gas filter with a pore size of 0.2 μm is used. To avoid clogging, it is mandatory to ensure that the gas outlet filter stays dry at all times.

Antifoam sensor: Mount the antifoam sensor and the antifoam stock bottle. The antifoam sensor acts as a level probe. As soon as foam reaches the sensor, the sensors' electrical conductivity changes, which leads to the activation of the antifoam pump. During the cultivation process, the formation of foam due to aeration, agitation, and the presence of foam-producing or foam-stabilizing substances like proteins, polysaccharides, or fatty acids can occur. Foaming can lead to a blockage of the outlet filter and gas lines, and has to be avoided. Antifoam agents, mechanical foam separators, or ultrasonic treatment can be used to reduce foam formation. However, surface tension-lowering antifoams are usually utilized to reduce energy consumption.

Peristaltic pumps: To achieve precise and reproducible media feeding and metering of acid, base, and antifoam, peristaltic pumps are used. These types of pumps are easy to handle and maintain. Via the dedicated control buttons, the rotation speed of a peristaltic pump can be regulated. It is noteworthy that the volumetric flow rate of a peristaltic pump depends on the internal diameter of the tubing. In preliminary measurements, the correlation between the rotation speed and the respective tubing must be identified. Thereby, a calibration of the favored flow rate range is possible. The required time for pumping a certain liquid volume into the reactor, e.g., 10 mL, has to be measured. Dividing the pumped liquid volume by the required time gives the desired flow rate (mL/min). The bioengineering base control unit offers several pump-support brackets to place the peristaltic pumps, so valuable bench space is saved.

Mass flow controllers: To supply the microorganisms with a defined volumetric gas flow, mass flow controllers are installed. Use gas tubing to channel the inlet gas between the gas lines and mass flow controllers inside the fermentation vessel. It is common to utilize a mixture of air and O_2 in aerobic fermentation processes. Airflow calibrators are recommended to enable an accurate and consistent measurement of the desired flow rate. Sterile filters (pore size 0.2 µm) are integrated to sterilize the gases.

As soon as the setup is completed, the functionality of the peristaltic pumps, mass flow controllers, gas inlet sensors, and the computer connection has to be checked once more. If all devices work properly, the sterilization process can be launched.

11.6.3 Sensor installation

During cultivation, pH, dissolved oxygen (pO_2), and temperature in the fermentation broth have to be measured and controlled.

pH sensor: Calibration has to be performed before autoclaving. Calibration buffers have to be selected according to the pH range of the fermentation. For standard fermentations, a pH 4.0 buffer and a pH 7.0 buffer are recommended. The calibration of the pH sensor (Fig. 11.6) has a two-point gauging system.

pO_2 sensor: A two-point gauging calibration is made after autoclaving while stirring. Zero point of calibration is N_2 saturated medium, while the slope calibration point is determined in an air saturated medium. Although pO_2 sensors usually compensate temperature effects automatically, it is preferable to calibrate the sensor at the fermentation temperature since the saturating oxygen concentration varies with temper-

Fig. 11.6: Technical drawing in three-quarter section of a pH sensor. The long shaft is for better handling and for reaching into the fluid. The structure at the upper end is designed to hold the sensor in the reactor opening and seal it tightly.

ature, pressure, and concentration of dissolved media substances. The maximal dissolved oxygen saturation concentration at 37 °C in water is 6.72 mg O_2/L.

Temperature sensor: A two-point gauging system is used to calibrate this sensor type. A temperature range similar to the fermentation process is recommended, e.g., 37 °C for cultivation and 121 °C for autoclaving. Heated water can be used for the calibration process. Please have safety arrangements for hot boiling liquids in mind.

A process control system is necessary to enable a fully automatized monitoring of the cultivation as well as the storage of all recorded online data. For this fermentation, a graphical software-based process control system is used. Monitoring is possible via in-built displays, Fig. 11.7. All sensors used to control and monitor the process need to communicate with the process control system. This is usually ensured by a supplier-designed communication bus system via analog or digital input/output signals.

11.6.4 Autoclaving the bioreactor

To ensure sterile working conditions, the fermentation vessel and all devices that will be in direct contact with the medium have to be autoclaved. Before starting the sterilization process, several specific preparations must be made. First, the fermentation vessel must be filled with the cultivation medium. Therefore, all unused ports must be closed with septa or screw caps. The liquid working volume of a bioreactor is usually 70% or 90% of its total volume to keep some headspace for foam formation; the volume increases by aeration and substrate feeding. In fact, the bioreactor for our example cultivation is only filled up to 50% at the beginning (batch phase), to leave sufficient space for the feed solutions during the fed-batch phase. With exception of the pO_2 sensor, each sensor has to be calibrated and the sensor connectors have to be protected by a cap. Each sensor should

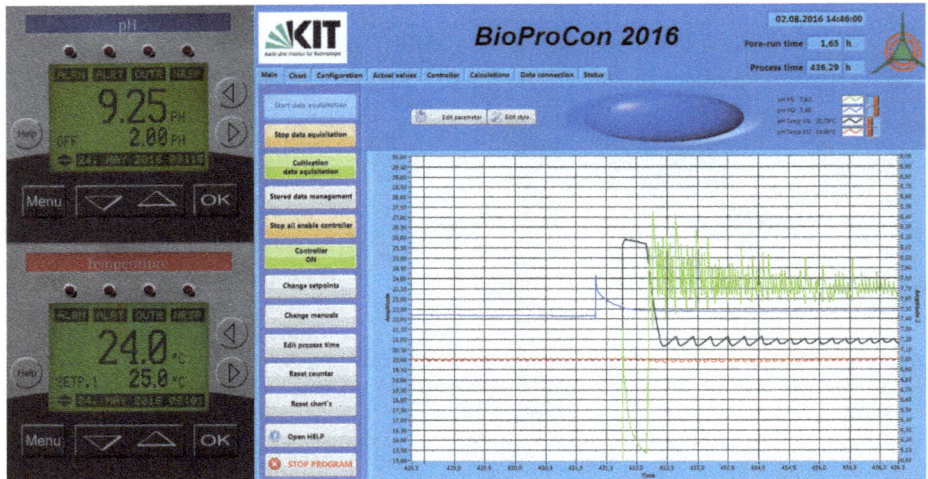

Fig. 11.7: Example pictures of an exemplary process control system "BioProCon 2016", a graphical process control software that was developed at the Karlsruhe Institute of Technology (KIT), Institute of Bioprocess Engineering (© KIT).

be disconnected from its cable during the sterilization process to avoid damage to the sensor cable. It is important to check once more if the overpressure security valve is installed in the reactor vessel. The gas outlet is to be closed at the beginning of autoclaving and will be opened later for pressure compensation during the autoclaving procedure. Other tubes being in contact with the liquid medium are to be closed. All storage bottles and pump lines have to be sterilized separately. Non dangerous liquids can be filled into the storage bottles and autoclaved directly. Acid, base, or liquid substances with high evaporation losses should be autoclaved separately. The utilization of pressure-resistant bottles is recommended. The attached storage bottles with pump lines can be autoclaved with some drops of water in them. After the sterilization, the sterile acid, base, or liquid substances with high evaporation losses is filled into the sterile storage bottles. Please make sure to use a ventilation line with a sterile filter, pore size 0.2 µm, connected to each storage bottle to allow pressure compensation. The liquid level in the bottles, which shall be sterilized in an autoclave, should not exceed two-thirds of the maximum bottle volume. A checklist to get ready for the autoclaving process is given in Tab. 11.8.

In our setup, an auto-sterile ceramic filter for manual sterilization is used. One main advantage of this system is the possibility of in place sterilization along with the reactor autoclavation process. The steam created in the vessel passes the opening in the filter housing, sterilizing the filter and the inlet piping. During fermentation, the filter housing is lifted to a position where the steam inlet is sealed by the reactor lid. However, other reactor systems often use other sterile filters instead of an auto-sterile filter for manual in-place sterilization. In that case, the gas inlet filter has to be autoclaved or

Tab. 11.8: Checklist to prepare the autoclaving process.

Tasks before autoclaving	Done
h) Calibrate the pH sensor.	
i) Check the functionality and pre-calibration of other sensors.	
j) Mount the reactor vessel, including the over pressure valve.	
k) Prepare the medium for batch phase and fill the reactor vessel.	
l) Disconnect all cables and protect the cable connectors with protection caps.	
m) Mount an autoclavable gas inlet and gas outlet filter of pore size 0.2 µm.	
n) Prepare storage bottles and tubing.	

sterilized externally, prior to autoclaving the reactor system. The installation of sterile filters has to be considered carefully for each reactor system individually.

Oil- and water-free air at a pressure of 1–3 bar is led into the medium via the sterile filter and the submerged stainless steel ring sparger (see also Section 11.6.1). Exhaust gas will pass through and exit the vessel. The exhaust gas will be moisturized by passing through the liquid phase. To prevent evaporation and blocking of the off-gas filter, water in the exhaust gas is condensed in the exhaust gas condenser and channeled away toward the vessel, while the gas can escape through the exhaust gas pipe.

Utmost attention is necessary during autoclaving, since all metal parts will get quite hot and the fermentation vessel will be under pressure. Safety glasses and thermal gloves are mandatory. During the sterilization procedure, different stages are passed through. Figure 11.8 shows the piping and instrumentation. The liquid reactor vessel content is heated with steam via the reactor jacket. The sterilization temperature has to be maintained constant for a certain period before the reactor vessel can be cooled down with water to cultivation temperature. To facilitate correct autoclaving, a proper heat exchanger network is mandatory.

At first, the temperature of the bioreactor system is increased from room temperature to 95 °C in the heating-up cycle. During this period, the auto-sterile filter for gas inlet is closed and the condensate drainage of the outgas condenser is open (In the case of other sterile filter techniques, the sterile filter is autoclaved externally). The air supply has to be closed during the whole heating-up process. When the temperature has reached 95 °C, the sterilization cycle for the gas outlet section is induced by closing the gas outlet valve to a minimum. Thereby, a minimal flow of hot steam passes through the gas outlet section, ensuring proper sterilization. The sterilization cycle is induced by increasing the reactor temperature to 121 °C. This temperature has to be kept for at least 20 min. According to the vapor pressure of water, an overpressure of 1 bar (2 bar absolute and vapor pressure of water at 121 °C) will build up in the reactor vessel at this temperature. However, due to the opened gas outlet valve, the overpressure is adjusted to 1 bar during the autoclavation process. The sterilization cycle is terminated by reducing the temperature to 100 °C. Once 100 °C is reached, the following steps have to be conducted: carefully open the gas outlet valve, activate the exhaust air cooling system,

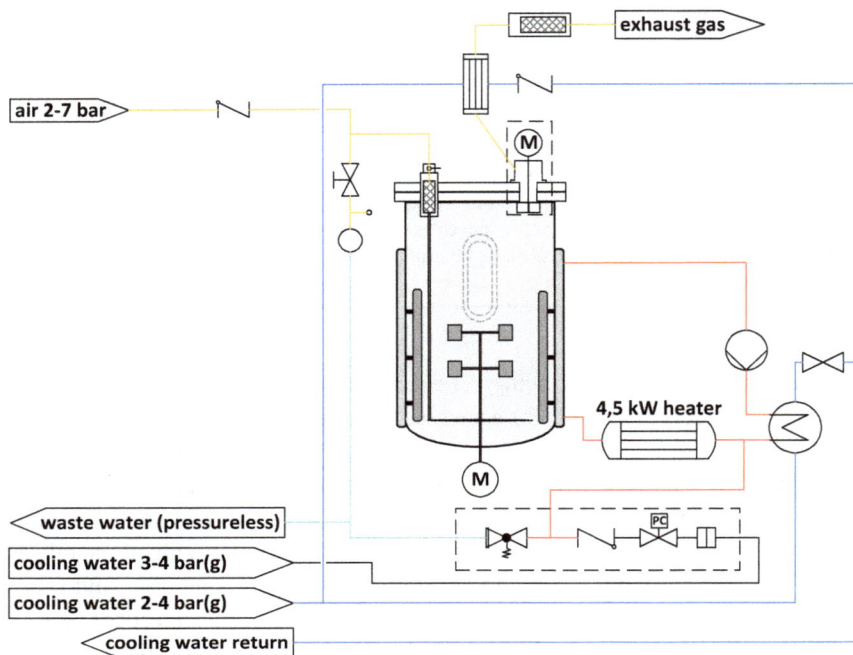

Fig. 11.8: Piping and instrumentation diagram to check the heating and cooling circuits.

activate sterile filter for gas inlet, open the gas supply, and close the condensate drainage of the outgas condenser (for auto-sterile filter only). Due to this procedure, the generation of a vacuum inside the fermentation vessel is prevented. By cooling down the reactor system to the cultivation temperature (37 °C), autoclaving is completed. Each peripheral unit, like storage bottles, tubing, and sensor cables, can be connected to the system as soon as the reactor vessel has a temperature of 75 °C or less. Table 11.9 gives a summary of the single steps of the autoclaving process in the form of a checklist.

Tab. 11.9: Checklist to perform the autoclavation process.

Tasks for autoclaving	Done
a) Close the exhaust gas cooling system and open the gas outlet section.	
b) Check if the air supply is closed, lower the sterile filter for gas outlet, and open the condensate drainage of the outgas condenser.	
c) Raise the temperature to 95 °C.	

Tab. 11.9 (continued)

Tasks for autoclaving	Done
d) When 95 °C is reached, close the gas outlet valve and the condensate drainage valve to a minimum.	
e) Raise the temperature to 121 °C.	
f) When the temperature reaches 121 °C, maintain it for at least 20 min.	
g) Lower the temperature to 100 °C.	
h) When 100 °C is reached, lift the sterile filter for gas outlet, close the condensate drainage valve, open the gas supply, and open the gas outlet valve.	
i) Lower the temperature to 75 °C.	
j) Connect all peripheral units: storage bottles, tubing and sensor cables.	
k) Adjust cultivation temperature (37 °C) and calibrate the pO_2 sensor.	

11.7 Managing the actual cultivation process

The fed-batch cultivation process is started by inserting the preculture into the fermentation vessel. Before inoculation, the agitator speed, gas flow rate, and the pH have to be set to the requirements of the experiment. All devices for measurement and control have to be connected to the system and must be functional. If not performed already, the pO_2 sensor has to be calibrated. Table 11.10 gives a checklist summarizing the tasks before inoculation.

Tab. 11.10: Checklist before inoculation.

Tasks before inoculation	Done
a) Reactor system at cultivation temperature.	
b) Connect each peripheral device to the system, if not already done, including storage bottles, tubing and sensor cables.	
c) Calibrate the pO_2 sensor.	
d) Adjust setpoints of the cultivation parameters (agitator speed, gas flow rate, pH, etc.) to the growth specification of the start conditions in batch mode.	
e) Inoculum is prepared and inoculation bottle is connected.	
f) Process control system (PCS): data recording and controller activated.	
g) Inoculation of preculture.	

Immediately after preculture inoculation, the first sample should be taken. To ensure a sterile sampling procedure, the tubing of the sample units has to be rinsed earlier. The washing procedure is similar to the sampling procedure. The forerun cultivation broth can be discarded. An appropriate sampling cycle is recommended considering the growth rate of the microorganism. For *Bacillus amyloliquefaciens*, a sampling cycle of once in every one to two hours is sufficient. During cultivation, increased foam formation can occur, therefore it is necessary to use an antifoam detergent or a mechanical foam separator. Both chemical and mechanical foam reducers can be controlled by implementing a level probe.

For the presented aerobic fed-batch cultivation of *Bacillus amyloliquefaciens*, a two-stage process is used to produce phytase. In the first stage, the batch phase, biomass is produced until the nutrient phosphate is almost depleted. In the second stage, the fed-batch phase, product formation is carried on by continuously feeding phosphate. Since the synthesis of phytase is enhanced under phosphate starvation, the fed-batch phase is necessary to guarantee an ongoing phytase production. An additional glucose feed is used to supply the bacteria with sufficient amount of carbon and energy to maintain metabolism.

Stage 1 – Batch: After a short lag phase, the microorganisms enter the exponential growth phase. By depletion of the phosphate source, the metabolic activity of the cells decreases. Increasing pO_2 value is a sign of reduced metabolic activity. At this point, the process is switched to the fed-batch mode.

Stage 2 – Fed batch: During fed-batch mode, the culture is supplied with glucose and phosphate. The feed rate of the phosphate source is adjusted to the target growth rate (μ). By the marginal supply of phosphate, the production of phytase is induced. The concept on fed-batch processes is given in Chapter 7.

The glucose feed strategy is to adjust a continuous volumetric feed rate to ensure a sufficient, nonlimiting glucose concentration in the fed-batch mode. An accumulation of glucose should be avoided and thus the supplied amount of glucose should be as high as the glucose consumption of the microorganisms. Furthermore, the feeding profile of glucose is related to the phosphate feeding profile and can be calculated by the related yield coefficient of the microorganisms for glucose and phosphate. By analyzing the biomass, the glucose, and the phosphate concentration during fermentation, these yields can be determined ($Y_{X,glucose} = g$ cell dry mass/g glucose; $Y_{X,PO4} = g$ cell dry mass/g phosphate). Tables 11.11 and 11.12 give an overview of the cultivation and feed parameters of the fermentation process.

Process parameters have to be ensured to be constant during the entire process time. The actual values for temperature, agitator speed, flow rate, and pH of the medium have to be checked continuously. Additionally, qualitative observations like color, turbidity, and smell of the cultivation broth, cell morphology, or possible contamina-

Tab. 11.11: Cultivation parameters of the fermentation process.

Parameter	Unit
Temperature	37 °C
Reactor type	NFL 19 L
Agitator speed	Controlled, range: 200–600 rpm
Aeration rate	2 vvm (16 L/min)
p_{O_2} setpoint	30%
pH setpoint	6.5
Filling volume	7,500 mL
Inoculation	500 mL, depending on $OD_{preculture}$

Tab. 11.12: Feed parameter of the fermentation process.

Parameter	Unit
Growth rate, μ_{set}	0.04 h
Start biomass, $c_{x,0}$	5.00 g/L
Yield, $Y_{x/s}$	12.59 (KH_2PO_4) g/g
Start concentration, $c_{s,in}$	5 (KH_2PO_4) g/L

tions have to be protocolled. For the determination of more cultivation parameters, samples are taken in defined sampling cycles via the sampling device. The samples have to be stored on ice to avoid further cell growth. By measuring the optical density in a spectrophotometer (UV/VIS), the cell growth and respective growth phases can be monitored quickly. For *Bacillus amyloliquefaciens,* it is sufficient to measure the sample at 550 nm using a culture medium or deionized water as a blank. If the optical density of the samples exceeds the linear measuring range, the samples have to be diluted. The measured values for optical density can be related to cell dry mass concentrations by a calibration curve. Here, the measured cell dry mass and optical density have to be correlated. The cell dry mass concentration can be determined gravimetrically by taking an appropriate sample volume. Cultivation parameters like glucose, phytase, or phosphate concentrations are estimated as well. The samples are centrifuged and the remaining cell pellets or supernatants are analyzed by photometric assays. An exemplary calibration curve for a photometric assay is given in Fig. 11.9.

However, analytical methods other than chromatography can be used as well to determine these parameters. High-performance liquid chromatography (HPLC), ion exchange chromatography (IEC), or gas chromatography (GC) are worth mentioning at this point. Figures 11.10 and 11.11 represent the exemplary fed-batch process with all determined parameters over the process time. Additionally, the plotted values are listed in Tab. 11.13. During the exponential phase within the first 8 h, cell dry mass increased from 0.2 to 6.0 g/L with a maximal growth rate of 0.7 1/h. After a process time of 8 h, the fed-batch process was started, reducing the growth rate to 0.04 1/h in the process inter-

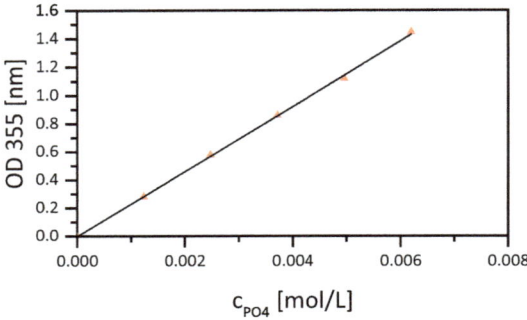

Fig. 11.9: Exemplary calibration curve for a photometric molybdate phosphate assay to determine the phytase activity.

val of 10–26 h. The maximal achieved cell dry mass was around 10 g/L. The phosphate concentration decreased from 6 to 1 mmol/L during batch phase and increased slightly to 1.5 mmol/L at the end of the fermentation. The phytase activity increased linearly after the fed-batch process was initialized till the maximum activity of 0.346 mmol/ (min L) was reached after 11 h. During the time from 16 to 26 h, phytase activity remained constant around at 0.2 mmol/(min L) due to the reduced cell growth.

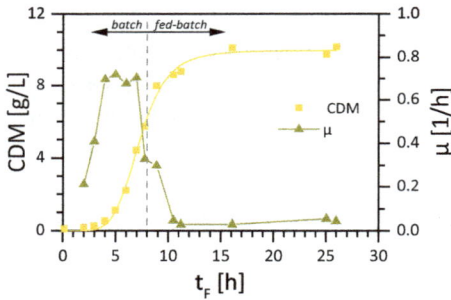

Fig. 11.10: Cell dry mass (CDM) in g/L and growth rate (μ) in 1/h of an exemplary two stage fed-batch process with *Bacillus amyloliquefaciens*.

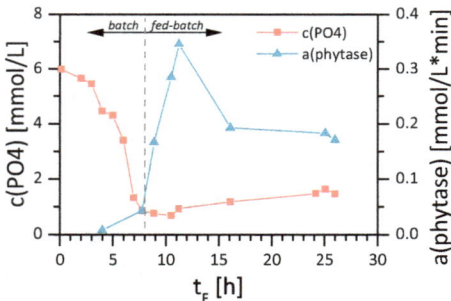

Fig. 11.11: Phosphate concentration in mmol/L and phytase activity in mmol/(L min) of an example two-stage fed-batch process with *Bacillus amyloliquefaciens*.

When the fermentation process is finished, the remaining samples have to be analyzed, the reactor system has to be demounted, and the vessel, including vessel components, storage bottles, and sensors have to be cleaned. It is recommended to decontaminate the culture broth by another autoclaving step. The empty reactor vessel and other peripheral parts have to be rinsed with cleaning agents like water, ethanol, H_3PO_4, NaOH, or HCl. For each fermentation device, a suitable cleaning agent has to be selected and applied in a defined amount and for a fixed time. Nearly all bioprocess fermentation devices require cleaning-in-place (CIP) operations. The necessary time for CIP should be considered for scheduling cultivation. A typical CIP sequence might consist of the following steps: washing with water, rinsing with an acidic solution, and washing with purified water.

In addition to fermentation results given in Tab. 11.13, an example protocol for sample acquisition is given in Tab. 11.14. It can be used as a template to create a protocol for the measured values of taken samples during cultivation.

Tab. 11.13: Offline measured biomass concentration and phytase activity of the exemplary two-stage fed-batch process with *Bacillus amyloliquefaciens*.

Process time (h)	Cell dry mass (g/L)	Growth rate (h^{-1})	Phosphate content (mmol/L)	Phytase activity (mmol/min L)
0.1	0.12	n.a.	5.99	n.a.
2.0	0.18	0.213	5.66	n.a.
3.0	0.27	0.409	5.46	n.a.
4.0	0.55	0.697	4.46	0.007
5.0	1.12	0.718	4.31	n.a.
6.0	2.20	0.677	3.40	n.a.
7.0	4.42	0.704	1.32	n.a.
7.8	5.73	0.329	0.85	0.043
8.9	7.98	0.299	0.75	0.167
10.5	8.60	0.047	0.67	0.285
11.2	8.79	0.029	0.92	0.346
16.1	10.12	0.029	1.17	0.193
24.2	n.a.	n.a.	1.47	n.a.
25.1	9.75	0.054	1.63	0.183
26.0	10.17	0.045	1.46	0.171

In this chapter, it was shown how to manage an actual cultivation process in a lab-scale fermentation unit for the production of phytase, from inoculation to data evaluation. It has to be mentioned that this is an exemplary fermentation; depending on the aim of fermentation, it is probably necessary to adapt the cultivation parameters and/or feed parameters. However, by following the general steps in this section and in the previous sections, it is possible to set up fermentation in a bioreactor from scratch. We want to encourage the use of this book chapter as a practical guide to performing a cultivation process, step by step.

Tab. 11.14: Example protocol for sample acquisition.

Sample no.	Date JJ:MM:DD	Time HH:MM	Process time (h)	Dilution (–)	OD 550 (–)	CDM (g/L)	PO$_4$ (mol/L)	Phytase activity (mmol/min/L)
0								
1								
2								
3								
. . .								

11.8 Sterilization – more than heating up

In this paragraph, the deeper sense and mechanisms of sterilization behind the to-dos will be highlighted. A distinction is made between the different processes and they are defined for clarity.

Sterilization: Remove living microorganisms, including their resting stages (e.g., spores) from materials and objects, including viruses, prions, plasmids, and other DNA fragments.

Disinfection: According to the German Pharmacopoeia (DAB), it is about bringing dead or living material into a state in which it can no longer infect. So, this is more of a medical term.

Axenic work: Elimination of all foreign germs, i.e., only the desired strains live in the reactor. Note the difference from sterilization (Greek: ξένος xénos, stranger)

Strict sterile working is often not possible on a large scale. Nevertheless, hygienic work means keeping contamination to a minimum by cleaning, disinfecting, careful working, etc. Hygienic design means avoiding weak points, right from the design stage.

11.8.1 Thermal sterilization – the most important approach in process technology

In order to design a thermal sterilization process, the first fundamental step is to consider the reaction of microorganisms to high temperatures. Corresponding experiments have shown that the number of living microorganisms decreases exponentially from a healthy culture after a temperature jump. That is quite surprising. Experience with plants, for example, shows that all specimens die more or less at the same time, holding out for a while after the temperature jump. This experiment also does not allow the organisms to adapt, which would also be interesting to assess possible threat from contaminating organisms. This behavior corresponds to a reverse growth process and can therefore be formulated in a completely analogous way:

$$\frac{dc_{\text{Cell}}}{dt} = -k_{\text{in}} \cdot N_{\text{Cell}} \rightarrow k_{\text{in}} = \frac{1}{t_{\text{in}}} \cdot \ln\frac{N_{\text{Cell},0}}{N_{\text{Cell}}} \qquad (11.1)$$

! The number of cells dying per time unit is therefore again proportional to the number of currently existing cells. It is therefore a statistical process, as it also occurs in the case of the thermal degradation of molecules or radioactive decay. Note that the exponential decrease does not predict a point in time at which all cells have actually been irreversibly deactivated.

The parameter k_{in} is called the inactivation rate and depends on the organisms and the temperature. Data for *Bacillus stearothermophilus* are listed in Tab. 11.15. This organism, which is particularly resistant to heat, as the name suggests, is used in food industry to prove the effectiveness of sterilization measures in production facilities.

Tab. 11.15: Inactivation constants for *Bacillus stearothermophilus* (spores).

T (°C)	k_i (h^{-1})	N/N_0 after 0.5 h
80	0.0065	0.997
100	1.16	0.56
110	12.6	1.8×10^{-3}
121	150	2.7×10^{-33}
130	1,038	4.0×10^{-226}
140	8,044	0.0

First, it is noticeable that all interesting temperatures are above 100 °C. Only above 110 °C does k_i increase sharply, and 121 °C is the temperature that is used as standard in practice. What does this mean for a bioprocess in which it is highly probable that all germs must be killed, whether before cultivation or afterwards, e.g., for the disposal of S2 organisms? A bioreactor with 1 m^3 working volume and a culture with 10^9 cells/mL contains 10^{15} cells. You would therefore have to maintain a temperature of over 110 °C and up to 121 °C for 1/2 h to be sure that no survivors can contaminate the next run. However, maintaining these high temperatures means an overpressure in the boiler that corresponds to the steam pressure, which is exactly 2 bars at 121 °C. Incidentally, this also explains why the table does not state 120 °C. This is because the boiler is designed for an even 2 bar and a corresponding safety valve. Theoretically, longer sterilization times would also be conceivable to guarantee that the heat reaches all nozzles and seals.

With the specific growth rate μ, the specification of doubling times is also common because it is more intuitive. For the k_i value, it is the D value (decimal) at which the bacterial count drops to 1/10. Some values are listed in Tab. 11.16.

Tab. 11.16: *D*-values (min) for different scenarios.

For different bacteria at 121 °C in water, medium, or humid heat			
Bacillus stearothermophilus	*Bacillus subtilis*	*Bacillus megaterium*	*Clostridium sporogenes*
1.5–4.0	0.4–0.7	0.04	0.8–1.4
For *Bacillus subtilis* at dry heat under atmospheric air conditions				
120 °C	140 °C	150 °C	160 °C	180 °C
30	3–5	2	1	0.22

The processes discussed so far correspond to the first row in the table – in the presence of water, i.e. the medium and/or steam that is fed through pipes and valves. Stainless steel parts such as probe holders or spatulas from the laboratory and medical tools can also be dry-autoclaved. However, this requires higher temperatures, as shown in the second line. However, this is not a problem because dry air can be brought to high temperatures without overpressure.

11.8.2 Other problems – and other solutions

To heat up the reactor, steam is passed through heat exchangers, such as the cooling jacket or internal pipes. In the first case, the heating process can take a long time; in the second case, the tubes (e. g. also smaller laboratory glass reactors) or even complicated shapes (large yeast reactors) must be cleaned after the cultivation, which is a cost factor. An alternative is the direct vaporization of the medium. However, this is not possible for all media. Depending on the application, other chemical and physical measures can also be taken to inactivate foreign germs. Incidentally, traditional cooking is also a sterilization process, in addition to increasing the bioavailability of food. The gentler and pressure-free heating process to 75–80 °C was invented by Louis Pasteur (1822–1895, France). Apart from experience, the real scientific achievement is to recognize that food spoilage is caused by microorganisms. Pasteurization is still widely used for food preservation of milk or fruit juices, for example. However, this does not deactivate any spores but also unwanted enzymatic activity.

At some point, someone came up with the idea that you should not pressurize an entire reactor to more than 2 bar, but that this is technically possible with thin pipes or small-plate heat exchangers. This means that the medium can be fed through a thin pipe, which is kept at 140 °C and 3.61 bar, for example. It is usually assumed that the k_i-values behave according to an Arrhenius equation with respect to temperature (eq. (2.1)). Sterilization times are then reduced so drastically that only a few minutes of holding time is required. The medium is therefore at ultra-high temperature (UHT), similar to milk. The short times are a major economic advantage. It therefore also allows a continuous sterilization process. However, there are fears that valuable ingredients in foods and complex media such as molasses will be lost. To investigate this assumption, Tab. 11.17 shows some values for the activation energy of some substances.

Tab. 11.17: Activation energies E_a (kJ/mol) for substances in biotechnological media.

B. stearothermophilus	Proteins	Vitamins
250–290	150–200	89–90

In this context, microorganisms are considered as macromolecules with very high molecular weight. The lower the molecular weight, lower is the activation energy. This means that the hotter the sterilization is performed, the shorter are the necessary times and the more selectively, the microorganisms are affected. Low molecular weight substances, on the other hand, tend to be spared.

However, there are also reactions between media components that are favored by high temperatures. The most important of these is the reaction between amino acids and sugars, the so-called "Maillard reaction," commonly known as caramelization. This not only causes an undesirable discoloration, but also a reduction in quality because many microorganisms, such as yeast, are unable to utilize the substances formed. One solution to the problem is the separate sterilization of media components. This is not a problem in the laboratory with synthetic media, but it is with complex media. Sterile filtration (steric separation principle) should always be carried out for completely dissolved and thermally sensitive components. Especially in food processing different approaches of playing with temperature and pressure are employed (Tab. 11.18).

Tab. 11.18: Overview of the various thermal processes for microbial inactivation.

Temperature (°C)	75–80	100	121	133	140
Dampfdruck (bar)	0.39	1.0	2.0	3.0	3.6
Application	Pasteurization; no spores	Boiling	Steam sterilization	Fermentation residues; prions	Ultra-high temperature (UHT)

The use of pressure has been suggested many times. However, more than 500 bar is needed to achieve a noticeable effect. This is also unthinkable in relation to the reactor, possibly as a flow method for the medium. The process is occasionally used on a small scale in the food sector. Chemical treatment (dosing of ozone, peracetic acid, and ethanol) is widely used in the laboratory (causing hygiene) or for water utilization, but not directly in bioprocess technology, as the cells and products should not come into contact with foreign chemicals, if possible. Ionizing radiation (γ-radiation and UV-C light) is effective, but technically limited (light is not miscible). Radiation is used to keep surfaces of clean benches aseptic and γ-radiation to reduce the contamination on surfaces of pepper or cereal grains. One application of UV-C radiation is the sterilization of plastic bag reactors in sensitive areas such as in pharmaceutical production, see Fig. 11.12. In contrast to conventional inactivation technologies, UV-C irradiation is particularly effective for small, non-enveloped viruses.

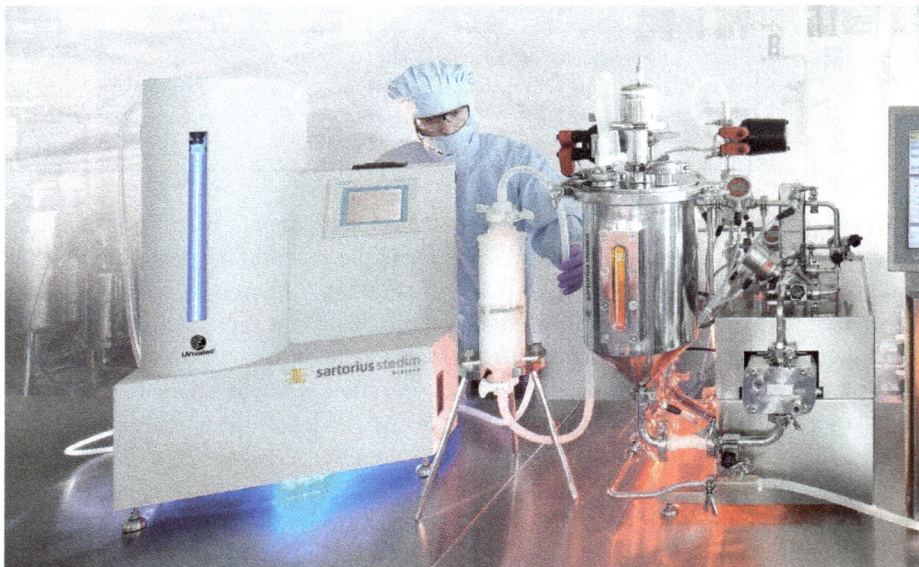

Fig. 11.12: Sterilization of a bag placed in a stainless steel container. UV-C radiation is used for this purpose (© sartorius). Source: Sartorius AG.

However, there are many cases where safe and completely fail-safe operation cannot be guaranteed. These can be open processes such as composting or algae production in open tanks. But large reactors in which a lot of natural material is converted are also a problem. In any case, the sterilization times could be absurdly long. For cost reasons, not all media can be sterilized or the reactor cannot be ideally processed. In countries with little infrastructure, even the production of yeast is a problem due to the availability of cooling water or energy for sterilization. Biological approaches can still help here. One possibility is the choice of extremophilic microorganisms. These can prevent low salt-, pH-, or temperature-resistant contaminants from occurring. The self-protection of yeast through low pH values and the production of alcohol protect the process from foreign organisms and made it possible for humans to bake bread and brew beer. The same applies to the cultural adaptation of lactic acid bacteria. The establishment of consortia is also a current topic. This can not only help to secure the supply of vitamins in the process, for example, but also to keep protective organisms alive, which in turn protect the production organisms.

11.9 Integration on the process level – how to include fermentation into a whole process chain

Closing the fermentation chapter, it should be reconsidered how the fermentation stage interacts with upstream and downstream steps. Many examples have already been mentioned in the previous chapters. There can be a direct physical integration of the process steps upstream, bioreaction, and downstream. Even if this is not the case, the stages influence each other in all directions, which the bioengineer should be aware of during the process development process. In the best case, these interactions can be anticipated to be of decisive advantage. Table 11.19 tries to give some order and previously mentioned examples. Some points are then outlined in a bit more detail.

Tab. 11.19: Aspects of integration on the process level and on the level of process design.

Stages	Aspect	Measures and examples
Microorganism and cultivation	Medium	Media for lab and production foaming
	Control of physiology	Process feeding strategy
	Robustness	
	No by-products	
Microorganism and downstream	Extracellular products	Targeting tags Cell wall permeability
	Cell size and shape	
	Aggregation	Auto/induced flocculation
Cultivation and downstream	Cell-recycle or -retention	High biomass concentration Wastewater
	Extracellular products	

Strain robustness is an important issue. Some strains are in use only based on historic development. Even simple things like shear stress resistance, stickiness, or lack of by-products could be targets. While, in the lab, complex media and supply of vitamins are ok, for production, cheap media are preferred. That has to be envisaged during strain selection and development. Finding additional pathways for certain sugars is one way of making strains fit for the rough production environment or to keep extracellular products in an inactive stable state. In other cases (e.g., microalgae), extremophiles can help suppress contamination. Many microorganisms with interesting products have been screened and isolated, but they are hard to cultivate. Sometimes, they live in symbiosis with other microorganism or work together in a conversion chain. Cocultivation is an option currently discussed. Recall the persistent problems of using *E. coli* or *Bacillus* as expression systems (auxotrophy, unnecessarily complex metabolism, undesired proteases, foaming, etc.). Are artificial or minimal cells delivered by systems biology, a solution? The idea of robustness also drives attempts to replace animal cells with phototrophic plant cells (moss and microalgae), which are more robust and can grow in pure mineral media. Intrinsic safety is also increased because no viruses are known that infect plants and humans. This is of course a long process, triggered by recent success in genetic engineering toward humanized protein products (glycosylation).

Extracellular products are a good idea to avoid intracellular inclusion body formation and to ease separation. Citric acid or penicillin are established examples as well as recombinant proteins, which may be naturally excreted or engineered with a targeting tag. In the case of citric acid, cell wall mutants are employed. Cell disruption or extraction also turn out to be problematic in some cases. Having a look at these unit operations with respect to, e.g., the cell wall, can be done during strain development. In situ extraction, during the cultivation, with an immiscible extractant has been tried out.

In the baker's yeast example, fed-batch processes in the cultivation stage are necessary to cope with the biological features of the Crabtree and Pasteur effects. Furthermore, the process strategy could influence product quality and, in the context of integration on the process level, filtration efficiency. As an aside, budding of yeast also influences the brewing process, be it bottom-fermented (sedimentation) or top-fermented (flotation). Back to baker's yeast, what else could be done during strain development, e.g., by genetic engineering (if it was allowed) to make the process simpler and more intensive? Yeast strains are the result of 100 years of strain development. This virtual glass bead game could include overexpression of the respiratory chain. Yeast can do that, in principle, as we have seen during aerobic growth on ethanol. Some strains cannot produce ethanol. Is it sensible to use such strains when some yeasts for single-cell protein (SCP) would come into question? This is obviously not expedient, as we need the fermentative pathway to let the dough rise. Thus, an inducible alcohol dehydrogenase (ADH) would be an option for switching on under the specific conditions in the dough. Similar considerations for other expressions systems are made.

Cultivation for a good cause can also be non-sterile and inexpensive (Fig. 11.13). This reactor concept can also be operated in multiple parallel operations by farmers near the rewetted moors as a secondary source of income.

Fig. 11.13: There is another way: moss cultivation for the renaturation of bogland in a 100 L water canister operated as bubble column (Reactor concept and picture © Maria Glaubitz).

11.10 Questions and suggestions

1. Draw a reactor scheme and indicate the parts of the reactor that have to be steamed, and mention which valves have to be opened/closed along with their sequence.
2. The oxygen transfer rate (OTR) is often the limiting step during cultivation. How can it be increased in a running process?
3. Antifoam agents often have a negative effect on cultivation. What it is the reason for this problem and what are some possible solutions?

Chapter 12
Modeling – the art and handcraft
of mathematically describing bioprocesses

Modeling is not an esoteric task that we can do at the end of process development as a cherry on the cake. In fact, it is an integral part of process design and has to be done iteratively from step to step during the design process. The willingness to understand the inner relationships of systems and to effectively design processes brings us inevitably to modeling. In this chapter, some general ideas about modeling of bioprocesses are collected. Specific features to model metabolic networks are one kernel of a bioprocess model. Modeling itself is a structured and integrated processes, following specific ideas, structures, and rules for the different aspects we meet in biotechnology. As generally applicable examples, aerobic, anaerobic, and phototrophic processes are given in some detail. The examples also give hints to different modeling techniques setting the follower in a position to build models for other or more complex bioprocesses. Finally, quantitative analysis of the model and model outcomes have to be computed, a process called simulation. No specific mathematics is needed; a short introduction will enable us to work with the models in the supplementary material.

12.1 Modeling – what it might be and what it is good for

All engineers work with models, sometimes without being fully aware of it. The imagination of electrical or gravitation fields are model representations, allowing us to quantitatively design technical devices. This concept has been applied very successfully, even knowing that the reality may be more complex. Behind all scientific terms stand structured ideas, making their impact on scientific thinking. Before being lost in detail, we can try to formulate an admittedly quite broad definition:

> **!** Models are descriptions of natural and artificial systems in a formal representation space.

Modeling is hence an integrative approach between different sciences or people who contribute their specific knowledge. More attributes filling this formal statement with life are made clearer in the next paragraphs. In our daily lives, we encounter models, often graphical models. Street maps can be taken as a common example. Comparing such a map with a satellite picture showing the same section of a landscape, the differences become clear. First, many bewildering details are left out, while the focus is on showing only the streets. So, simplification or better said, abstraction, is one major issue in setting up models. The different streets, like highways or country roads, are

https://doi.org/10.1515/9783110773354-012

kept in different colors. Of course, in reality, streets appear in gray. This difference between a graphical model and the reality is a tribute to applicability, the second basic issue. Orientation on usefulness defined by a model builder or model applicant is a pragmatic consequence. In biology, graphical models are also a classic method. A drawing of a cell with several details is based on many microscopic pictures, none of them alone showing all the details of the drawing. Analog models can be useful means of communication. An example in bioprocess engineering is the bottleneck model. Behind the purpose of every model, stand people who work with the model. This has to be considered by the model builder. Liebig's barrel model is a good example of user-oriented graphical modeling.

In engineering, graphical representations are also a common method, as process flow sheets prove. To go more into details and to describe the invisible relations between parts of a system, mathematics is the language of choice. Consequently, in the context of this book, we are interested in mathematical models, especially process models. This leads us to the third issue: complexity. The decision – which and how much information is lumped into the model – has its limits, on the one hand in known details and on the other hand, in the purpose of the model. Figure 12.1 gives an overview of models of different complexity, based on data and knowledge.

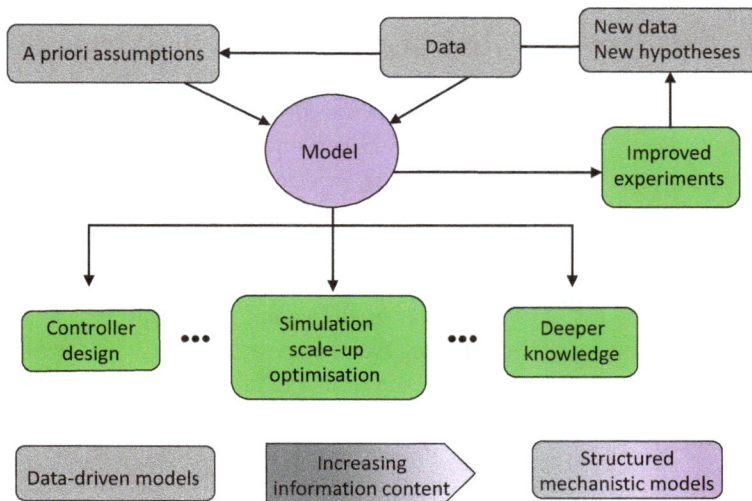

Fig. 12.1: Flow sheet to show generation of structured models, from simple ones to increasing information content, by cyclic running through the modeling process.

In control theory, information is needed to adjust parameters of the controller, e.g., PID control. Often, there is not enough knowledge to directly calculate dynamic responses, e.g., to disturbances. In such cases, simple-step response functions are measured and are used to set up basic system representations as P, PI, or PID systems.

Such simple models are called data-driven models. Despite the fact that no mechanistic assumptions about the system are made, such models are good enough to fulfill the valuable task of control. Neural networks are another more complex example. In multifactorial experimental design, a quadratic cost function is the simple model background on which experiments are planned. The inclusion of more information leads to moderately structured models. The simple growth model we used until now is an example of this. It includes some assumptions about uptake kinetics and growth yield but without overstressing any details. Putting more and more knowledge into the model will presumably lead to more precision and more reliability of such a mechanistic model. The highest meaning of a model, although it is assembled from already known bricks, is to help get new insight into a system and to gain better understanding of the system as a whole.

Models in the context of bioprocess engineering are meant to be employed for process development. This includes plant simulation, optimization, media design, and scale-up. Changing process policy, e.g., from batch to fed-batch needs precise understanding of the interaction between the medium and the cell. Exact kinetics for most of the medium compounds are necessary. Simple heuristic models, valid only for a specific case, are not adequate. The intracellular structure also has to be considered in order to understand the constraints and flexibility of the metabolism, because intracellular bottlenecks and stoichiometry will dominate the whole process. Models for successful process development are therefore structured models. Validity, as the next important issue, increases with the inclusion of mechanistic items. In the best case, a model is equipped with clearly defined limitations of validity.

The framework to start (Fig. 12.2) is definitely conservation laws as they are standard in chemical engineering, in general. This holds, especially for reactor equations. Known stoichiometry, and macromolecular and elemental composition will be added in the next step on the cell level. The model building process is a structured procedure with formal meta-rules on different levels and model aspects. Facing the complexity of a cell and the whole process, it is clear that model components will always be missed. Depending on the purpose of the model, we can be satisfied or look for further relevant information. Even here, the model developed so far can be useful. It supports model-based experimental design to bring out as much information as possible to allow for a quantitative description of the missing parts. Finally, the model can be regarded as a complete description of the process.

What happened until now in this book? Process models have already been set up intuitively. But, most important, is also that a structure has been assigned to the problem of model set up. The system was composed of three subsystems, namely gas phase (as described in Chapter 5 in some detail), the liquid phase, and the "bio-phase" (Fig. 12.3). All three phases, of course, have mutual mass transfer and other ways of exchange. In the liquid phase, usually no or well-known reactions take place, allowing us to apply deterministic rules of mass and energy balances.

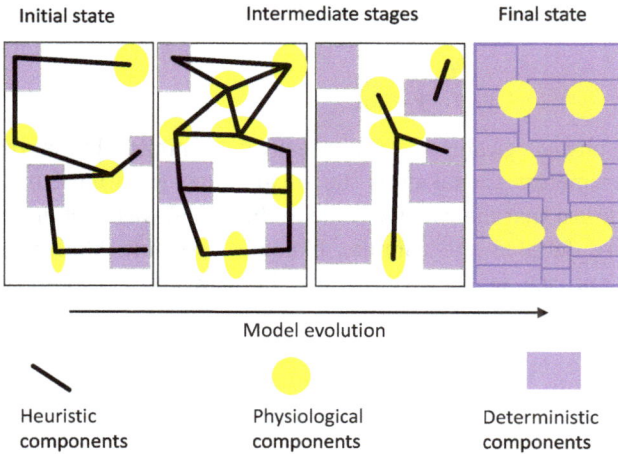

Fig. 12.2: Framework for building up a model from different model building blocks with different deterministic, mechanistic, or statistical background.

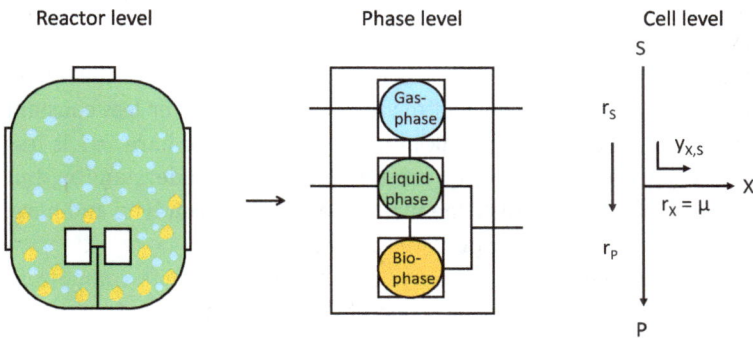

Fig. 12.3: Breaking down of a reactor model to submodels, here the physical phases; a metabolic submodel of the biophase is further indicated.

These equations (12.1a), we called the reactor model:

$$\frac{d\boldsymbol{c}}{dt} = \pm \boldsymbol{q} \times \boldsymbol{c} \pm \boldsymbol{r}(\boldsymbol{c}) \tag{12.1a}$$

This well-known reactor model (here, given in bold as vector representation), considering accumulation, transport, and reaction, has to be amended with a model for the bio-phase. These were based basically on observations. For substrate uptake, a saturation curve is observed, which we interpreted, maybe a bit prematurely, as enzyme kinetics. Reducing several substrate-uptake systems of many microorganisms to only one is here a simplification. Further data show that growth is linearly dependent on substrate uptake, which was interpreted as metabolic energy balance. In compact vector form (12.1b) reads:

$$\boldsymbol{r}_{\text{kin}} = f(\boldsymbol{c}); \ r = \boldsymbol{y} \times \boldsymbol{r}_{\text{kin}} \tag{12.1b}$$

Information processed in the physiological model is not really deterministic but more heuristic and the related parameters are taken as statistical values from measurements. There remains the concern that the way we came to the model was a bit arbitrary and would fail in more detailed and complex biological systems. Modeling should be a well-structured approach, which we want to go into now. We start with a classic example from ecology.

12.2 Predator-prey model

In the beginning of the last century, the Hudson's Bay Company observed that the number of sold furs of snowshoe hares and lynxes was affected by strong fluctuations over several years. Concerned about the severe impact on economics, they tried to understand the phenomenon, following the work of the chemist Alfred Lotka. He worked on oscillation in chemistry, society, and "physical biology." To follow his ideas, we first have a look at the data (see Fig. 12.4) and find the reasons for the fluctuations. Observations are that the frequency of the oscillations is more or less constant as well as the mean values. Further, the peak number of furs of the hare sets is less than that of the lynxes. These observations are not enough for understanding but may act as a benchmark after we uncover the inner relations between lynx and hare: basically, this is the relation between the predator and the prey.

The first steps are to define the system boundaries and to list the most important variables involved. In our example, it is the scope of Canadian northern forests and the number of lynxes $N_l(t)$ and hares $N_h(t)$. The dataset represents an inherently discrete time process with discrete values. Nevertheless, we know that the number of animals is very high and that they also propagate during the year, so we are encouraged to use a time-continuous model. Remember that a similar transition has been done already describing microbial growth. Here, we keep the letter "N" understanding that is a real value with the meaning of an areal population density. In the next step, the interrelations between each of the variables with themselves as well as with the other variables have to be fixed. Therefore, we formulate some model hypothesis and find mathematical representations for it:

The undisturbed growth of the hare population follows the same idea as the growth of microorganisms. The "undisturbed" growth of the lynxes is characterized by slow dying, in case they have no feed. We define the decay rate positively, but will consider the decay with a minus sign in the final balance:

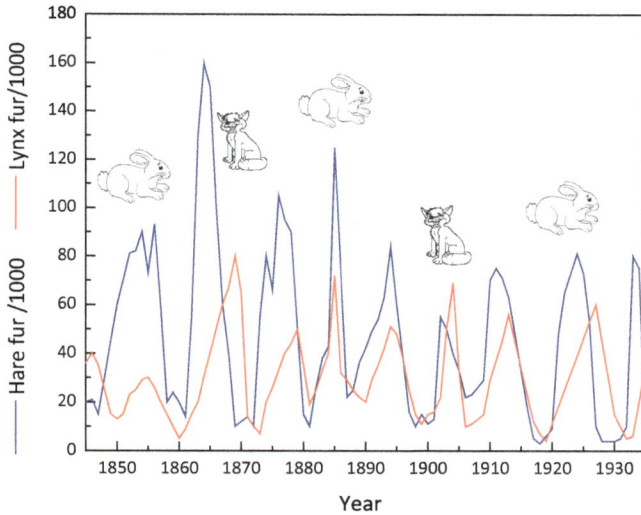

Fig. 12.4: Original data set from Hudson's Bay Company showing the number of lynx and hare furs sold.

$$\frac{dN_H}{dt}\bigg|_{undis} = R_{H,growth} = r_{H,growth} \cdot N_H; \quad \frac{dN_L}{dt}\bigg|_{undis} = R_{L,death} = r_{L,death} \cdot N_L \quad (12.2)$$

The structure of this simplest possible linear system, which includes feedback from the state variable to its derivative is common in nature and technology, at least in simplifications. Other examples are radioactive decay, chemical reactors with a positive temperature coefficient (can even explode!), or the Lambert-Beer law (flux of absorbed photons per path length is proportional to photon flux).

Something happens only if lynx meet hare, which is the first interplay. The probability is higher if the population densities of the respective species are higher. This can be fixed by introducing a probability coefficient, $w_{H,L}$. With a given "efficiency" in hunting by the lynx, in the model represented by a yield coefficient $y_{-h,l}$, this ends up with the death of the hare. The other way around, the lynxes need a given amount of caught hare for reproduction, noted as $y_{l,h}$:

$$\frac{dN_H}{dt}\bigg|_{meet} = R_{H,meet} = y_{-H,L} \cdot w_{H,L} \cdot N_H \cdot N_L; \quad \frac{dN_L}{dt}\bigg|_{meet} = R_{L,meet} = y_{L,H} \cdot w_{H,L} \cdot N_H \cdot N_L \quad (12.3)$$

Take notice of the analogy to the mass action law, where the same probability argument leads to a comparable formulation. The parameters y and w appear only as a product and are therefore linearly dependent. Further, they cannot be determined from other observations, so we lump them in this model to the new parameter k, similar to the reaction constants in chemical reaction technology. Remember this lumping process also in the derivation of the $k_L a$ value. We can now set up a complete population balance for both species as the sum over all of their single rates, Σ_R:

$$\frac{dN_H}{dt} = R_H = r_H \cdot N_H - k_H \cdot N_H \cdot N_L; \quad \frac{dN_L}{dt} = R_L = k_L \cdot N_H \cdot N_L - r_L \cdot N_L \qquad (12.4a, b)$$

A simulation example is given in Fig. 12.5. Indeed, the periodicity and the other features of the measurements are fairly well represented.

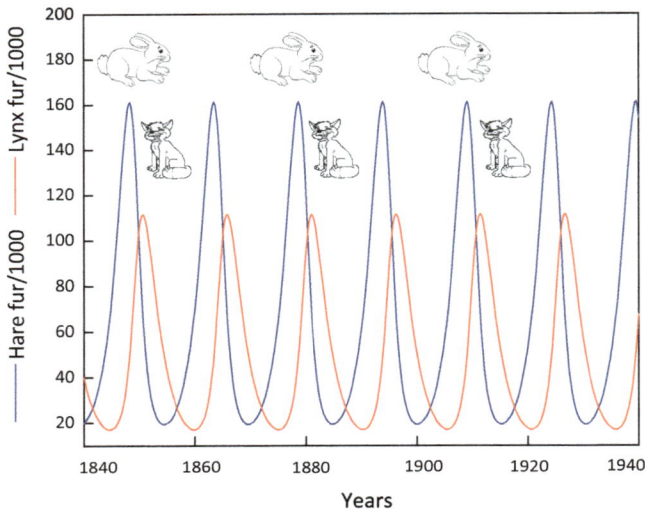

Fig. 12.5: Simulation of the predator–prey model with the parameters, $r_{H,growth} = 0.5$, $r_{L,death} = 0.4$, $w_{H,L} = 0.01$, $y_{-H,L} = 1.0$, $y_{L,H} = 0.6$, $N_{H,0} = 20$, and $N_{L,0} = 40$ (given in thousands).

A simulation program is given in the supplementary material. Such population models are also called Lotka-Volterra models, according to the already mentioned Lotka and the Italian mathematician Vito Volterra, who later described similar observations for fish populations in the Mediterranean Sea. The impact of such models goes beyond population dynamics. Activator-inhibitor systems follow the same pattern. Such systems are, for example, biceps and triceps working against each other to allow a stable movement of the arm, a principle called antagonism. Pairs of hormones often work according to this principle – like insulin and glucagon or adrenalin and noradrenalin. Such activator–inhibitor systems may be stable or unstable. A continuous cultivation can also be understood in this way. The substrate is the activator while "undisturbed" feed leads to substrate accumulation in the reactor. This supports increased substrate uptake and growth and finally more biomass. Biomass, as inhibitor, leads to negative feedback on the substrate concentration. We have already seen that for substrate inhibition, the sign of feedback changes to positive, potentially leading to an unstable situation. Other examples of cyclicality are business or climate cycles. In most cases, oscillations are not so desirable like damping factors. More complex predator-prey systems tend to be more stable. Lynxes may have other prey than hares and vice versa, which turns out to be a stabilizing factor. This makes it understandable that oscillations are observed in nature, mainly in species-poor regions. Mathematically speaking, stable cycles require a nonlinear element, like the product of the two species, and a positive feedback loop. The undisturbed growth equation is such a loop, as more hares gives birth to even more hares, in analogy to microbial growth.

Did we really find a basic description of reality in population dynamics and is our model hypothesis proven? Doubts are definitely appropriate. First of all, the number of furs is, in the best case, proportional to the real population density and, if not, with the same factor for both species. It is indeed unbelievable that, on average, only three times more hares exist as lynxes and that the predator can live by catching only three preys per year. Furthermore, human hunters may have an influence in reducing the number of both species with different efficiency. Could an external influence control the systems, like solar cycles (sunspots)? Some people found the time to exclude this specific idea. As many experts think now, the real predator–prey system is the one between grass and hares, with hares now being the predator. The number of lynxes follows the cycle passively, so much for model validity.

12.3 Linear networks – finding stationary flux equations

After we have identified activator-inhibitor systems as structures of possible interest to describe biological systems, we seek other general patterns. Many systems have a network structure. We know them as electrical or computer networks. They may be traffic systems like railways or streets. Also, rivers, water supply networks, or leaf veins fall into this category. Here, the question of interest is fluxes (transport) along edges, like the number of cars per second driving along a street or the water flux in a river. These fluxes interact with each other and the points where that happens are called nodes of the network. A node can be the mouth of a small river entering a bigger one. The interconnection would simply be an addition of both water flows upstream to the one downstream. A more complicated view could be the temperature or pollution load of the rivers. Typically, these nodes are connected only by a few edges (arcs, pathways, etc.) with other nodes, thus forming meshes. This structure makes final modeling easier than dealing with a lot of variables, all of which are interconnected. Another simplification is to ignore all kinds of storage. A parking place at a highway will not change much the number of cars per time unit between a motorway access and the next exit.

Metabolic networks are of highest interest in the context of bioprocess engineering. Beyond a graphical metabolic structure showing metabolic pathways and their mutual links, the metabolic fluxes along the pathways should be calculated and made visible. We did this already on a strongly simplified basis by calculating the specific turnover rates. Accumulation of metabolites may have biological meaning but are ignored, for now. This stationary view leads to sets of algebraic equations. The simple approach, called "physiological model," consisting of Michaelis-Menten kinetics for substrate uptake and of a Pirt equation was the first attempt, based on observation. It is now time to find a stringent and unified procedure for setting up such models for more complicated cases. As the first example, the aerobic metabolism, as depicted in Fig. 12.6, is chosen.

! For complex metabolic networks, it has to be set up a structured approach for finding suitable balance boundaries and setting up the balance equations from network analysis. Here are important rules:
- Each node to be considered has to be located at least inside one system boundary. Otherwise, the stoichiometry of this node is not represented in the model. In the example, three nodes are defined for the catabolism K_{cat}, the respiratory chain K_{resp}, and the anabolism K_{Ana}. Consequently, three system boundaries can be identified, shown here as ellipses (brown).
- The number of system boundaries equals the number of metabolic nodes. The definition of an additional system, e. g., around the whole cell may be useful for later model validation but does not give additional information beside the balances around the individual nodes.
- Each metabolic path has to be intersected (at least) by one balance border. Otherwise, it will not appear in the respective balance equations, and therefore not in the model.
- The maximum number of linear independent balance equations $n_{max,knot}$ for one node cannot reach or exceed the number of nodes n_{knot}. Otherwise, the node would determine itself completely and would be decoupled from the remaining part of the network and the outer world.

These rules are also applied to networks with several hundred nodes. Besides, weaker linear relations are also observed in living cells. These are not stoichiometrically determined directly but have their origin in intracellular energy or material balances, which are subject to change by the cells, according to environmental conditions. Metabolic fluxes interact in specific ways, corresponding to specific approaches to finding model equations. This will be exercised now for the aerobic metabolism.

12.4 Aerobic growth – setting up the first general model

The first step is to look at the metabolism and to select important metabolic pathways. For this example, the metabolic structure in Fig. 12.6 is chosen. In the balance, fluxes are counted positively if they go inside a node, and negatively, if going outside. That does not mean that they go in reality (always) in this direction. The numerical value of a flux will then have a negative sign.

Substrate (glucose) as C6 body is taken up and converted via glycolysis to pyruvate (C3), which stands here as representative for all metabolites of the TCC. The two ATP produced are neglected here. The same holds for $NADH_2$, as it compensates more or less with redox demand in growth; the remaining parts are small, compared to redox production in respiration. So, the virtual metabolite "Cat" has the same degree of reduction as glucose. The brown circles represent system boundaries for setting up balance equations. All parts having to do with respiration, including production of

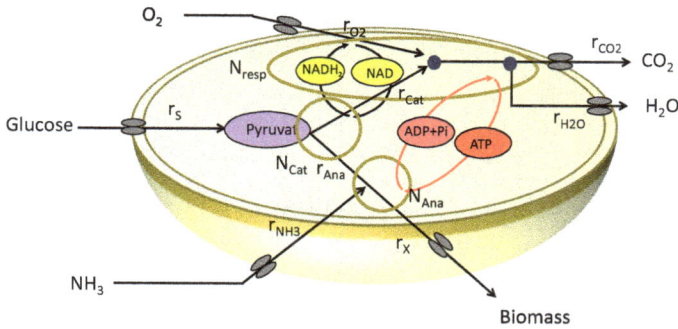

Fig. 12.6: Metabolic structure of a typical aerobic microorganism; the circles are possible subsystem boundaries.

redox equivalents and CO_2, as well as oxygen uptake and ATP generation, are lumped in one subsystem. The anabolic pathways include biomass formation from metabolites in the TCC and nitrogen uptake, also as a model for other nutrients.

The first approach is setting up material balances and stoichiometric relations. Elemental balances are a specific form of mass balances. Balance equations can be set up either by looking at mass or at molar fluxes. The latter option has the advantage that some side reactions like CO_2 or H_2O addition can be ignored for a while. On the other hand, a conversion factor, from the molar growth rate to real biomass, has to be found. We start here with mass flux balances as they link to the representation used in the other chapters. For the mass flux balance around the TCC, all three fluxes have to be summed up. All fluxes are given in a condensed representation. The first one is the flux-through glycolysis. In the TCC, precursors for growth are produced (r_{Ana}) and redox equivalents (r_{Cat}) for respiration. Outgoing fluxes, r_{Ana} and r_{Cat}, for growth and respiration are counted negatively, leading to $r_S - r_{Cat} - r_{Ana} = 0$; pyruvate does not accumulate. This is actually the first row in eq. (12.5). Note that no additional knowledge about yields is required. A second equation could be allowed but would result in redundant information. Each elemental balance, for example, would lead to a linearly dependent equation. From a biochemical and physiological viewpoint, it is clear that the organisms can allot the glucose arbitrarily into respiration as energy source or into growth as carbon source, according to the actual needs. Anaplerotic reactions in the TCC help them to do so:

$$
=
\begin{bmatrix}
\text{mass} - \text{balance } N_{\text{Cat}} \\
\text{mass} - \text{balance } N_{\text{Ana}} \\
N - \text{balance } N_{\text{Ana}} \\
C - \text{balance } N_{\text{Resp}} \\
O_2 - \text{stoich } N_{\text{Resp}} \\
H_2O - , CO_2 - \text{stoich } K_{\text{Resp}} \\
\text{ATP} - \text{balance}
\end{bmatrix}
$$

$$
=
\begin{bmatrix}
1 & 0 & -1 & -1 & 0 & 0 & 0 & 0 \\
0 & 0 & 0 & 1 & 1 & -1 & 0 & 0 \\
0 & 0 & 0 & 0 & \frac{M_N}{M_{NH_3}} & -e_{N,X} & 0 & 0 \\
0 & 0 & \frac{M_C}{M_{Cat}} & 0 & 0 & 0 & 0 & \frac{-M_C}{M_{CO_2}} \\
0 & \frac{1}{M_{O_2}} & \frac{-3}{M_{Cat}} & 0 & 0 & 0 & 0 & 0 \\
0 & 0 & 0 & 0 & 0 & 0 & \frac{1}{M_{H_2O}} & \frac{-1}{M_{CO_2}} \\
0 & 0 & \frac{16}{M_{Cat}} & 0 & 0 & -y_{ATP,X} & 0 & 0
\end{bmatrix}
\times
\begin{bmatrix}
r_S \\
r_{O_2} \\
r_{Cat} \\
r_{Ana} \\
r_{NH_3} \\
r_X \\
r_{H_2O} \\
r_{CO_2}
\end{bmatrix}
$$

$$
=
\begin{bmatrix}
0 \\
0 \\
0 \\
0 \\
0 \\
0 \\
0
\end{bmatrix}
\tag{12.5}
$$

In the same manner, the mass flux balance for the metabolism is set up (second row), meaning that the anabolic mass flow plus the nitrogen source (representative for other inorganic salts) delivers the final biomass flux, which is actually the specific growth rate. Assuming constant biomass composition, nitrogen uptake is regarded as proportional to the anabolic flux; compare the medium calculation in Chapter 3, leading to the third row. We now have two equations for the metabolic node with three unknown fluxes, which is sufficient, and the maximum, according to the rules. Respiration is a completely defined process without any degree of freedom. So, three additional equations (rows 4 to 6) for four unknowns, based on oxidation stoichiometry, can be set up. The carbon balance is set up first, based on the fraction of C in Cat and CO_2. Stoichiometry can be best formulated on a molar basis. So, fluxes are divided by their molar mass (M). The virtual metabolite Cat (C3 body) needs three O_2, to be completely oxidized. Row 6 says that for each mole CO_2 one mole H_2O is produced. Now, six equations for eight unknown fluxes have been set up.

Basically, the allotment of glucose to respiration or growth is one degree of freedom. As a biological hypothesis, we assume that the cell will use all ATP generated in respiration for growth. On the one hand, growth requires a thermodynamically given amount of energy, while on the other hand, ATP from respiration will not be wasted. So, setting up the ATP balance sounds sensible. Complete oxidation of glucose is fairly well known to deliver 32 ATP/mol. We therefore set 16 mol ATP per mole Cat. ATP demand for growth can only be roughly estimated from references, so we use a yield coefficient, $y_{ATP,X}$, meaning the amount of ATP in moles to allow for building up 1 g of biomass. We now have seven equations – being the maximum for the whole cell. Only one possible equation is left to describe the interaction of the cell with the environment, in the form of kinetics. In the example, it could be the substrate, oxygen, or nitrogen. In the following, we look at the substrate-uptake kinetics as the limiting step. All other fluxes then follow the glucose flux. If the nitrogen source or oxygen is the limiting factor, the cell has to downregulate the substrate uptake correspondingly. The metabolism of the aerobic cell is indeed highly stoichiometrically determined.

The matrix describing the intracellular linear dependencies is called the system matrix or, more specifically for our biotechnological purpose, the stoichiometry matrix Y. The dimensions of the matrix are n_{equ} rows (number of equations) and n_{rates} columns (number of fluxes). Of course, $n_{equ} < n_{rates}$. In our example, for aerobic growth, $n_{rates} = 8$ and $n_{equ} = 7$. The difference (here 1) is the number of degrees of freedom the cells react to the environment. This is measured with the rank(Y) of the matrix. The calculation (see virtual supplement) in the example case delivers 7, so one degree is left, meaning one additional equation can be set up. It can also turn out that in complex networks, the matrix is not of full rank. In that case, the rank is lower than the number of equations. This might happen in cases where the metabolic network under investigation has equilibrium reactions or anaplerotic sequences or circumvents the rigid stoichiometry. During setting up the model equations, these cases may not be detectable intuitively. So, looking at the rank of the stoichiometry matrix delivers a valuable outcome to understand better the physiological behavior of the cell. To sum up, a generic stoichiometric model can be written as

$$Y \times r^T = 0 \qquad (12.6)$$

This contains the whole information with respect to the intracellular stoichiometry. The degrees of freedom $n_{free} = n_{rates} - \text{rank}(Y)$ can be filled up with kinetic equations conveying between intracellular and extracellular fluxes.

Finally, the model is used to find a value for the a priori unknown ATP demand of growth. From numerous observations, we know that approximately only half of the glucose is used for growth and the other half is respired, which leads to $y_{X,ATP} = 18/90 \approx 1/y_{Cat,ATP}$. In Fig. 12.7, the flux distribution for our example is shown as a bar graph. The values are normalized for $r_S = 1$. This is possible as the relations do not change with increasing or decreasing substrate uptake.

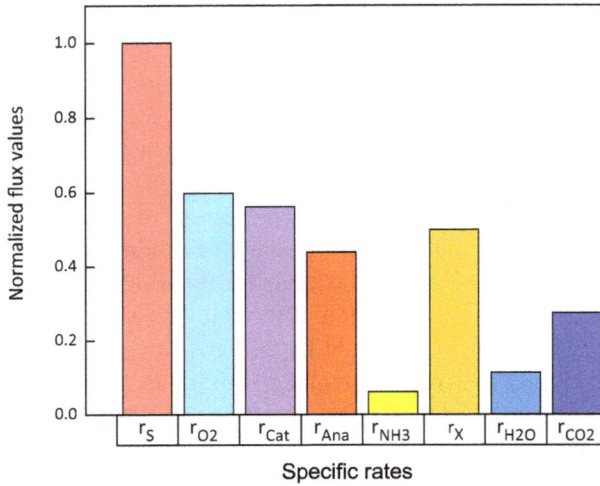

Fig. 12.7: Flux distribution in an aerobic cell; the calculation is based on the stoichiometry matrix given in eq. (12.4).

12.5 Anaerobic growth – traps and pitfalls

In anaerobic pathways, things look a bit different but simpler on the first glance, as shown in Fig. 12.8. The substrate is degraded via glycolysis, delivering two ATP per glucose. Note that two ATP per glucose are needed firstly to form phosphorylated compounds, while each one of the two resulting C3 bodies produce two ATP per mole. Furthermore, two redox equivalents per mole of glucose are generated. We now feel encouraged to simplify the situation by neglecting the possible byproduct formation. As the amount of available ATP is much lower than in the aerobic case, only a small amount of metabolite "Cat" can be used for growth. The remaining part cannot be simply excreted because $NADH_2$ is produced in excess. Only a small amount, according to the low growth rate, is needed. The predominant part has to be transferred to "Cat" again, and ethanol is finally excreted.

The stoichiometric matrix (12.7) can now be set up from mass balance around the central metabolism, followed by the energy and redox balance:

$$\begin{bmatrix} \text{mass balance } N_{Cat} \\ \text{Redox balance } N_{Cat} \\ \text{ATP balance } N_{Cat} \end{bmatrix} = \begin{bmatrix} 1 & -1 & -1 \\ \frac{1}{M_{Cat}} & -\frac{1}{M_X} & -\frac{1}{M_{Ferm}} \\ \frac{1}{M_{Cat}} & \frac{-18}{M_X} & 0 \end{bmatrix} \times \begin{bmatrix} r_{Cat} \\ r_X \\ r_{Ferm} \end{bmatrix} = \begin{bmatrix} 0 \\ 0 \\ 0 \end{bmatrix} \qquad (12.7)$$

This results in three equations for three unknown fluxes. This is forbidden as over-determination, as no degree of freedom is left to react to the environment. But there is a way out of this dilemma, looking at the rank of the matrix. Rank (Y_{anaer}) can be 2

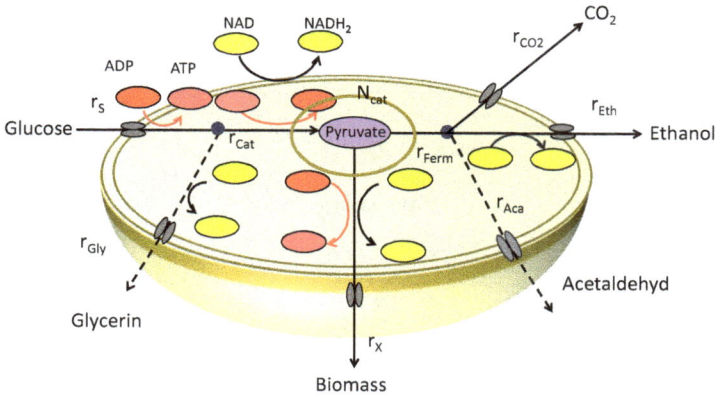

Fig. 12.8: Simplified structure of an anaerobic metabolism. The dotted fluxes are switched on or off by the cell, depending on the present needs.

when the redox demand of the cell exactly matches the produced amount from the metabolites allotted to growth. In the matrix, it can be seen that the first and the second rows are linearly dependent. The two rows can be transformed into each other by a constant factor for the case that $M_{Cat} = M_X$ and M_{Ferm}. During model simplification, this was supposed as C3 bodies with the same degree of reduction. This means, in particular, that the organism has to have the same degree of reduction as the substrate glucose. This is, of course, physiologically speaking, not sensible. Lipid accumulation would, for example, not be possible. In any case, the cell needs further degrees of freedom. For anaerobically growing yeast and anaerobic bacteria, nature has given us this degree of freedom via acetaldehyde and glycerol excretion. Per mole of glycerol, one $NADH_2$ can be "disposed." As a phosphorylated compound (dihydroxy-acetone-phosphate) is used for this purpose, the process is at the cost of one ATP. The usually high maintenance leads to a change in ATP/$NADH_2$ ratio, making balancing via byproducts necessary. Furthermore, acetaldehyde can freely diffuse through the cell membrane, leaving $NADH_2$ behind. Diffusion depends on the concentration gradient over the cell membrane, making acetaldehyde flux not only dependent on the growth rate but also on the biomass concentration and process feeding strategy with feedback on glycerol excretion. Byproduct pattern is therefore difficult to understand by simply looking at the data. What we learned from this model is that byproduct formation in anaerobic processes is not only necessary to give beer a good taste but is a biological necessity. In more complex situations, the only way to get insight into the behavior of cells is using models.

The examples we investigated until now give rise to the impression that cells are highly chained by all kind of compulsions, as indicated in Fig. 12.9. Thermodynamics, here in the appearance of $NADH_2$ and ATP balance, or mass conservation in the form of elemental balances, form a tight corset. That allowed us as practical spinoff to assign flux values, also to intracellular fluxes. Based on modeling, it is possible to look one step deeper into the metabolism than is possible based only on extracellular measurable metabolic fluxes.

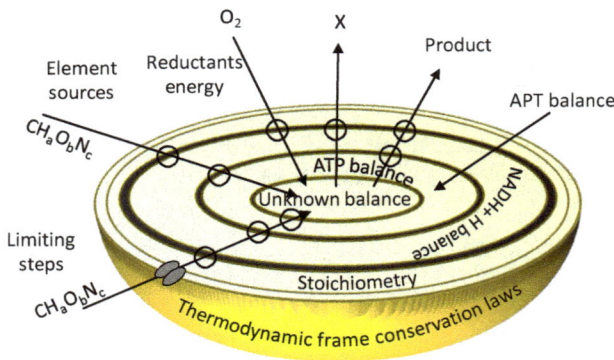

Fig. 12.9: Different material fluxes into and out of the cell are strictly constrained by thermodynamic conservation laws and stoichiometries.

But life is much more flexible. We already looked at examples of anaplerotic sequences and switching on/off additional pathways. In other cases, cells have two parallel pathways for different purposes, but possibly compensating each other. Geneticists are sometimes astonished by how good some microorganism can compensate knockouts. Glucose can be degraded via glycolysis or via the pentose phosphate pathway. As there is no rigid stoichiometry, a prediction about the fluxes through these two pathways is not possible. An approach from metabolic analysis to measure such fluxes directly is using C^{13}-labeled substrate compounds. The principle is shown in Fig. 12.10.

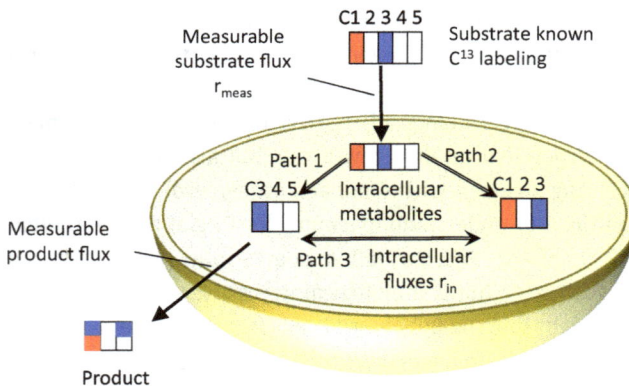

Fig. 12.10: Principle of flux measurement using labeled carbon sources.

As a substrate example, a C5-compound is shown at the top of the picture; the carbon atoms are counted from 1 to 5. The two-colored carbons (C1 and C3) are labeled. The colors (no difference in reality) are used to follow their fate on the way through the metabolism. In pathway 1, C1 and C2 are cleaved, while in pathway 2, the first three carbons, including the labeled ones, are kept. Both remaining C3 bodies are in chemical equilibrium, allowing no stoichiometric statements. It is now possible to measure the labeled carbons in the product using NMR technology. The result is that C1 of the product (mixture of the former C1 and C3) is completely labeled while C3 (the former C5 and C3) is only labeled to 50%. This allows the conclusion that half of the substrate is degraded via pathway1 and the other half via pathway2. This information about unmeasurable intracellular fluxes is a powerful tool to be included in linear flux models or to verify them.

12.6 Back to baker's yeast – more degrees of freedom

In Chapter 7 about fed-batch processes, we got to know something about Pasteur and Crabtree effects. Under oxygen limitation or above a certain growth rate, yeast is able to produce ethanol, parallel to respiration. According to the spirit of modeling, it is not enough to formulate these relationships as an "if . . . then . . . else" rule. Rather, we want to find a mechanistic reason either on the reactor (oxygen transport) or on the metabolic (limited respiratory capacity) level. In order to approach the problem from the outside, a dataset of a continuous cultivation (X/D-diagram) is given in Fig. 12.11a and b.

For small dilution rates, the curves for measured concentrations look as expected. Above a critical growth rate, $D = r_X > r_{X,\text{crit}}$, ethanol formation sets on. It is understandable that ethanol production is at the cost of substrate and biomass, but otherwise, the curves look a bit unusual. On the right-hand side, the calculated specific turnover rates are plotted. The onset of ethanol formation is also visible by the increasing respiratory quotient (RQ value) and coincides with the maximum of the specific oxygen-uptake rate. Considering this maximum could be an idea to follow up when setting up a model. In fact, a bottleneck is a good working hypothesis to start with.

Yeast catabolism, shown in Fig. 12.12 as a strongly simplified graphical model, is a combination of the aerobic and the anaerobic cases. From the viewpoint of modeling, we now have a problem. There are four metabolic fluxes, but only two equations can be formulated. One degree of freedom is necessarily used by the cell for the link between growth and energy formation. This leaves one degree of freedom for the cell to channel carbon into respiration or the fermentation pathway.

What makes the yeast cell produce ethanol with low energy gain instead of only respiration? We cannot follow all intracellular control mechanisms on the genetic or epigenetic level, which would be the pure mechanistic view. A successful approach to cope with this situation is to anticipate the control goals of the cells and model the

(a)

(b)

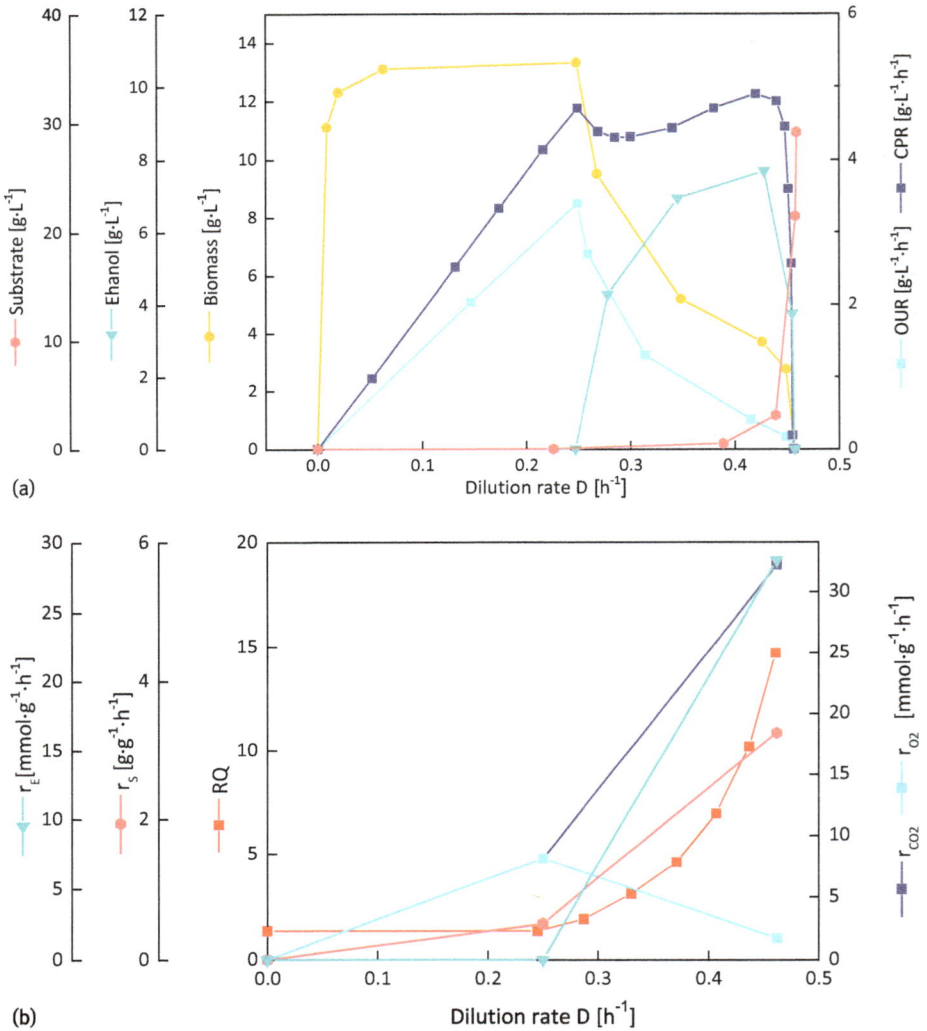

Fig. 12.11: *X/D* diagram showing yeast production in continuous cultivation: (a) measurements of concentrations and volumetric gas turnover rates and (b) calculated specific rates.

resulting effect. In continuous cultivations, it is not necessary to consider control dynamics. In several published metabolic flux models, the control goal has been stated quite generally as: "The cell utilizes all degrees of freedom to maximize growth." The formal notation (12.8) reads:

$$\max\left\{r_X\,|\,\boldsymbol{Y}\times\boldsymbol{r}=0,\ r_S\le r_{S,\max}\cdot\frac{c_S}{k_S+c_S},r_X\le r_{X,\max}\right\} \tag{12.8}$$

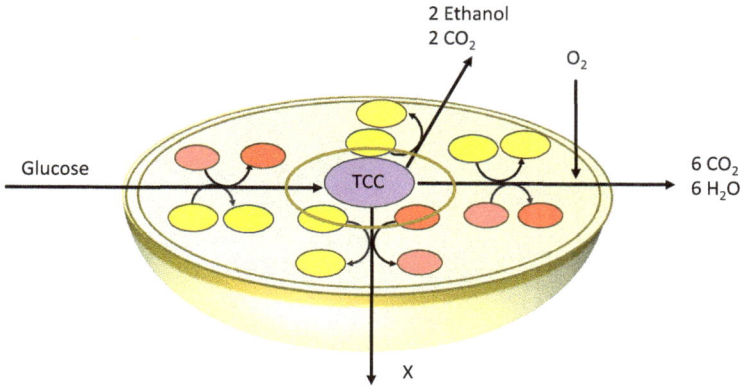

Fig. 12.12: Metabolic scheme of yeast metabolism as an additional flux ethanol production is considered.

The problem description is interpreted in the following way: The cells try to optimize their individual specific growth rates, r_X. This is first a linear optimization problem. Without a specific cost for increasing μ, which is the case here, it would go to infinity. But the cell is subjected to the linear constraints given by the stoichiometric matrix and the possible substrate-uptake rate. The less than or equal sign is used here to allow the cell to downregulate the substrate uptake in case it brings an advantage, e.g., to prevent a glucose overflow. Additionally, we are not sure that substrate uptake can go to its maximum before another intracellular step becomes the bottleneck, e.g., the speed of DNA replication.

This is easily written down but how is this approach evaluated? To answer this question, we have to take a trip down linear optimization, an approach known as "linear programming." The kernel of the approach is a linear cost function to be minimized or maximized. Linear means, in principle, it can go to infinity. This is prevented by linear constraints, given either as linear equations, in the yeast case stoichiometries, or as inequalities to formulate the upper or lower limits. A general form for x being the vector of the "adjusting screws" is

$$\max\left(c^T \times r\right) \text{ subject to} \left(Y \times r^T \leq b\right) \text{ and } \left(r \geq 0\right) \tag{12.9}$$

So, a linear combination of unknowns is also possible to be chosen as the lowest cost or maximum profit function. The parameter b was 0 in our examples but could contain other fixed terms like maintenance energy. The situation is visualized for two unknown variables in Fig. 12.13.

Now comes the interesting point from theory: All constraints, together, span a multidimensional polygon. The maximum is always reached at exactly one edge of this polygon and not somewhere in the middle. Optimization means, therefore, to calculate the linear cost function for all possible edges and find the best one out of the

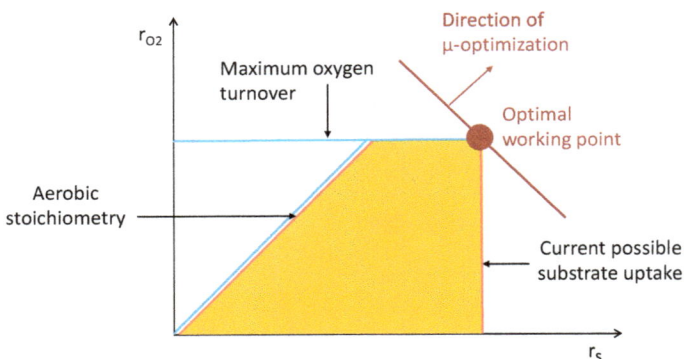

Fig. 12.13: Pictorial representation of linear programming for the example of substrate and oxygen uptake as independent variables. The profit function can be maximized until it finds constraints at the upper right bound.

list of results. In simulation packages, there are of course more clever routines to solve the linear programming problem. We now have a look at the results for yeast growth (Fig. 12.14).

The simulation reflects the data quite well (which has to be proven). For low dilution rates, the normal situation for aerobic growth is reproduced again, especially the nearly constant biomass concentration. At about $D = r_X = 0.23$ h^{-1}, the specific oxygen-uptake rate reaches its maximum. To grow faster, the yeast cells have to produce ethanol, visible from the increasing ethanol concentration and the increasing specific ethanol-formation rate. The curve for specific substrate uptake increases more steeply from this point, while biomass concentration sharply drops. This is debited to the lower ATP yield of the fermentative pathway. Finally, at about $r_X = 0.42$ h^{-1}, the specific uptake rate also reaches its maximum. Further increase of D is not possible and the simulation shows the washout case. However, the real yeast data show a difference. At high dilution rates, the specific oxygen-uptake rate even decreases. This worsens the Crabtree effect further, which is of course not good for the yield. We can, at this point, only speculate. Oxygen diffusion could be limited but this is unlikely, as there is no visible reason why that should be the case. Another assumption is that the cell "intentionally" reduces the formation of enzymes in the respiratory chain. The background could again be an intracellular optimization, where these complex enzymes are more "expensive" to produce than kinases for sugar transport, or that the mitochondria do not have enough membrane space at high growth rates, or that the expression rate is somehow limited. Note that each cell optimizes its local cost criterion. This is not automatically good for the population. (Is this an analog model for rise in population of mankind? Another optimization idea for microorganisms could be efficiency during growth at strong substrate deprivations.

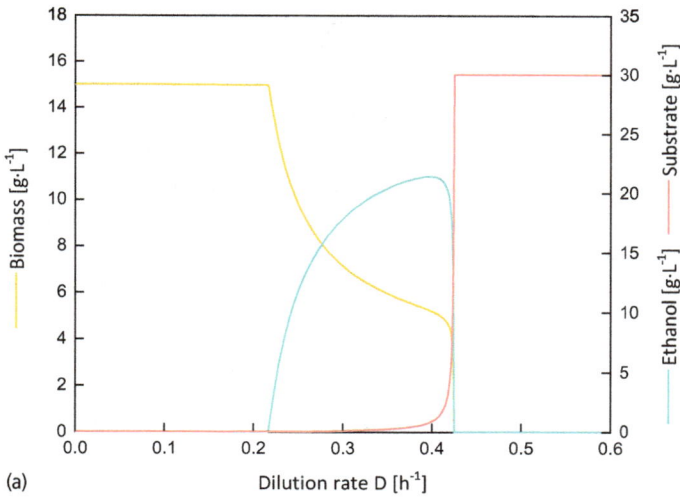

(a)

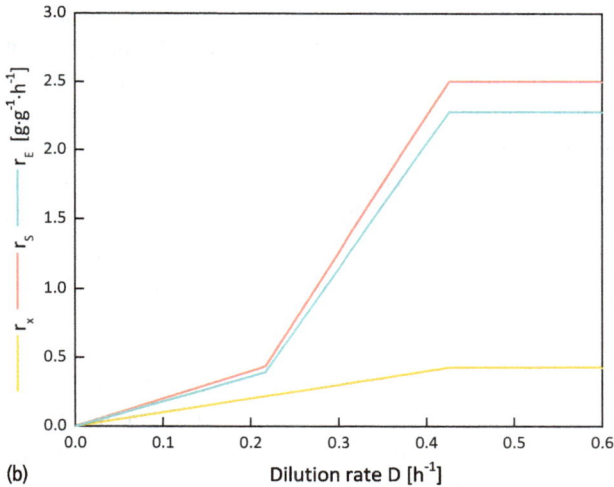

(b)

Fig. 12.14: Simulation of a continuous yeast cultivation; parameters are $c_{S,f} = 30$, $M_{Cat} = 90$, $e_{N,X} = 0.1$, $r_{S,max} = 2.5$, $k_S = 0.1$, $y_{ATP,Cat} = 16/M_{Cat}$, $y_{X,ATP} = 18/90$, and $y_{ATP,Eth} = 2/90$. Further details can be taken from the model in the virtual appendix. Note that the simulator does not deal with units, so consistent conventions should be used.

The optimization approach is also a useful tool in complex cultivation patterns, where the cells run through different limitations. Until now, substrate uptake was insinuated as limiting. In cases where only one limiting step is expected, the limitation could switch between substrate, nitrogen, or oxygen limitation during the process. Setting all three kinetics as possible constraints for the optimization approach intrinsically delivers the switching point between the possible physiological states. Under limitation conditions, like nitrate or phosphate limitation (no protein and nucleic acid formation is possible), some microorganisms do not reduce substrate uptake or produce byproducts to fulfill the stoichiometric requirements, but they produce intracellular storage compounds. Examples are PHB production in bacteria or starch and oil accumulation in microalgae (see Fig. 2.15). In such cases, the optimization approach is also useful to simulate this behavior as the simulation examples (virtual material) prove. In such cases, cells may change their physiological behavior. Considering this in representing the different physiological cell types leads to the so-called segregated model. An example for yeast growth, where yeasts in the different cell cycle states play a role, is given in the virtual material.

12.7 Back to microalgae – more spatial and hierarchical structure for subsystem definition

As already stated above, a process model can be decomposed into single stages, and the bioreactor itself into three physical phases. The "bio-phase," consisting of potentially different cells, can be subdivided into submodels for different cell types, leading to segregated models (different cell types from one species) or population models. But, inside the cells, hierarchical structuring is also applicable. A cell is not an ideally mixed reactor. Rather different reactions take place in different part of the cell, often in organelles as distinct compartments. Models considering this structure are called "compartment models." Until now, we considered mainly rigid metabolic pathways for cells with a given composition. Depending on environmental conditions, it is entirely possible that the cells change their macromolecular composition. To do so, the cell has a complex network of sensors, actuators, and controllers on the genetic, epigenetic, and enzymatic level. In Fig. 12.15, this is marked as the "control level," one hierarchical level above the metabolism. It is, in principle, not possible to model the control level in detail, so we summarized it in the optimization paradigm. In the following paragraph, an example of a spatial and hierarchical model is elaborated where microalgae and especially the chloroplasts play the main role.

Photosynthesis, located in the chloroplasts, consists of light absorption, ATP and $NADH_2$ generation, and carbon fixation. The dependency of growth, light intensity, and light absorption is condensed in the PI curve (Fig. 8.10, eq. (8.8)). We now have a closer look at the different steps to distinguish between light absorption and the consecutive reaction steps. Firstly, the most important specific turnover rates must be listed before using them in the model. Light capturing by light absorption, denoted as $r_{hv,abs}$ as a passive physical transport step can theoretically go to infinity. The consecutive steps on photosynthesis have, of course, a limited capacity; so, depending on growth conditions, not all photons can be used. This situation is aggravated as the cells cannot "close their eyes" where a bacterium can downregulate the substrate uptake. So, we have to split the photons into used ones, $r_{hv,use}$, and radiated ones, $r_{hv,NPQ}$.

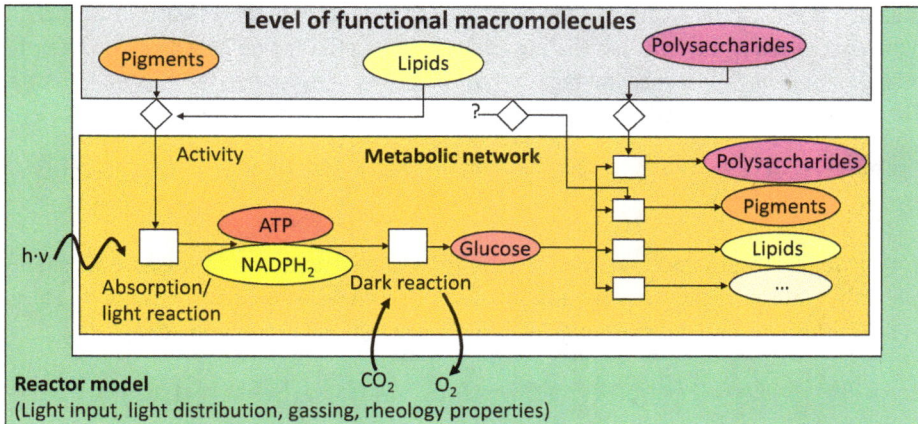

Fig. 12.15: Spatially and hierarchically structured graphical model for phototrophic growth. Ellipses are metabolic compounds, rectangles are material converters (metabolic pathways), and rhombus are intracellular controllers.

The energy and redox balance can be set up only for the chloroplasts, based on the gross equation (8.3) in Chapter 8. As primary product of photosynthesis starch is accounted for, formally setting up both balances would lead to an overdetermined set of equations, as in the anaerobic case. To balance both, the cell can adjust the relation of redox and energy generation by shifting between linear and cyclic electron flux. This allows us to lump the process directly into starch production, r_{Starch}, from a given amount of photons, $r_{hv,use}$. Further, CO_2 uptake, r_{CO2}, is needed as well as oxygen formation, r_{O2}. The final vector of specific turnover rates in the chloroplasts reads:

$$\boldsymbol{r}_{Photo} = \begin{bmatrix} r_{hv,abs} & r_{hv,NPQ} & r_{hv,use} & r_{CO_2} & r_{O_2} & r_{Starch} \end{bmatrix} \tag{12.10a}$$

Setting up the system matrix is now comparatively easy, see eq. (12.10).

It contains the photon balance in the first row, photon-starch yield in the second row, and the stoichiometry of starch (here, calculated as glucose) from CO_2 and O_2 formation in the last two rows:

$$\boldsymbol{Y}_{Photo} = \begin{bmatrix} 1 & -1 & -1 & 0 & 0 & 0 \\ 0 & 0 & y_{ATP,hv} & 0 & 0 & -y_{Starch,ATP} \\ 0 & 0 & 0 & M_{CO_2}^{-1} & 0 & -6 \cdot M_{Starch}^{-1} \\ 0 & 0 & 0 & M_{CO_2}^{-1} & M_{O_2}^{-1} & 0 \end{bmatrix} \tag{12.10b}$$

The consecutive steps on photosynthesis have, of course, a limited capacity. Maximum starch production depends on limitations of different reaction steps, either somewhere in water splitting, noted here as $r_{hv,use}$, or in CO_2 availability. This latter one is predominant in nature. In sum, this leads again to overdetermination, as not all photons are used. The way out for the cell and for modeling is the release of absorbed light energy from the light-harvesting complexes as photons or heat (NPQ). The optimization approach includes the fact that the cell tries to minimize the wasted photons. These aspects can now be formulated in a submodel holding for the chloroplasts:

$$\max\{r_{\text{Starch}}|\mathbf{Y}_{\text{Photo}}\times r_{\text{Photo}}^T = \mathbf{0}, r_{\text{hv,abs}} = \sigma_X \cdot I_{\text{hv}}, r_{\text{CO}_2} \le f\left(c_{\text{CO}_2}\right), r_{\text{hv,use}} \le r_{\text{hv,use,max}}\{\}\}$$

(12.10c)

For the heterotrophic part, we can assume the same structure as in the aerobic example, only that the substrate is the self-produced starch, as written in eq. (12.10.d). Here, to be on the safe side, maximum growth for other reasons is also considered:

$$\max\{r_X|\mathbf{Y}_{\text{aerob}}\times r_{\text{aerob}} = \mathbf{0}, r_S \le r_{\text{Starch}}, r_{\text{NH}_3} \le f_{\text{kin}}\left(c_{\text{NH}_3}\right), r_X \le r_{\text{X, max}}\{\}\}$$ (12.10d)

The consequences for the PI curve become clear from this model reflection and are visualized in Fig. 12.16. For low light intensities, the linear branch depends only on absorbed photons, from which all are used for growth. No additional limitation occurs. The saturation branch, however, is the consequence of different limitations and not necessarily from the light reaction. For carbon limitation inside the chloroplast, starch production is reduced – having direct impact on growth. This is actually a manifestation of Blackman kinetics (Chapter 4). NPQ should rise under this condition. Further down in the anabolism, nitrogen limitation may occur. In fact, we did not consider any feedback to photosynthesis. The consequence is then an ongoing starch production. That is actually the case in many microalgae, where nitrogen deprivation leads to starch or to lipid accumulation. A simulation example is available in the supplementary material.

Fig. 12.16: Light-response curve (PI curve) for different limitation conditions.

For quantitative analysis, the P curve has now to be applied to photobioreactors, where no homogenous light conditions prevail. Reactor equations are not simple mass balances as in the ideally mixed heterotrophic case. The specific growth rate, μ, might be constant at μ_{max} in the bright front of the reactor. At an a priori, not exactly known point inside the reactor, it will decay according to local light intensity and kinetics. As a comparatively simple case, a flat cuboid geometry is chosen, where an exponential decay of light intensity is observed along the light path.

As we practically cannot measure μ along the light path, a formal deduction of the macroscopically apparent average growth rate $\mu_{av}(I_{hv,0})$ has to be developed. The most direct case is to virtually divide the total light path length; here, the thickness or depth of the reactor D_R, into infinitesimal slices dl_{path} and to sum up the respective local $\mu(I_{hv})$ over all slices. This is to integrate $\mu(I_{hv})$ along the light path from the side of incident light to the dark remote side of the reactor, as formally deduced:

$$\mu_{av,\mu}(I_{hv,0}) = \frac{1}{D_R} \int_0^{D_R} \mu \cdot I_{hv}\left(l_{path}\right) \cdot dl_{path} \tag{12.11}$$

This approach is called "μ-integration," indicated here with the index μ. In fact, a formal integration is not possible for all possible kinetics. Some examples are given in the supplementary material. A simulation for different σ_X, comparing ideal local kinetics $\mu(I_{hv})$ with observations, μ_{av}, is plotted in Fig. 8.19 as a function of the incident light intensity $I_{hv,0}$. Also given is $\mu_{hv,I}(I_{hv})$, assuming that the cells can somehow store the absorbed light and grow finally according to the mean value of the light distribution in the reactor as follows:

$$\mu_{av,I}(I_{hv,0}) = \mu\left(\frac{1}{D_R} \int_0^{D_R} I_{hv}\left(l_{path}\right) \cdot dl_{path}\right) \tag{12.12}$$

This approach is called "light integration." Simulation results for the resulting μ versus I_0 curves are simulated in Fig. 12.17.

In the case where all cells experience this light, we observe the ideal kinetics. As expected, light integration leads to a linear increase not reaching the maximum growth rate. For μ integration, with increasing I_0, an increasing part of the reactor is bright, leading to light saturation and therefore not to higher growth rates. This looks very similar to Monod kinetics and is therefore sometimes misinterpreted. Real values can lie a bit higher, as microalgae can indeed store activated states for some ns and ATP level for some ms. Fast mixing with fast changes between dark and light parts of the reactor, from the view of a single cell, supports this energy storage, leading to the so-called intermittent (or flashing) light effect.

With the optimization approach, it is possible to cover some aspects of the changing macromolecular composition of a cell, especially accumulation of storage compounds. Closing up modeling considerations, we now collect some examples for visible impact of the control level onto flux distribution and macromolecular composition. Reduction of maximum respiratory capacity in yeast with increasing glucose availability is already the first example on the enzymatic level. Having the choice between different substrates like glucose and fructose, many species can downregulate the uptake of all but one. High temperature induces chaperone production in many bacteria, a behavior exploited in recombinant protein production. Microalgae avoid losing photons by NPQ, so most of them are able to reduce chlorophyll content under

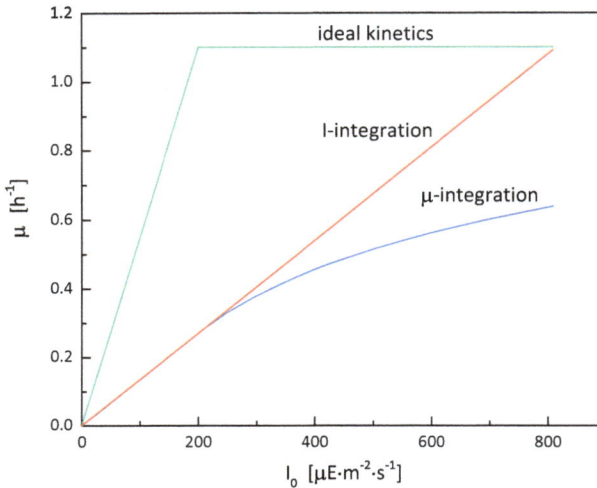

Fig. 12.17: Average specific growth rate as a function of incident light intensity for different assumptions.

high radiation conditions with feedback to flux distribution. With polysaccharides, microalgae influence their environment in a controlled manner. While flux models for bioprocesses are usually defined as stationary, control reactions on the level of macromolecules show time constants in the range of minutes or hours. So, dynamic modeling has to be applied in these cases. Make notes about other cases that you get to know!

Even in a late stage of development, bioprocess models contain only weakly known relations or unknown parameters. These should be addressed by model-based experimental design. Here is finally in this chapter, a saying of Norbert Wiener (who coined the term "cybernetics"): "The best material model of a cat is another, or preferably the same, cat."

12.8 Integration of products and processes into society – the final proof of meaning

All the attempts to develop bioprocesses are only sensible if somebody needs and wants the product. That sounds self-evident but has different characteristics in different fields of application. For medical applications, cost-benefit ratio is rated differently as for food, cosmetics, or bulk chemicals. Typically, competing chemical products can be found in the latter case, while therapeutics are often natural or nature-identical substances, where microorganisms have huge advantages as producers. Not only do economic aspects play a role, but in addition and in mutual dependency, ecological, ethical, societal, or simply practical aspects can be decisive for acceptance in the market. This also affects the process as such. Use of renewable resources and the avoidance of organic solvents are critical points. Labels like "bio," "without fossil carbon," or "without animal additives" count in the market. Things like diatomaceous earth as filter aid are not in public awareness but are worth considering, from the ecological view as well. It should not be forgotten that there are people running the process in a factory. Process design to allow daily or weekly production cycles mark only one consequence. Worker skill and education is a concern in process development. Finding and paying a bioengineer could be a knockout criterion for a sophisticated bioprocess with microorganisms, especially for small companies. The mentioned aspects, together with the wages for the workers, are criteria for the decision on the country in which a production process is established. In the previous chapters, some examples have been outlined for different aspects of integration into the society. These are ordered in Tab. 12.1, not for the sake of completeness but to encourage reconsidering of the examples in this book and making your own thoughts. As bioengineers, we are called to integrate ourselves in scientific networks and the fast developing vertically and horizontally structured scientific communities.

Tab. 12.1: Aspects of integration on the societal level.

Field	Aspect	Measures and examples
Direct impact on humans	Healthcare	Antibodies (mammalian cell culture), recombinant proteins (*E. coli*)
	Food processing, food supply	Fermentation (*Lactobacillus*), microalgae (*Chlorella*)
	Cosmetics, wellness	
	Military use	"The Dirty Dozen"
Working world	Occupational safety, product safety	Reactor design, GMP, PAT
	Work rhythm, shift work	Process duration
	Existing infrastructure	Biorefinery, closed material cycles
	Education and training	
Environment	Environmental protection	Wastewater treatment
	Renewable resources	No organic solvents
	Biodegradable products	Polyhydroxybutyrate (PHB)
Economics	Production costs	Process intensification
	Market price	Specific advertising
Political, cultural, and ethical aspects	Public perception	Gene technology, animal welfare

Tab. 12.1 (continued)

Field	Aspect	Measures and examples
	National regulations	"Bio"-label
	Tradition	Cheese
Sciences	Integration of sciences	Extensive networking, education
	Art and design	Architecture

What will be future of bioprocess engineering? As the future is per se not predictable, asking questions about current trends may be sufficient here. Will synthetic biology provide standard organisms for all products or even artificial cells making process development superfluous? Modern enzyme processes with enzyme cascades and co-factor regeneration could make the employment of living cells obsolete as well. Will better understanding of biology make scale up from single-cell technology over micro-titer plates to production-scale possible, again with less engineering input? Or is it imaginable the other way around: that more and more engineering thinking is needed to understand intracellular thermodynamic processes? Further interlocking between biology and bioprocess engineering is mandatory. Food processing is traditionally the field of most interferences between technical production process and public perception. With respect to bioengineering, fermented food with genetically engineered microorganisms is an example of this, or the newest developments in making meat substitutes with animal cell cultures. The science fiction author Isaac Asimov dared a prediction at Expo 1964: "Ordinary agriculture will keep up with great difficulty and there will be 'farms' turning to the more efficient microorganisms. Processed yeast and algae products will be available in a variety of flavors . . . but there will be considerable psychological resistance to such an innovation." This statement could be made in these days in the same way. Scientific targets changed with changing economic constraints; in this case, from making yeast from oil to making oil from yeast. In addition, taste, smell, and color can still be improved. Raw material from microorganisms for food production is still a small sector; microalgae could make the breakthrough. But also, products from plastics made from or with microorganism could be a giant amount, in the range of 7% of fossil raw materials used today.

For large quantities of product, large quantities of organic substrates must, of course, also be available, for which there is not yet much use. This could be straw, wood chips, or other plant waste with a high lignocellulose content. Only fungi can really cope with this material. As everybody can see, leaves need a whole year to be degraded in nature. There are only a few economic approaches, e.g., cultivation of fungi on straw for production of enzymes. Vera Meyer shows one way to the coupling of a big problem with a big opportunity in biotechnology. She combined the need for building materials with the production of fungal mycelium on lignocellulosics (see Fig. 12.18).

Vera Meyer is professor for fungal biotechnology and materials research in Berlin, and combines this with her second passion as visual artist. She wants to give the science she does, an image. Her followers call her "the fungi prophetess." Her view of the integration of bioprocesses into society is best described in her own words: "What fascinates me about fungi is their diversity, versatility and, in principle, this oscillation between friend and foe. We could not live our daily lives without the benefits that fungi give us. But they also threaten our health and our nutrition. And that's why research on fungi is a very rewarding field, because you can do research for society, for humanity."

Fig. 12.18: MY-CO SPACE is a habitable sculpture made of wood and fungi designed by the Berlin ArtSci Collective MY-CO-X. The roofing is made by a fungal-plant composite material. Hemp shives are inoculated with mycelium of the tinder fungus *Fomes fomentarius*. It is used for thousands of years to make fire, already from "Ötzi." Here, it converts the plant biomass into its own mycelium, forming a fungus-plant composite. This is then dried at 60 °C (© Vera Meyer, Wolfgang Günzel).

Integration into society, therefore, means not only developing sustainable processes, but also promoting them in society. By the way, words and concepts can be models just as much as formulas, and the value of a model also lies in enabling communication. Only the appreciation and subsequent action generated in this way, can lead to the product and process being accepted and thus established in the long-term. The collaboration between science and design/art for mediation up to the personal union continues to develop to the point of cross-fertilization. To conclude this chapter and the book, a quote from Vera Meyer says: "At the moment, it is good for us to think visionary. In view of the climate crisis, we need to find answers now. And nature offers us many."

Dear reader, now the reading has come to an end, and you can recreate in a fungi igloo to become fit for your own ideas.

12.9 Questions and suggestions

1. Add into the metabolic scheme, the number of carbons in the metabolic flux schemes and the number of $NADH_2$ and ATP. Adjust the stoichiometric matrices accordingly. As these are linear operations, the simulation results should not change.
2. Now go step-by-step through the simulations in the virtual material and try to verify the models in this chapter.
3. In the microalgae example, there is a serious oversimplification, making an adjustment of the heterotrophic submodel necessary. Starch contributes to dry weight, but not to light absorption and the growth machinery. Decompose growth into functional (active) biomass $c_{X,func}$ and intracellular starch.

Acknowledgments – dedicated to all the people who supported me

Writing this book was mainly fun. It helped to clear up the own thoughts on scientific aspects and to think about how best to convey it to the reader by text, tables, pictures, and (sorry to say that) mathematical formulas. But it means also quite some efforts. That concerns many practical things, editorial work, and data retrieval. But also professional scientific support turned out to be necessary, to fill the different topics with living information. That was only possible with the help of many people, who supported me with tireless work and creative ideas. The most important day-to-day support came from my coworkers at the time in terms of exploring content and creating graphics. Even colleagues from all over Europe and their employees also found time and pleasure in supporting the book project. It is not possible to express all the shades of gratitude of feel for those people, and certainly not to weigh their contribution. But I feel the need letting the reader know who it was and what he or she did, thus giving a little bit back. Please find below some small remarks to them, and their names are listed in alphabetical order.

My coworker Dr. Meike Dössel (KIT, Germany) is chemist and contributed with the drawing of chemical structures and expert advice on the medium chapter.

My former coworker Prof. Dr. Gözde Gözke (University of Yalova, Turkey) gave substantial support in setting up the reactor chapter.

Among the students, Eva Heymann (student of bioengineering, KIT) was noticed by her photographic skills; all the reactor and product pictures are her work.

The specialty of my coworker Dr. Ioanna Jakob (KIT) is bioprocesses development, where her knowledge on inorganic particle formation found input in the continuous cultivation chapter.

Also Mirco Katzenmeyer (KIT) is coworker with chemical engineering background; he delivered flow sheets and other inputs to the bioreactor and the fermentation chapters.

Dr. Viktor Klassen (Bielefeld University, Germany) is an expert in microbiology, especially microalgae in biogas production. His huge contribution in the form of pictures and text blocks was indispensable for the biosystems chapter and the corresponding paragraphs in the continuous culture and microalgae chapters.

https://doi.org/10.1515/9783110773354-013

My secretary Lisa Rest (KIT) cared for the correct sequence of figure and table numbers and in general for all formal issues of the manuscript. Further, she invented the nice generic product labels.

My former lab engineer Lena von Riesen (KIT) carried out the bioreactor runs to gather native data and pictures.

Dr.-Ing. Rosa Rosello (KIT) was something like a general editor of the writing process, from formal things to logical progression.

I know Dr. Jens Rupprecht (Sartorius, Germany) from a former project in Australia. Now he is involved in reactor development and shared his insight by discussing current trends in bioreactor development with me.

An immense of work and ideas came from Selima Sayah (student of bioengineering, KIT) by creating most of the pictures in ppt, handling all the data plots, and giving feedback from the perspective of a student.

The enlightening simulations of the Sun owe their existence to Kira Schediwy (KIT), coworker at my institute.

The support of the institute in all engineering questions is Christian Steinweg (bioprocess engineer, KIT). He delivered the CAD reactor drawings with much engagement, detailed view, and perfection.

My dear friend Renate Suda (IIT, Germany) was a bioscientific documentalist. She performed literature retrieval, also including more specific issues.

Michael Taebling (alias Käsemichl, Schönegger Käsealm, Germany) from my regional market brought to me the idea of cheese making as a significant bioprocess and shared with me some secrets about it.

Dr. Andreas Trautmann (KIT, Lonza, Switzerland) supervised with great dedication the fermentation course of my institute and condensed his experiences in the fermentation chapter.

Dr. Ines Wagner (KIT, OHB, Germany) worked as a scientific coworker in my group and made her PhD in the field of microalgae, and the photosynthesis part in this book is a highly welcome spin-off.

My former college Prof. Dr. Şems Yonsel (Okan University, Turkey) worked several years in a baker's yeast factory and knows the process from the theoretical and the practical side, making him to a perfect advisor for the running yeast example.

Special thanks go also to Marc Bisshops (Director SLS Biopharm, Pall Life Sciences), who discussed with me the pros and cons of continuous processing and wrote the precise statement in the chapter.

My special thanks go also to the designers/artists/scientists from the field of biomedia, who were happy to get involved in this book project and were willing to go into the discourse about ideas and to provide exhibits. Marie Jamroszczyk drew my attention to her beautiful drawings from the world of microalgae alongside her master's thesis. She enthusiastically devoted herself to the idea of combining the technical aspect of filtration with biology in a drawing. Prof. Vera Meyer, although already heavily involved in leading her institute and creating bio-based installations, found time to instruct me and to propose the two exhibits. Pitched by our interest in microalgae, Anthea Oestreicher performed a practical project at my lab, inspired and impressed me with her deep understanding of symbiosis and the importance of watery ecotones Dr. Tim Otto Roth contributed already to the first edition. During a common excursion to his atelier and an installation in the Black Forest, we found enough idea for further cooperation. I hope this experiment of art in a technical textbook will be a small step but a step nonetheless toward mutual inspiration, living together, and integration.

Last but not least, I want to say "Thank You" to Ria Sengbusch and Karin Sora from De Gruyter publisher for their encouragement to publish this book, practical help, advice with the manuscript, and their professional patience.

To express my sincere gratitude: THANK YOU SO MUCH!
Clemens Posten

Copyrights – pictures provided with courtesy and accepted with thanks

The following persons and companies delivered pictures and kindly copyrights to be printed in this book.

Figure 1.1a	Meine Pestoria, Grötzinger Straße 42–44, 76227 Karlsruhe, Germany.
Figure 1.1b	Schönegger Käse-Alm GmbH, Steinwies 20, 86984 Prem, Germany, with special thanks for Matthias Köpf and Michael Taebling (Michel's Rollender Käseladen, Eggweg6, 86989 Steingaden).
Figure 1.11	With gratitude from my former co-worker Christian Steinweg.
Figure 2.2a,b	Olga Blifernez-Klassen, personal com. by Viktor Klassen.
Figure 2.3a	From Hutchison, Clyde A. et al., "Design and synthesis of a minimal bacterial genome." In: Science (New York, NY) 351 (6280), aad6253. DOI: 10.1126/science. aad6253. Reprinted with permission from AAAS. License Number: 3953251395268, License bought by Viktor Klassen.
Figure 2.3b	Courtesy to CDC/ National Escherichia, Shigella, Vibrio Reference Unit at CDC.
Figure 2.4a	Viktor Klassen, KIT, Culture provided by Jan-Philipp Schwarzhanz, University of Bielefeld.
Figure 2.4b	Permission achieved by communication (research gate) from book chapter "Sample Preparations for Scanning Electron Microscopy – Life Sciences" both authors: Patchamuthu Ramasamy, Mogana Das, Courtesy of EM Unit, University Sains, Malaysia.
Figure 2.5a	Permission achieved by communication (e-mail) with author Marian Petre, taken from "Environmental Biotechnology for Bioconversion of Agricultural and Forestry Wastes into Nutritive Biomass". In: Marian Petre (Hg.): Environmental Biotechnology – New Approaches and Prospective Applications: InTech.
Figure 2.5b	Permission achieved by communication (e-mail) with authors Ms. Wardah. Mogana Das Murtey; Patchamuthu Ramasamy, coutesy of Ms. Wardah, taken from "Sample Preparations for Scanning Electron Microscopy – Life Sciences. Modern Electron Microscopy in Physical and Life Sciences". InTech.
Figure 2.6b	Courtesy of Juliane Steingröver, TU-Dresden, Germany.
Figure 2.7a	Courtesy of Eric Gottwald, KIT-IFG.
Figure 2.7b	Courtesy of Iznewton (Own work) [CC BY-SA 4.0 (http://creativecommons.org/licenses/by-sa/4.0)], via Wikimedia Commons.
Figure 2.8a	Courtesy of Michael Wagner, KIT-IFG.
Figure 2.8b	Courtesy of Ursula "Uschi" Obst, KIT-IFG, redrawn and modified by author.
Figure 2.13a	Licensed from science photo, StockFood GmbH, Munich, Germany, communicated by Martina Braun, with special thanks to Mike Allen, Plymouth Marine Laboratory, UK, who provided some of the pictures.
Figure 2.16	With courtesy and kind regards from Anthea Oestreicher.
Figure 5.13	Redrawn from material of CyBioTec MicroReactor.
Figure 5.14	With courtesy from my successor Alexander Grünberger with best wishes for further success from me.
Figure 5.15	With kind permission of Eppendorf AG, Bioprocess Center Jülich, Germany, communicated by Ulrike Becken.

https://doi.org/10.1515/9783110773354-014

Figure 5.16	General electric company, GE Healthcare, reproduced by permission of the owner, with thanks to Yana Ostrovskiy and Stijn Vercammen.
Figure 5.17	With kind permission of Sartorius-Stedim, Germany, communicated by Jens Rupprecht.
Figure 5.18	With kind permission of Infraserv-Hoechst, Frankfurt am Main, Germany, communicated by Jasmin Graf.
Figure 5.24	With kind permission of the authors/developers Björn Frahm, University of applied sciences, Ostwestfalen-Lippe and Helmut Brod, Bayer Technology Services, Leverkusen, Germany. Described at "Improving bioreactor cultivation conditions for sensitive cells by dynamic membrane aeration" Cytotechnology, 2009, 59, 17-30.
Figure 5.25	With kind permission and courtesy of Vera Meyer, Berlin, after personal communication, image credit: Martin Weinhold.
Figure 6.6	Picture and permission achieved from GEA Wiegand GmbH, communicated by Norbert Strieder with many thanks for the interesting discussion.
Figure 6.15	Permission achieved by communication (e-mail) with author Jian Ju, thanks to her, taken from the book chapter "Generation and Utilization of Microbial Biomass Hydrolysates in Recovery and Production of Poly-(3-hydroxybutyrate)".
Figures 6.16 and 6.17	With courtesies of Matthias Köpf, Schönegger Käse-Alm GmbH, Steinwies 20, 86984 Prem, Germany.
Figures 7.7a and 7.23	Taken at the plant of algae for future a4f in Pataias, Portugal, communicated by Edgar Santos, greetings to Vitor Verdelho Vieira.
Figure 7.12	Permission achieved by grateful communication (e-mail) with author Chan-kyu Park, taken from "Green fluorescent protein as a scaffold for high efficiency production of functional bacteriotoxic proteins in Escherichia coli".
Figure 7.18	Taken at TUM-AlgaeTec Center, Ottobrunn, Germany, provided by Dirk Weuster-Botz, Technical University Munich, Germany.
Figure 7.21	Taken at the algae pad of KIT from Mirco Katzenmeyer, Germany.
Figure 8.19	With courtesy from Marie Jamroszczyk especially drawn for this book.
Figure 9.11	Taken by Martina Nolte, Creative Commons CC-by-sa-3.0 de (http://creativecommons.org/licenses/by-sa/3.0/de/legalcode), communicated by Viktor Klassen.
Figure 9.12	Redrawn Axel Schippers, Federal Institute for Geosciences and Natural Resources, Hanover, Germany after Johnson et al., Curr. Opin. Biotechnol., 2014.
Figure 9.13	Licensed by Science Photo.
Figure 9.15a	REM picture taken by Nikolay Krumov, KIT, Lonza.
Figure 9.15	ESEM picture taken by Frank Friedrich, KIT.
Figure 10.18	With courtesy of Tim Otto Roth in cooperation with Prof. Dr. Rüdiger Hardeland, imachination projects, Oppenau, Germany.
Figure 11.1	From printed user material of bioengineering bioreactor with permission of Bioengineering AG, Wald, Switzerland.
Figures 11.4 and 11.5	Taken at KIT owned bioreactor, with permission of Bioengineering AG, 8636 Wald, Switzerland, communicated by Claudia Kälin.
Figure 11.12	With thanks and courtesy from Maria Glaubitz.
Figure 2.18	With kind permission and courtesy of Vera Meyer, Berlin, after personal communication, image credit: Wolfgang Günzel.

Index

https://doi.org/10.1515/9783110773354-015

www.ingramcontent.com/pod-product-compliance
Lightning Source LLC
Chambersburg PA
CBHW080705220326
41598CB00033B/5310